SYSTEM MODELING AND OPTIMIZATION

IFIP – The International Federation for Information Processing

IFIP was founded in 1960 under the auspices of UNESCO, following the First World Computer Congress held in Paris the previous year. An umbrella organization for societies working in information processing, IFIP's aim is two-fold: to support information processing within its member countries and to encourage technology transfer to developing nations. As its mission statement clearly states,

> *IFIP's mission is to be the leading, truly international, apolitical organization which encourages and assists in the development, exploitation and application of information technology for the benefit of all people.*

IFIP is a non-profitmaking organization, run almost solely by 2500 volunteers. It operates through a number of technical committees, which organize events and publications. IFIP's events range from an international congress to local seminars, but the most important are:

- The IFIP World Computer Congress, held every second year;
- Open conferences;
- Working conferences.

The flagship event is the IFIP World Computer Congress, at which both invited and contributed papers are presented. Contributed papers are rigorously refereed and the rejection rate is high.

As with the Congress, participation in the open conferences is open to all and papers may be invited or submitted. Again, submitted papers are stringently refereed.

The working conferences are structured differently. They are usually run by a working group and attendance is small and by invitation only. Their purpose is to create an atmosphere conducive to innovation and development. Refereeing is less rigorous and papers are subjected to extensive group discussion.

Publications arising from IFIP events vary. The papers presented at the IFIP World Computer Congress and at open conferences are published as conference proceedings, while the results of the working conferences are often published as collections of selected and edited papers.

Any national society whose primary activity is in information may apply to become a full member of IFIP, although full membership is restricted to one society per country. Full members are entitled to vote at the annual General Assembly, National societies preferring a less committed involvement may apply for associate or corresponding membership. Associate members enjoy the same benefits as full members, but without voting rights. Corresponding members are not represented in IFIP bodies. Affiliated membership is open to non-national societies, and individual and honorary membership schemes are also offered.

SYSTEM MODELING AND OPTIMIZATION

Proceedings of the 22nd IFIP TC7 Conference held from July 18-22, 2005, in Turin, Italy

Edited by

F. Ceragioli
Politecnico di Torino, Torino, Italy

A. Dontchev
AMS and University of Michigan, Ann Arbor, USA

H. Futura
Kansai University, Osaka, Japan

K. Marti
Federal Armed Forces University, Munich, Germany

L. Pandolfi
Politechnico di Torino, Torino, Italy

Springer

System Modeling and Optimization
Edited by F. Ceragioli, A. Dontchev, H. Futura, K. Marti, and L. Pandolfi

p. cm. (IFIP International Federation for Information Processing, a Springer Series in Computer Science)

ISSN: 1571-5736 / 1861-2288 (Internet)

ISBN 978-1-4419-4103-9 **e-ISBN 978-0-387-33006-8**

eISBN: 10: 0-387-33006-2
Printed on acid-free paper

9 8 7 6 5 4 3 2 1
springeronline.com

Contents

Preface

We publish in this volume the plenary talks and a selection of the papers on numerics, optimization and their applications, presented at the 22nd Conference on System Modeling and Optimization, held at the Politecnico di Torino in July 2005. The conference has been organized by the Mathematical Department of the Politecnico di Torino.

IFIP is a multinational federation of professional and technical organizations concerned with information processes. It was established in 1959 under the auspices of UNESCO. IFIP still mantains friendly connections with specialized agencies of the UN systems. It consists of Technical Committees. The Seventh Technical Committee, established in 1972, was created in 1968 by A.V. Balakrishnan, J.L. Lions and G.I. Marchuk with a joint conference held in Sanremo and Novosibirsk.

The present edition of the conference is dedicated to Camillo Possio, killed by a bomb during the last air raid over Torino, in the sixtieth anniversary of his death. The special session "On the Possio equation and its special role in aeroelasticity" was devoted to his achievements. The special session "Shape Analysis and optimization" commemorates the 100th anniversary of Pompeiu thesis.

All the fields of interest for the seventh Technical Committee of the IFTP, chaired by Prof. I. Lasiecka, had been represented at the conference: Optimization; Optimization with PDE constraints; structural systems optimization; algorithms for linear and nonlinear programming; stochastic optimization; control and game theory; combinatorial and discrete optimization. identification and inverse problems; fault detection; shape identification. complex systems; stability and sensitivity analysis; neural networks; fractal and chaos; reliability. computational techniques in distributed systems and in information processing environments; transmission of information in complex systems; data base design. Applications of optimization techniques and of computational methods to scientific and technological areas (such as medicine, biology, economics, finances, aerospace and aeronautics etc.).

Over 300 researchers took part to the conference, whose organization was possible thanks to the help and support of the Department of Mathematics of the

Politecnico di Torino. We would like to thank Istituto Boella, Fondazione Cassa di Risparmio Torino, Unicredit Banca and the Regione Piemonte for financial support. We also acknowledge support from Politecnico di Torino and INRIA.

We would like to thank the following colleagues, who organized special and invited sessions, greatly contributing to the success of the conference:

V. Agoshkov, O. Alvarez, G. Avalos, A. Bagchi , U. Boscain, A.V. Balakrishnan, M. Bardi, F. Bucci, J. Cagnol, P. Cardaliaguet, P. Cannarsa , M.C. Delfour, D. Dentcheva, A. Dontchev, H. Furuta, K. Juszczyszyn, P. Kall, A. Kalliauer, R. Katarzyniak, D. Klatte, B. Kummer, M. A. Lòpez, V. Maksimov, K. Marti, J. Mayer, S. Migorski, Z. Naniewicz , N.T. Nguyen, J. Outrata, B. Piccoli, S. Pickenhain, M. Polis, E. Priola, A. Ruszczynski, I.F. Sivergina, H. Scolnik, J. Sobecki, J. Sokolowski, G. Tessitore, D. Tiba, F. Troeltzsch , S. Vessella, J. Zabczyk, T. Zolezzi, J.P. Zolesio.

L. PANDOLFI

Contributing Authors

S.I. Aihara Tokyo Univiversity of Science, Nagano, Japan

M. Arakawa Kanazawa University, Kanazawa, Japan

G. Avanzini Politecnico di Torino, Turin, Italy

A. Bagchi University of Twente, Enschede, the Nederlands

A.V. Balakrishnan University of California, Los Angeles, U.S.A.

T. Birsan Gh. Asachi University, Iaşi, Romania

O.A. Brezhneva Russian Academy of Sciences, Moscow, Russia

P. Cannarsa Università di Roma "Tor Vergata", Rome, Italy

L. Dedè Politecnico di Milano, Miland, Italy

A. De Rossi Università di Torino, Turin, Italy

A. De Santis Università di Roma "La Sapienza", Rome, Italy

X. Ding University Shangai for Science Technology, Shangai, China

R. Fletcher University of Dundee, Dundee, Scotland, U.K.

A. Forsgren Royal Institute of Technology, Stockholm, Sweden

D.M. Frangopol University of Colorado, Boulder, U.S.A.

H. Furuta Kansai University, Osaka, Japan

U. García-Palomares Univ. Sìmon Bolìvar, Caracas, Venezuela

J.P. Gayon INPG, Grenoble, France

P.R. Graves-Morris University of Bradford, Bradford, U.K.

W. Hager University of Florida, Gainsville, U.S.A.

R. Hochreiter University of Vienna, Vienna, Austria

T. Kameda Kansai University, Osaka, Japan

F. Karaesmen Koç University, Istambul, Turkey

S. Kitayama Kanazawa University, Kanazawa, Japan

G. Kulvietis Vilnius Gediminas Technical University, Vilnius, Lithuania

M. Liu Formerly, Univ. of Colorado, Boulder, U.S.A.

Y. Matsui Kanazawa University, Kanazawa, Japan

T. Matsumori Kanazawa University, Kanazawa, Japan

J. Mayer University of Zürich, Zürich, Switzerland

E. Mijangos University of the Basque Country, Bilbao, Spain

S. Minkevicius Institute of Mathematics and Informatics, Vilnius, Lithuania

W. Ogryczak Warsaw University of Technology, Warsaw, Poland

M. Opolska-Rutkowska Warsaw University of Technology, Warsaw, Poland

L. Örmeci Koç Universitesi, Istanbul Turkey

G. Ch. Pflug University of Vienna, Vienna, Austria

R. Pytlak Military University of Technology, Warsaw, Poland

A. Quaini Politecnico di Milano, Miland, Italy

A. Quarteroni Politecnico di Milano, Miland Italy

G. Rozza Ècole Polyt. Fèdèrale, Lausanne, Switzerland

I. Talay-Degirmenci Duke University, Durham, U.S.A.

T. Tarnawski Military University of Technology, Warsaw, Poland

D. Tiba Romanian Academy, Bucarest, Romania

A.A. Tret'yakov Russian Academy of Sciences, Moscow, Russia

E. Venturino Università di Torino, Turin, Italy

R. Voss DLR Institute of Aeroelasticity, Göttingen Germany

D. Wozabal University of Vienna, Vienna, Austria

K. Yamazaki Kanazawa University, Kanazawa, Japan

M. Yoshimura Kyoto University, Kyoto, Japan

H. Zhang University of Florida, Gainsville, U.S.A.

THE LEGACY OF CAMILLO POSSIO TO UNSTEADY AERODYNAMICS

R. Voss,[1]
[1]*DLR Institute of Aeroelasticity, Goettingen, ralph.voss@dlr.de* *

Abstract First a brief overview is given of Camillo Possio's short but outstanding and fruitful career. This is followed by an outline of the state of the art in flutter and unsteady aerodynamic research, and the challenges and problems like high-speed flight that arose in aircraft development at that time. Possio's first publications on gas dynamic and supersonic problems are reviewed. The main focus is on the 1938 report on unsteady subsonic compressible 2D flow that became famous and was named after him, because he was the first person to developed an unsteady compressible aerodynamic theory, which was urgently needed in those years. The theory, which is based on Prandtl's acceleration potential is briefly outlined. Some discussions and comments that took place in Germany and other countries at that time highlight the importance of this work for the scientific community. Early solutions of Possio's integral equation developed by himself and later ones developed by other researchers are presented, as well as approaches that extended the theory to 3 dimensional flows before the war, like Kuessner's theory, which was probably influenced by Possio. Finally Camillo Possio's later scientific contributions to wind tunnel interference and to hydrodynamics are described. A summary of some developments of the 2nd half of the 20th century demonstrate that Camillo Possio created a milestone for modern aircraft research during his very short career.

keywords: Aeroelasticity, Unsteady Aerodynamics, Flutter, Integral Equation, Possio.

1. Introduction

Camillo Possio was born on October 30. 1913 and died on April 5. 1945. He was killed by a bomb during the last air-raid over Turin. He received the Laureate Degree in Industrial Engineering and in Aeronautics, both in Turin in 1936 and 1937. He was a pupil of Modesto Panetti and Carlo Ferrari. He was Assistant Professor and then Professor at Politecnico di Torino. His work is well known in the world's aeronautic community, particularly his method

*The author would like to thank Prof. H. Foersching for many fruitful discussions.

Please use the following format when citing this chapter:
Author(s) [insert Last name, First-name initial(s)], 2006, in IFIP International Federation for Information Processing, Volume 199, System Modeling and Optimization, eds. Ceragioli F., Dontchev A., Furuta H., Marti K., Pandolfi L., (Boston: Springer), pp. [insert page numbers].

of calculating the unsteady airloads of harmonically oscillating wings in 2D compressible flow, the key equation of which has been named after him. But his fruitful scientific carreer also covers general problems of gas dynamics and supersonic flow, 3D wing oscillations in incompressible flow, the influence of free surfaces on hydrodynamics, as well as the influence of wind tunnel walls on measurements. His work was published in about 16 papers, many of which were translated from Italian to German and English. When he died at the age of 31, he was one of the most promising experts of aerodynamics in Italy. He possessed the special skill of using high-level mathematics to obtain results of practical interest.

2. Aeroelastic and high-speed flight challenges

During the fast development of aircraft at the beginning of the last century, the significance of aeroelastic effects soon increased. These phenomena resulted from the interaction of the elastic structural system with a surrounding airflow. Airplanes have to be built extremely lightly and their structure is flexible. Elasticity of an aircraft increases with size more than proportional. Fig. 1depicts the wing deformations of modern aircraft wings in flight.

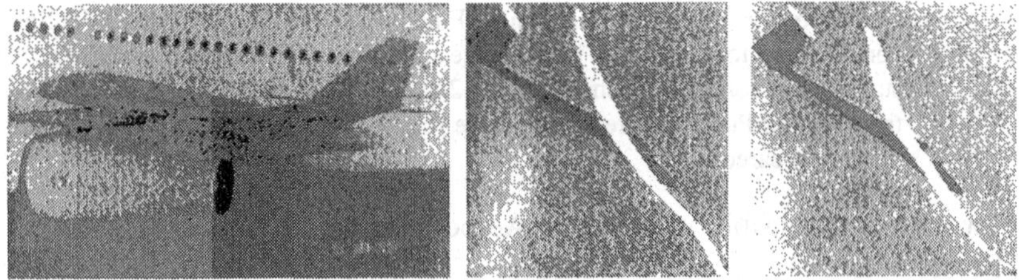

Figure 1. Deformation of a transport aircraft wing (l) and of a sailplane (r)

As long as strength requirements are fulfilled, structural flexibility itself is not necessarily objectionable. But the static and dynamic deformations of lifting surfaces generate steady and unsteady aerodynamic reactions, that in turn alter the deformations. These interactions may lead to several different aeroelastic problems with far-reaching technical consequences for flight. Of special importance is the flutter stability of wings and control surfaces. Particularly during the First World War pilots often reported of heavy oscillations of the wings and tail planes, that resulted in many fatal accidents. Back then aircraft were usually biplanes and the most frequent aeroelastic problems did occur not on the wings but with tail flutter. One of the first documented cases was the tail flutter problem of the British Handley-Page 0/400 bomber. Biplane constructions had a high torsional wing stiffness due to the interplane bracing. This was lost when cantilever monoplanes were developed, and flutter became even more dangerous. A remarkable example is the destruction of a big Ju90 aircraft in 1938 by bending-torsional tail-plane flutter. Although these accidents were not recognized to be a stability problem at that time, it seemed clear from the beginning that a theoretical investigation of these destructive vibrations required knowledge of both the elastic oscillatory behaviour of the lifting surfaces and the oscillatory motion-induced unsteady airloads. This presented a great scientific challenge. A systematic investigation of wing flutter and unsteady aerodynamics began about 1920 in Goettingen, Germany and in Turin, Italy. In general three different forces interact and generate aeroelastic effects, namely elastic forces E, aerodynamic forces A and inertial or mass forces I. They are distributed continuously over an aircraft and are in equilibrium :

$$E(q) + A(q) + I(q) = Q \tag{1}$$

Q denotes a known external force, which is independent from the aeroelastic system, rising for example from: atmospheric turbulence, gusts, landing gear impacts. E, A, I are functions of the elastic geometrical deformations q. If $Q = 0$, the equation describes the general aeroelastic stability problem. The analytical solution of this problem thus requires the calculation of the involved system forces as a function of local elastic deformations for a complete aircraft or for its components and the solution of the equilibrium equation.

The governing equations of fluid dynamics, the Navier-Stokes equations, have been known since the 19th century. But with the exception of a few limiting special cases there was no chance of finding analytical solutions. Only the development of numerical CFD (Computational Fluid Dynamics) methods since 1970 opened the way to computing particularly compressible and viscous flow problems around complex 3D configurations. When powered flight began, mainly experimental observations were used together with simplifications of the basic equations in order to derive mathematical models of the flow around an aircraft. In 1918 L. Prandtl developed his idea of a lifting line to represent the effect of a lift generating wing. Since the mathematical value of the circulation is directly related to the lift, the wing is replaced by a vortex line along the span at 1/4 chord of the wing, with constant strength Γ. Flow visualisations first in water tunnels for 2D wing sections for a starting motion showed the presence of a bound vortex and of a free starting vortex of opposite sign, which is swept down from the trailing edge to infinity by the main flow. These two vortices compensate, thus fulfilling the vorticity conservation law. If the fluid is incompressible and free of rotation - except for the bound and free vortices -

a potential function exists for the velocity vector and fulfills the Laplace equation. This follows from mass conservation, while a relation between pressure and velocity follows directly from momentum equation :

$$rot\vec{v} = 0 \rightarrow \vec{v} = \nabla\phi \;\; and \;\; \frac{\partial \rho}{\partial t} + \nabla(\rho\vec{v}) = 0 \;\; \rho = const \rightarrow \nabla\vec{v} = \Delta\phi = 0 \tag{2}$$

In 1922 Prandtl extended his concept to a theory of unsteady oscillating 2D lifting surfaces. Every time the bound vortex strength γ - and thus the lift - changes, a small free vortex with the opposite strength ϵ of this change is created to fulfill the vorticity conservation law, and is carried downstream from the trailing edge by the main flow. This model of unsteady free and bound vortices was applied by Prandtl's student W. Birnbaum, who developed a mathematical singularity method for computing unsteady airloads induced by an oscillating 2D wing [17]. He showed that aerodynamic lift and moment on the oscillating surface lag behind the forcing motion, and that this phase shift as well as the magnitude of unsteady lift and moment strongly depend on a new similarity parameter, the reduced frequency, see fig. 2.

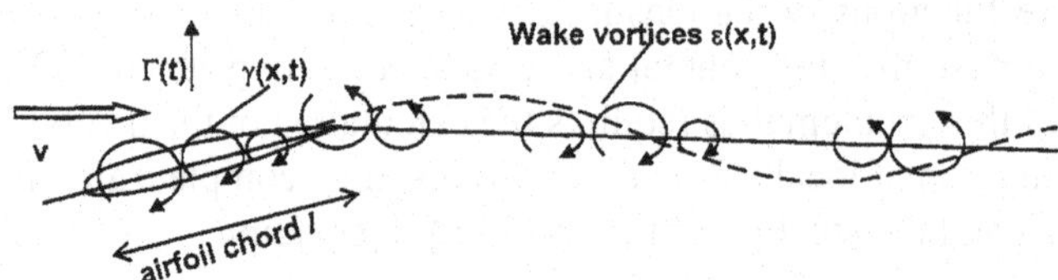

Figure 2. Vortex generation in the wake of an oscillating airfoil

He further remarked that modeling the unsteady problem requires a chordwise vorticity distribution instead of one bound vortex. His work was key to understanding the physical mechanism of flutter as a dynamic aeroelastic stability problem. This could be solved by assuming all forces including the unsteady aerodynamics to be harmonic and then treating eq.(1) as a complex Eigenvalue problem. Birnbaum died in 1925 at the age of only 28.

Structural dynamic modeling was either provided by a simplified elastic beam theory or from experimental ground vibration tests. These soon reached a high standard. In contrast the prediction of unsteady airloads remained a key problem, due to the mathematical complexity. Measurements of unsteady motion-induced aerodynamics did not exist before 1938, and therefore the flutter test, either in a windtunnel or in flight, provided the only -indirect - validation of unsteady aerodynamic models.

Birnbaum's theory was later expanded by Kuessner and Schwarz in Germany to include also control surface oscillations. Derivations of other solutions for the same problem were performed in Italy by Possio's colleague Cicala and in the US by Theodorsen, adopting mathematical singularity models or conformal mapping. As their results agreed, the development of unsteady incompressible 2D flow around airfoils was completed about 1930. The efforts that followed focussed on 3D and compressible flow.

When flight speeds increased, new problems were encountered, especially when sonic speeds were approached. A significant amount of flight stability was lost, control surfaces lost their efficiency or even reversed their effect and wings and rudders started heavy vibrations. The only chance to understand these phenomena was to investigate the influence of the compressibility on unsteady flows. The special complexity of compressible flow was already known at the beginning of the last century. If flow velocity v was no longer much smaller than the speed of sound a, the speed of disturbance propagation in the fluid began to play an essential role, and the governing velocity potential equation for unsteady flow became a wave equation, see fig. 3

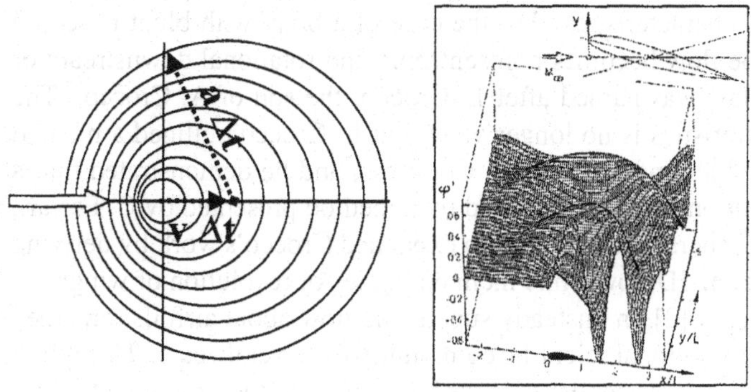

Figure 3. Disturbance wavefronts in a compressible flow (l) and spatial waviness of the velocity potential value (r)

The priciples of compressible flow, including supersonic flows and shock waves, had already been elaborated around 1900 as gasdynamic theories. The only practical application of this knowledge was in the design and analysis of steam turbines and in the research of ballistic projectiles. Propeller blade tips had already reached these velocities and after 1930 several airplanes reached velocities higher than 600 km/h, among them the German Messerschmidt 109 and the first jet airplane He178VI and the Italian waterplane MC72. For practical aircraft applications Prandtl and Glauert derived that lift and pressure on wings in compressible flow depend on Mach number by the factor $\beta = \sqrt{1 - Ma^2}$. Mach number $Ma = v/a$ denotes the ratio between flight velocity and isentropic speed of sound. A major milestone on the path of compressible flow research was the 5th Volta Congress of the Royal Italian Academy of Sciences in 1935 in Rome. This conference with the topic of "High Velocities in Aviation" had been initiated by Arturo Crocco, a general and aeronautical engineer. The most famous aeronautical scientists of that time were invited to present state of the art in high-speed flow research. Several topics adressed at this conference soon influenced the work of young scientists like Camillo Possio in the following years.

3. First publications until 1937

In [1] and [2] Camillo Possio investigated supersonic flow problems. The choice of this topic was certainly influenced by the Volta Congress. In supersonic flow about bodies of revolution, like ballistic projectiles, often a curved shock wave occurs, which abruptly decelerates the flow from supersonic to subsonic and produces a strong drag on the body. Behind non-curved shocks the flow around a prescribed sharp nosed geometry could already be calculated by the method of characteristics. For the case of a body with blunt nose and a curved shock the flow becomes nonisentropic and rotational downstream of the shock. This law was named after L. Crocco, the son of A. Crocco. The method of characteristics is no longer valid, but L. Crocco outlined a method to overcome this dilemma at the Volta Congress, and he demonstrated that a mathematical solution could be obtained by a method presented by C. Ferrari, one of Possio's teachers. Camillo Possio expanded Crocco's work by deriving a concept for general 3D flows, like inclined bodies of revolution or wings.

Possio's key paper [2] on unsteady supersonic flow about airfoils remained unknown in the international scientific community for several years. This might be due to the fact that it was published in the "Pontificia Accademia Scientiarium Acta", and that its summary was written in Latin. Possio started his theory by deriving the velocity potential of an acoustic source of pulsating strength that moves rectilinear with supersonic velocity. This method was later also adopted by Kuessner for his General Lifting Surface Theory. Possio derived the induced velocity component normal to the airfoil which was represented by a flat plate. He combined the effects of two different sources, with an infinitesimal small distance to the upper and lower side of the thin plate, and thus obtained the effect of a doublet singularity with pulsating strength. The whole unsteady 2D supersonic flow was then modeled by a doublet distribution along the chord of the plate, and the doublet strength distribution was calculated from the normal velocity which in turn was forced by the airfoil oscillation, the so-called downwash w.

4. Possio's integral equation

In 1938 24-year old Camillo Possio outlined the very first theory for unsteady compressible flow around 2D oscillating airfoils [3]. This was a complicated problem, because for the unsteady compressible wave equation the use of vortices as elementary solutions was not possible and the Prandtl-Glauert rule did not hold for unsteady flow. The wave equation for the velocity potential is derived from mass- and momentum conservation, and by linearly superimposing steady and unsteady flow velocity components $\vec{v} = u_\infty + \nabla\varphi \rightarrow \phi = \varphi + u_\infty x$ the following form is obtained :

$$(1 - Ma^2)\frac{\partial^2\varphi}{\partial x^2} + \frac{\partial^2\varphi}{\partial y^2} - 2\frac{Ma^2}{v}\frac{\partial^2\varphi}{\partial x\partial t} - \frac{Ma^2}{v^2}\frac{\partial^2\varphi}{\partial t^2} = 0 \tag{3}$$

As for isentropic flow the density is a function of pressure only, the acceleration vector has a potential ψ too, which is a function of pressure only and thus has no discontinuity in the flow except at compression shocks and at the lifting surface, but not in the wake. Prandtl had presented the concept of the acceleration potential at the Volta Congress:

$$\rho\frac{Dv}{Dt} = -\nabla p \rightarrow \frac{Dv}{Dt} = -\nabla\int\frac{dp}{\rho} = \nabla\psi \tag{4}$$

Adopting a Galilei-Transformation $x = \bar{x} - vt, y = \bar{y}, t = \bar{t}$ Possio derived a corresponding wave equation and fundamental solution $\psi_D(x, y)$ in form of a doublet singularity of strength $A(\xi, \eta)$.

$$\frac{\partial^2\psi}{\partial\bar{x}^2} + \frac{\partial^2\psi}{\partial\bar{y}^2} - \frac{1}{a^2}\frac{\partial^2\psi}{\partial\bar{t}^2} = 0 \tag{5}$$

$$\psi_D = Av^2\frac{4i}{\beta}e^{i(\omega t + \frac{\omega^* Ma(x-\xi)}{\beta^2 l/2})}\frac{\partial}{\partial\eta}H_0^{(2)}\left(\frac{\omega^* Ma}{\beta^2}\frac{\sqrt{(x-\xi)^2 + \beta^2(y-\eta)^2}}{l/2}\right) \tag{6}$$

The reduced frequncy is defined by $\omega^* = \omega b/v$ and $b = l/2$ is half chord. If the oscillating flat plate is modeled by doublet distribution of varying strength along the chord, the total acceleration potential is computed by an integral with a singularity at $(x, y) = (\xi, \eta)$. Evaluation of Cauchy's mean value, yields a relation between doublet strength and pressure jump across the plate, and the doublet strength can be calculated from the boundary condition, which prescribes the disturbance velocity w in y-direction induced by the airfoil oscillation. Performing differentiation and the limiting process $y \rightarrow 0$ and $\eta \rightarrow 0$

finally yields Possio's equation

$$\bar{w} = w(x,t)e^{-i\omega t} = \frac{\omega}{\rho v^2} \int_{-l/2}^{+l/2} \delta p(\xi) K[Ma, \omega^* \frac{(x-\xi)}{l/2}] \tag{7}$$

$$\begin{aligned} K &= \frac{1}{4\beta}\{e^{iMa\Xi}[iMa\frac{|x-\xi|}{x-\xi}H_1^2(\Xi) - H_0^2(\Xi)]\} \\ &+ \frac{i\omega^*}{4\beta}e^{-iMa\Xi}\int_{-\infty}^{x-\xi} du H_0^2(\Xi)e^{i\frac{\omega^* u}{b\beta^2}} \end{aligned} \tag{8}$$

The kernel K depends on Mach number and reduced frequency and contains cylindrical Hankel functions. As Possio did not find an analytical expression for the kernel, he numerically calculated tables of K for Ma = 0, 0.25, 0.50, 0.75 and for a limited range of values $\Xi = \omega^*|x-\xi|Ma/(\beta^2 b)$. Using these values he solved the integral equation by a series expansion for the unknown pressure jump δp according to Birnbaum and Ackermann

$$\delta p(\theta) = A_0 \cot\frac{\theta}{2} + \sum_{n=1}^{\infty} A_n \sin n\theta \quad \cos\theta = \xi/l \tag{9}$$

Possio applied only 3 terms of this series and correspondingly solved his integral equations by fulfilling it for w at 3 chorwise positions. This approximation was sufficient for rigid heave and pitch motions, and the unsteady components of lift and moment coefficients read

$$c_L = \frac{L}{\rho l v^2} = \frac{\pi}{2}[A_0 + \frac{1}{2}A_1] \quad c_M = \frac{M}{\rho l^2 v^2} = \frac{\pi}{16}[A_1 + \frac{1}{2}A_2] \tag{10}$$

These results are valid for sinusoidal oscillations and depend linearly on the oscillation amplitude and on its time derivative around the mean position. Possio presented his results in a form that was common in Italy and the UK, see fig. 4. The paper was translated in Germany in 1939, and the diagrams were transferred to the German standard definition of coefficients, which differed from Possio's. This work supported engineers in Germany with unsteady compressible airloads in a form directly applicable for flutter computations. For more complex oscillation modes, like trailing edge flaps, which encounter a discontinuous behaviour at the hinge axis, more terms would have to be calculated. In 1942 Dietze suggested an iterative scheme that is based on the development of the kernel function in terms of its simpler incompressible counterpart, which had been presented by Possio in his paper too.

The outstanding character of Possio's key paper was mentioned soon by Kuessner [18] in 1940 : "The oscillating wing for $0 \leq Ma \leq 1$ has been investigated

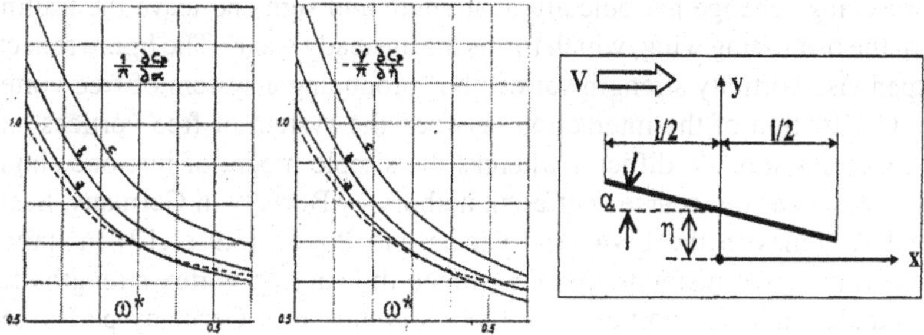

Figure 4. Results from Possio's original paper

up to now in just a single investigation for the two dimensional problem" and in 1981 by Garrick and Reed [23]: "There followed shortly afterward two short outstanding contributions by Camillo Possio in Italy. In 1938 he applied the acceleration potential to the two-dimensional nonstationary problem and arrived at an integral equation (Possio equation)".

5. Further unsteady aerodynamic contributions

With increasing flight speed v, the aerodynamic loads acting on an aircraft structure increase like v^2. Thus high elastic bending moments appeared for wings with large aspect ratio (ratio of wing span to chord). The need to investigate wings with low ratios was also driven by the requirements of more agile military aircraft and by the role of tail planes in flutter. As a result the effects of the wing tips became important and unsteady 3D theories had to be developed. In 3D flow the bound vortex line on the wing, together with the free vortices behind the tip and the starting vortex far downstream form a closed vortex line. For 3D unsteady flows around an oscillating wing the 2D unsteady model of Birnbaum and the 3D steady model of Prandtl were refined by Sears, see fig. 5

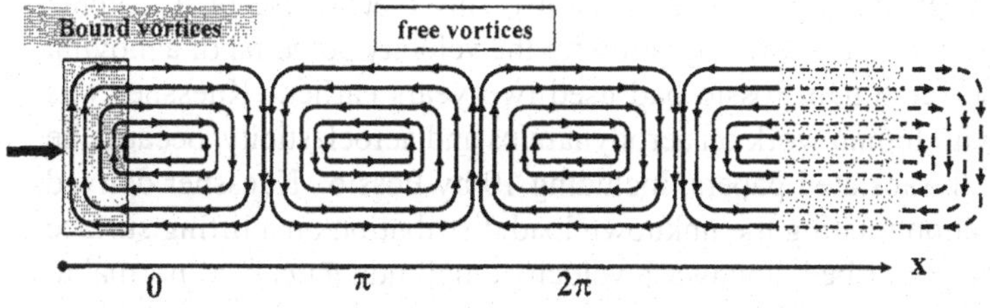

Figure 5. Lines of constant vorticity in the wake of a 3D oscillating wing

Vortex rings change periodically in strength and sign and leave the trailing edge of the oscillating wing, which forms the unsteady wake. The figure reflects that spanwise vorticity strength varies, thus producing a system of free vortex lines. Calculation of the interaction between the system of free vortices and bound vortices was too difficult whereas the simple model of just one single bound vortex was too coarse. Different authors - v.Borbely in Germany, Sears in the UK, Jones in the USA, and Cicala and Possio [8,9,11,12] in Italy - developed different methods to approximate the varying vortex strengths for both bounded and free vortices. In 1943 Kattenbach in Germany performed a comparison of the different theories for a 3D wing of elliptical planform in heavy oscillations with different reduced frequencies in incompressible flow. Remarkably he spent much more effort in reviewing Possio's theory in his paper than the other ones. After the war unsteady incompressible 3D problems were solved as a special case of Kuessners General Lifting Surface Theory, as soon as computers became available.

Several high-speed wind tunnels, several of them in Germany and in Guidonia near Rome, as well as both flutter and unsteady aerodynamic tests in high speed flow were planned in the 1930s, but tests were not realized before the end of the war. One of the problems with test results is that the model is tested in an airstream, which is bounded either by the free atmosphere or by the walls of the test section. Camillo Possio was among the first ones to investigate the influence of tunnel walls on oscillating models [15]. He first demonstated, that tunnel wall disturbances decreased with frequency like $1/\omega$. Then he calculated the effects of the above mentioned closed walls and free jet conditions. He found that esp. the imaginary part of unsteady lift coefficient was significantly changed by the wall effects. In [6], [7] and [8] Possio investigated the problem of general unsteady motions in 2D and 3D flows, namely non-harmonic impulsively started motions.

6. Influence on the development of 3D theories

In Germany H.G Kuesnner in 1940 published his General Lifting Surface Theory, which was the first German paper dealing with unsteady compresssible flow. The strategy he chose for derivation of his theory, use of a Lorentz Transformation to derive a wave equation for the velocity potential of a moving source, was the same that Possio had used two years earlier. Kuessner was very familiar with Italian work on aerodynamics and aeroelasticity, because he reviewed several translated papers, including 10 by Possio. Kuessner derived an integral equation relating the unknown load distribution on a lifting surface $\delta p(\xi, \eta)$ and the prescribed downwash velocity amplitude $\bar{w}(x, y, z)$ normal to the surface, by means of a new Kernel function K.

$$\bar{w} = \int\int d\xi d\eta \frac{\delta p}{4\pi\rho v} e^{i\omega\frac{\xi - x}{v}} \frac{\partial^2}{\partial z^2} \int_{-\infty}^{x-\xi} d\lambda \frac{e^{i\omega(\lambda - Ma\sqrt{\lambda^2+\beta^2(y-\eta)^2+\beta^2 z^2})}}{\sqrt{\lambda^2 + \beta^2(y-\eta)^2 + \beta^2 z^2}} \tag{11}$$

The theory is valid for lifting thin surfaces of an arbitrary planform in inviscid and irrotational flow, but not in transonic flow. Kuessner demonstrated that all other known theories of that time were special cases of this theory, but the Kernel was left in the form of a highly singular integral, whose solution could only be found for special cases like Possio's 2D theory. Since no analytical solution was found before the war, Possio's theory of 2D strips along the span of wings and tail planes was applied for the flutter analysis in high-speed flight.

7. Work in non-aerodynamic fields

The topic of lateral firing from an airplane was probably chosen in 1939 due to Camillo Possio's military service. The spinning axis and path of a projectile fired from board of an aircraft in a direction different from the aircraft's flight direction are not parallel. This yields an effective aerodynamic incidence angle, and therefore aerodynamic side forces. Possio calculated the airloads and the projectile motion and showed that due to the spinning forces the axis of the projectile soon turns to the direction of the path [10]. In 1941 Possio focused on hydrodynamic problems. He investigated the influence of the free water surface on moving underwater airfoils [13], [14]. He calculated forces on a 2D hydrofoil, the motion of which is governed by the Laplace equation for the velocity potential. The effect of the free water surface is assumed to be a linear perturbation. Possio used the constraints, that disturbance velocity vanishes at infinity and that pressure on the water surface has a constant value. Modeling both the hydrofoil and the disturbance velocity by a vortex distributions Possio derived the result that lift and drag on the hydrofoil depend on the Froude number $Fr = v/\sqrt{gl}$ and on the depth in which the hydrofoil of chord length l moves with velocity v, g denotes the gravitational constant.

8. Extension of Possio's method after the war

Before the war 2D compressible unsteady flow based on Possio's integral equation was the best method available to model unsteady airloads in flutter calculations. 3D unsteady methods existed only for incompressible flow, and it was not before the 1960s that 3D unsteady compressible flow could be calculated by solving Kuessners equation. The Doublet Lattice Method (DLM)[19] became the standard method for the years that followed. Similar to Possio's 2D theory, the flow is modeled by a doublet distribution on the lifting surfaces. The unknown doublet strength is discretized on trapezoidal panels. This powerful method yields large linear systems of equations and fully populated matrices

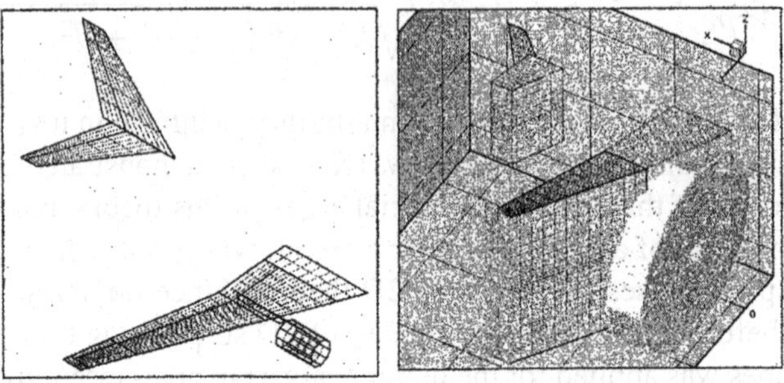

Figure 6. Discretization of an aircraft in DLM (l) and in TDLM (r)

A transonic extension, the so-called TDLM (Transonic Doublet Lattice Method) has been developed later on [20]. In this method the unsteady transonic flow is modeled as a perturbation of a mean transonic steady flow and is governed by a nonhomogeneous wave equation for the acceleration potential. In contrast to purely subsonic flow additional source singularities have to be distributed in the flow field, the strength of which depends on the velocity potential of the transonic steady flow. Thus the DLM model of doublet panels is extended by volume elements of constant source strength, see fig. 6.

$$\frac{\partial^2\psi}{\partial x^2}+\frac{\partial^2\psi}{\partial y^2}+\frac{\partial^2\psi}{\partial z^2}+\lambda^2\psi=(\frac{\partial}{\partial x}+i\frac{\omega^*}{\beta^2})\,\sigma[\frac{(\gamma+1)Ma^2}{\beta^2}\frac{\partial\phi^0}{\partial x}] \tag{12}$$

Flutter became a problem in turbomachines when attempts to increase efficiency and to reduce noise and emissions made the size of aircraft engines and thus the blade size grow. Additional geometrical parameters as well as blade-to-blade interactions play a significant role. A method developed in 1973 Carstens [21] extended Possio's integral equation to the case of a 2D cascade blade row, which is a model of a surface cut through a fan at constant radius. The kernel of the integral equation depends not only on Mach number and reduced frequency but also on stagger angle λ, interblade distance τ and interblade phase angle θ. In [21] efficient numerical methods are developed both for this kernel and for the original Possio kernel. The computational results show the interesting effect of blade resonance. For constant values of the Mach number and the reduced frequency the lift coefficient of a reference blade varies strongly when the interblade phase angle is changed, and even drops to zero for specific resonance values. This effect is typical for unsteady compressible flows, and appears when disturbance waves propagating in a compressible fluid are reflected between boundaries within a time period fitting the oscillation period of the boundary motion. This effect was also observed in wind tunnel tests on oscillating models. Fromme and Golberg [22] reformulated Possio's kernel for a 2D airfoil in order to include the boundary conditioins of wind tunnel walls and were able to compute the effects of wind tunnel wall resonance.

These examples are demonstrating, that fast analytical methods are still of high value, if the influence of several parameters in unsteady aerodynamics has to be scanned. The performance of similar studies with modern CFD methods requires a tremendous effort.

9. Conclusion

Comparing today's capabilities of unsteady aerodynamic computation with those of 60 years ago, much progress has been made due to the CFD development and the enormous growth of computer power. The first unsteady nonlinear transonic flow simulations with complex shock motions were obtained around 1975. They were based on nonlinear potential equations, and were soon followed by solutions of the inviscid Euler- and the viscous Reynolds Averaged Navier-Stokes (RANS) equations. All of these methods adopt finite volume or finite difference methods to model the conservation laws of mass momentum and energy together with gas equations of state and different turbulence models. Today they allow for computations of 3D flows around complete aircraft configurations with strong shock waves and flow separation in the whole speed range. Flutter computations are beginning to be performed by directly integrating the equations of motion of the structure in time, together with computing the interaction between structural deformations and fluid dynamics. Such flutter simulations usually take many hours of CPU, while calculations of the flutter boundary as an Eigenvalue problem with Possio's aerodynamics needs just seconds or just minutes if TDLM aerodynamics is chosen.

We can conclude that Possio's legacy is manifold. First his mathematical models for unsteady aerodynamics are still valuable and his papers are still being cited. Secondly he paved the way for Kuessner's theory and for later 3D unsteady compressible methods like DLM and TDLM in addition to sophisticated analytical methods in turbomachinery. Third he has shown, how fruitful analytical methods can be for understanding of physical mechanisms and for providing engineers with a feeling for the importance and magnitude of different parameters, both aspects of which are sometimes neglected today.

References

[1] C. Possio. Sul moto razionale dei gas. *Atti Accad. Naz. Lincei, Rend.* VI. S. 25, 455-461, 1937

[2] C. Possio. L'azione aerodinamica sul profile oscillante alle velocita ultrasonore. *Pontificia Accademia Scientiarium Acta* 1, 93-106, 1937

[3] C. Possio. L'azione aerodinamica sul profilo oscillante in un fluido compressibile a velocita iposonora. *L'Aerotecnica* 18, 441-459, 1938

[4] C. Possio. L'azione aerodinamica su una superficie portante in moto oscillatorio. *Atti Accad. Naz. Lincei, Rend.* VI. 28, 194-200, 1938

[5] C. Possio. Determinazione dell'azione aerodinamica corrispondente alle piccolo oscillazioni del velivolo. *L'Aerotecnica* 18, 1323-1351, 1938

[6] C. Possio. Sul moto non stazionario di una superficie portante.*Atti Accad. Sci. Torino* 74, 285-299, 1939

[7] C. Possio. Sol moto non stazionario di un fluido compressibile *Atti Accad. naz. Lincei* Rend. VI, 29, 481-487, 1939

[8] C. Possio. L'azione aerodinamica su di una superficie portante in moto vario *Atti Accad. Sci. Torino* 74, 537-557, 1939

[9] C. Possio. Sulla determinazione dei coefficienti aerodinamici che interessano la stabilita del velivolo *Comm. Pontif. Acad. Sci.).* 3, 141-169, 1939

[10] C. Possio. Sullo sparo di fianco da bordo di un aereo. *Atti Accad. Sci. Torino* 74, 1939

[11] C. Possio. Sul problema del moto discontinuo di un ala. Nota 1. *L' Aerotecnica* 20, 655-681, 1940

[12] C. Possio. Sul problema del moto discontinuo di un ala. Nota 2. *L' Aerotecnica* 21, 205-230, 1941

[13] C. Possio. Sulla teoria del moto stazionario di un fluido pesante con superficie libera. *Ann. Mat. pura appl.* IV, 20, 313-329, 1941

[14] C. Possio. Campo di velocita creato da un vortice in un fluido pesante a superficie libera in moto uniforme. *Atti Accad. Sci. Torino* 76, 365-388, 1941

[15] C. Possio. L'interferenza della galleria aerodinamica nel caso di moto non stazionario. *L' Aerotecnica* 1940-XVIII. 1940

[16] C. Possio. The influence of the viscosity and thermal conductibility on sound propagation. *Atti Accad. Scienze Torino* 78. 274-292. 1943

[17] W. Birnbaum. Das ebene Problem des schlagenden Flügels. *ZAMM* 4. 277-292. 1924

[18] H.G. Kuessner. Allgemeine Tragflächentheorie. *LuFo* 17. 370-378. 1940

[19] E. Albano, W. Rodden. A doublet lattice method for calculating lift distributions on oscillating surfaces in subsonic flow. *AIAA Journal* 1969-7. 279-285. 1969

[20] S. Lu, R. Voss. TDLM - a Transonic Doublet Lattice Method for 3D potential unsteady transonic flow calculation and its application to transonic flutter prediction. *Proc. IFASD 1993* 77-95. AAAF 1993

[21] V. Carstens. Berechnung der instationaeren Druckverteilung an harmonisch schwingenden Gittern in ebener Unterschallstroemung. *DFVLR-IB* 73-J06 and 75-J02. 1973 and 1975

[22] J.A. Fromme, M.A. Golberg. Aerodynamic interference effects on oscillating airfoils with controls in ventilated windtunnels. *AIAA Journal* 18. 417-426. 1980

[23] I.E. Garrick, W.H. Reed III. Historical development of flutter. *Journal of Aircraft* 18. 897-912. 1981

THE POSSIO INTEGRAL EQUATION OF AEROELASTICITY: A MODERN VIEW

A.V. Balakrishnan[1]
[1] *University of California, Department of Electrical Engineering, Los Angeles , CA, USA*
*bal@ee.ucla.edu**

Abstract A central problem of AeroElasticity is the determination of the speed of the aircraft corresponding to the onset of an endemic instability known as wing 'flutter'. Currently all the effort is completely computational:wedding Lagrangian NASTRAN codes to the CFD codes to produce 'Time Marching' solutions. While they have the ability to handle non-linear complex geometry structures as well as viscous flow,they are based approximation of the p.d.e. by o.d.e., and restricted to specified numerical parameters.This limits generality of results and provides little insight into phenomena. And of course are inadequate for Control Design for stabilization. Retaining the continuum models,we can show that the basic problem is a Boundary Value/Control problem for a pair of coupled partial differential equations,and the composite problem can be cast as a nonlinear Convolution/Evolution equation in a Hilbert Space. The Flutter speed can then be characterized as Hopf Bifurcation point, and determined completely by the linearised equations. Solving the linearised equations is equivalent to solving a singular integral equation discovered by Possio in 1938 for oscillatory response.In this paper we examine the Equation and its generalizations from the modern mathematical control theory viewpoint.

keywords: Possio Equation, AeroElasticity, Instability, Wing Flutter

1. Introduction

The genesis of the Possio Equation and its role in the Aeroelasticity theory of the 1950's has been amply documented in [1]. This paper presents the current outlook on this equation, including generalizations, from the vantage point of recently developed control theory for partial differential equations [2].

A central problem of AeroElasticity is the stability of the wing structure in air flow. Much of the interest is in subsonic compressible flow. This can be formulated (see[3]) in the Time Domain as a nonlinear convolution/evolution equation in a Hilbert Space,and the instability ('Flutter') speed as a Hopf Bi-

*Research supported in part under NSF Grant No. ECS-040073.

Please use the following format when citing this chapter:
Author(s) [insert Last name, First-name initial(s)], 2006, in IFIP International Federation for Information Processing, Volume 199, System Modeling and Optimization, eds. Ceragioli F., Dontchev A., Furuta H., Marti K., Pandolfi L., (Boston: Springer), pp. [insert page numbers].

furcation point which by the Hopf theory is completely determined by the linearised equations about the undisturbed flow. The linearised equations are of the Neumann boundary type, and hence can be cast equivalently as an Integral Equation – this is the Possio Equation,with a singular kernel.

We may also place it in the context of the currently fashionable numerical computation schemes-indeed almost all the work in AeroElasticity today is computational. In essence the partial differential equations are approximated by ordinary differential equations – both Structural Dynamics and AeroDynamics,and the most subjective part-often mysterious even-is the wedding of the Lagrangian Structure Dynamics to the Eulerian AeroDynamics. This is exactly where the Possio Equation would come in,if the full continuum models are retained.

We begin in section 2 with the Wing Structure model,where we need to calculate the aerodynamic loading In section 3 we consider the AeroDynamic flow model-the Euler Full Potential Equation with aeroelastic boundary conditions for attached flow, and the Kutta-Joukowsky conditions. The linearization of the equations is in section 4. Finally in section 5 we study the role of the Possio Equation.

2. Structure Model

The simplest model – a uniform rectangular beam, endowed with two degrees of freedom, plunge and pitch – goes back to Goland [4] in 1945 (too late for Possio!). Let the projection of the flow velocity be along the positive $X-$ axis,with x denoting the chord variable, $-b \leq x \leq b$. Similarly with y denoting the span or length variable, along the $Y-$ axis, $0 \leq y \leq l$, let

$$X(y,t) = \text{Column}\,(h(y,t), \theta(y,t)),$$

where $h(\cdot)$ is the plunge or bending along Z-axis; and $\theta(\cdot)$ is the pitch or torsion angle about the elastic axis located at $x = ab$. Then the structure dynamics equation is:

$$M_S\,\ddot{X}(.,t) + KX(.,t) = \text{Column}\,(L(.,t), M(.,t)),$$

where M_S is the Mass/Inertia matrix and K is the stiffness differential operator:

$$\text{Diagonal}\,(EI\frac{\partial^4}{\partial y^4},\ -GJ\frac{\partial^2}{\partial y^2}),$$

$L(\cdot)$ is the aerodynamic lift and $M(\cdot)$ the moment about the elastic axis, with boundary conditions: a) Cantilever

$$h(0,t) = h'(0,t) = 0 = \theta(0,t) = \theta'(l,t) = h'''(l,t) = 0 = h''(l,t).$$

b) Free-Free

$$\theta'(0,t) = \theta'(l,t) = h'''(0,t) = h'''(l,t) = h''(0,t) = h''(l,t) = 0.$$

See [5,6] for a Hilbert space formulation. The functions $L(\cdot)$ and $M(\cdot)$ have to be determined from the Aerodynamic model.

3. AeroDynamic Model

The aerodynamic lift and moment (per unit length) are given by:

$$L(y,t) = \int_{-b}^{b} \delta p \, dx$$

$$M(y,t) = \int_{-b}^{b} (x-a)\delta p \, dx$$

$$\delta p = p(x,y,0+,t) - p(x,y,0-,t), 0 < y < l; |x| < b$$

where $p(x,y,z,t)$ is the aerodynamic pressure,which along with the velocity vector $q(x,y,z,t)$, the density $\varrho(x,y,z,t)$ are the basic aerodynamic variables of interest. Under some simplifying assumptions (see[8]), we can show that the velocity is curl-free and is then characterized by the velocity potential $\phi(x,y,z,t)$ which satisfies the (Euler) Full Potential Equation:

$$\frac{\partial^2 \phi}{\partial t^2} + \frac{\partial(\nabla\phi.\nabla\phi)}{\partial t} = a_\infty^2 \nabla.\nabla\phi(1 + \frac{(\gamma-1)}{a_\infty^2}(\frac{(q_\infty.q_\infty)}{2} - \frac{\partial\phi}{\partial t}$$
$$-\frac{(\nabla\phi).(\nabla\phi)}{2})) - \nabla\phi.\nabla(\frac{\nabla\phi.\nabla\phi}{2})$$

where q_∞ is the undisturbed far-field velocity, a_∞ is the far field speed of sound, ϱ_∞ is the far field density.

$$M(\text{Mach Number}) = (\frac{|q_\infty|}{a_\infty}) \leq 1.$$

The pressure is given by

$$p = \varrho_\infty^2 \frac{a_\infty^2}{\gamma}\left(1 + (\frac{(\gamma-1)}{a_\infty^2})(\frac{(q_\infty.q_\infty)}{2} - \frac{\partial\phi}{\partial t} - \frac{(\nabla\phi.\nabla\phi)}{2})\right)^{\gamma/(\gamma-1)}$$

It is assumed that the far field potential is given by

$$\phi_\infty = q_1\, x + q_2\, y + q_3\, z$$

where

$$q_i = |q_\infty| \cos\alpha_i$$

AeroElastic Boundary Conditions.

The aeroelastic boundary conditions are:

a) Attached Flow

$$\frac{\partial \phi}{\partial z}|_{z=0} = \frac{\partial \phi_\infty}{\partial z}|_{z=0} + w_a(x,y,t)$$

where $w_a(\cdot)$ is the normal velocity of the structure, and is given by:

$$\begin{aligned} w_a(x,y,t) &= -\dot{h}(y,t) - (x - a\,b)\,\dot{\theta}(y,t)) - \frac{\partial \phi(x,y,0,t)}{\partial x}\,\theta(y,t) \\ &- \frac{\partial \phi(x,y,0,t)}{\partial y}\,(h'(y,t) + (x-a)\theta'(y,t)). \end{aligned}$$

b) Kutta-Joukowsky Condition:

$\delta p = 0$, off the structure and at the trailing edge (goes to 0, as $x \to b_-$),

where δp is the pressure jump :

$$\delta p = p(x,y,0+,t) - p(x,y,0-,t)$$

We do not have an existence theorem for this problem as yet!

4. Linearization

Because of the lack of existence theorem and other reasons it is customary to simplify the Full Potential Equation to the Transonic Small Disturbance Potential (TSD) equation which is quasilinear and yet retains sufficient non-linearity to yield shocks – see [8]. Here however we go straight to the linearzation. Thus defining

$$\varphi = \phi - \phi_\infty$$

we have:

$$\begin{aligned} \frac{\partial^2 \varphi}{\partial t^2} + 2U(q_1\frac{\partial^2 \varphi}{\partial x \partial t} + q_2\frac{\partial^2 \varphi}{\partial y \partial t} + q_3\frac{\partial^2 \varphi}{\partial z \partial t}) &= a_\infty^2((1 - M^2 q_1^2)\frac{\partial^2 \varphi}{\partial x^2} \\ +(1 - M^2 q_2^2)\frac{\partial^2 \varphi}{\partial y^2} + (1 - M^2 q_3^2)\frac{\partial^2 \varphi}{\partial z^2}), \end{aligned} \tag{1}$$

where now

$$U = |(q_\infty)|; \quad q_i = U \cos \alpha_i; \quad M = \frac{U}{a_\infty}$$

The boundary conditions are

$$\frac{\partial \varphi}{\partial z} = w_a(x,y,t), \quad 0 < y < l; \; |x| < b, \tag{2}$$

where

$$\begin{aligned} w_a(x,y,t) &= -\dot{h}(y,t) - (x - a\,b)\,\dot{\theta}(y,t)) - q_1\theta(y,t) \\ &\quad -q_2\,(h'(y,t) + (x - a\,b)\,\theta'(y,t)). \end{aligned}$$

With ψ denoting the linearised acceleration potential

$$\psi = \frac{\partial \varphi}{\partial t} + q_1 \frac{\partial \varphi}{\partial x} + q_2 \frac{\partial \varphi}{\partial y} + q_3 \frac{\partial \varphi}{\partial z}$$

the Kutta-Jukowski conditions become:

$$\delta\,\psi = \psi|_{z=0+} - \psi|_{0-} = 0, \qquad \text{off the structure,} \tag{3}$$

$$\delta\,\psi \to 0 \text{ as } x \to b-, \quad 0 < y < l. \tag{4}$$

These are the 3-D linear subsonic Compressible flow conditions with the aeroelastic boundary conditions – see [8] for more details.

5. The Possio Integral Equation

Let us begin with a statement of the Possio Integral Equation — actually this is a generalization of the original equation bearing his name which was 2-D, zero angle of attack, Fourier Transform (sinusoidal response) version. We state it for the 3-D case, in terms of the Laplace Transform variable λ, Re $\lambda > \sigma \geq 0$, because the integrals defining the equation will be convergent (which is not the case for $\lambda = i\omega$, as in the original formulation). Let

$$\begin{aligned} \widehat{w}_a(x,y,\lambda) &= \int_0^\infty e^{-\lambda t}\, w_a(x,y,t)\,dt \\ A(x,y,t) &= -\frac{2}{U}\,\delta\,\psi \\ \widehat{A}(x,y,\lambda) &= \int_0^\infty e^{-\lambda t}\, A(x,y,t)\,dt \end{aligned}$$

To reduce complexity, we shall take

$$q_1 = 1 \text{ (zero angle of attack)}$$

see [8] for the case $0 < q_1 < 1$. Then the equation is (see [9])

$$\widehat{w}_a(x,y,\lambda) = \int_0^l \int_{-b}^b \widehat{P}(x-\zeta, y-\nu, k)\,\widehat{A}(\zeta,\nu,\lambda)\,d\zeta\,d\nu \tag{5}$$

where $|x| \leq b$, $0 \leq y \leq l$ and

$$k = \frac{\lambda}{U}$$

and the spatial Fourier Transform of the kernel $\widehat{P}(\cdot,\cdot,k)$ is

$$\begin{aligned}\widehat{\widehat{P}}(i\omega_1, i\omega_2, k) &= \int_{-\infty}^{\infty}\int_{-\infty}^{\infty} e^{-i(\omega_1 x + \omega_2 y)}\,\widehat{P}(x,y,k)\,dx\,dy, \\ &= \frac{\sqrt{D(i\omega_1, i\omega_2, k)}}{2(k+i\omega_1)}, \quad -\infty < \omega_1, \omega_2 < \infty, \qquad (6)\end{aligned}$$

where

$$D(i\omega_1, i\omega_2, k) = M^2k^2 + 2kM^2\, i\omega_1 + (1-M^2)\omega_1^2 + \omega_2^2. \qquad (7)$$

We prefer the succinct form of the spatial Fourier Transform in contrast to the $\widehat{P}(\cdot,\cdot,k)$ which is too long to specify see [10,13]. It has a singularity at the origin so that we have a singular integral equation [9]. Assume that (5) has a solution. Then the solution of the linearised potential equation (1) specialized for $q_1 = 1$,q_2,q_3 both zero , is given by:

$$\begin{aligned}\widehat{\widehat{\varphi}}(i\omega_1, i\omega_2, z, \lambda) &= \frac{\widehat{\widehat{A}}(i\omega_1, i\omega_2, \lambda)}{k+i\omega_1}\, e^{-z\sqrt{D(i\omega_1, i\omega_2, k)}}, \quad z>0, \\ &= -\widehat{\widehat{\varphi}}(i\omega_1, i\omega_2, -z, \lambda), \quad \text{for } z<0\end{aligned}$$

where $-\infty < \omega_1, \omega_2 < \infty$ and

$$\widehat{\widehat{\varphi}}(i\omega_1, i\omega_2, z, \lambda) = \int_{-\infty}^{\infty}\int_{-\infty}^{\infty} \widehat{\varphi}(x,y,z,\lambda)\, e^{-(i\omega_1 x + i\omega_2 y)}\,dx\,dy,$$

$$\widehat{\varphi}(x,y,z,\lambda) = \int_{-\infty}^{\infty} e^{-\lambda t}\,\varphi(x,y,z,t)\,dt,$$

$$\widehat{\widehat{A}}(i\omega_1, i\omega_2, \lambda) = \int_0^l \int_{-b}^{b} e^{-(i\omega_1 x + i\omega_2 y)}\,\widehat{A}(x,y,\lambda)\,dx\,dy.$$

This is essentially a formula due to Kussner, an early German pioneer (see [1]). We note that the existence of solution to (5) is still an open problem, despite early work on the problem [11].
par To obtain the original $2D$ version of Possio we need to specialize to the 'airfoil' case – or, 'high-aspect-ratio' wings where

$$\frac{l}{b} \approx \infty$$

so that we may neglect the dependence on the y-coordinate . With q_1 equal to unity, this becomes

$$\frac{\partial^2\varphi}{\partial t^2} + 2U\frac{\partial^2\varphi}{\partial x\partial t} = a_\infty^2((1-M^2)\frac{\partial^2\varphi}{\partial x^2} + \frac{\partial^2\varphi}{\partial z^2}). \qquad (8)$$

And correspondingly (5) becomes:

$$\widehat{w}(x,y,\lambda) = \int_{-b}^{b} \widehat{P}(x-s,k)\widehat{A}(s,\lambda)\,ds, |x| \le b \tag{9}$$

where setting ω_2 in (5) to be zero, we have, for $\omega \in (-\infty, \infty)$,

$$\widehat{\widehat{P}}(i\omega,k) = \int_{-\infty}^{\infty} \widehat{P(x,k)}e^{-i\omega x}dx = \frac{\sqrt{k^2M^2 + 2kM^2 i\omega + (1-M^2)\omega^2}}{2(k+i\omega)}, \tag{10}$$

where we have discarded the subscript 1.Ê In this case it becomes actually a Mikhlin multiplier – see [12].

Second we need to consider the case of 'oscillatory' response-Fourier Transform in the time-domain; formally putting $i\omega$Ê for λ everywhere. In this case, the corresponding kernel function becomes rather involved and too long to specify [13]; further, the integrals in the kernel function also require special interpretation as in [10].

The importance of the Possio equation is that it links directly the structure velocity-the "input' in the problem to the 'output' Ê– the pressure jump which is all that is needed in the aeroelastic problem. We do NOT need to solve for the potential everywhere. On the other hand the potential can be determined from the pressure jump – this is the formula of Kussner (8). Thus solving the Possio equation is equivalent to solving the boundary value problem for the potential. It is true that this holds only for the linearised equations-we don't have yet a 'non-linear' Possio Integral equation. But if stability – or Flutter speed – is the prime concern,then all we need is the solution to the Possio equation! Given this,it is surprising there is hardly a mention of this equation in recent Texts [15]. Indeed, a systematic use of the Possio equation would have reduced the size of the classic text [13]. Finally we note at present the existence/uniqueness of solutions to the Possio Equation is known only for the 'air-foil' case and even at that only for $M = 0$ and $M = 1$, (see [14]). See [7] for some approximations. Otherwise the problem is open.

6. References

[1] R. Voss. The Legacy of Camillo Possio to Unsteady AeroDynamics. Proceedings. This Conference.

[2] I. Lasiecka and R. Triggiani: Control Theory for Partial Differential Equations: Continuous and approximation theories. Cambridge University Press 2000.

[3] A.V. Balakrishnan. NonLinear AeroElasticity: Continuum Theory : Flutter/Divergence Speed: Plate Wing Model. Journal of AeroSpace Engineering,

2005, To appear.

[4] M. Goland: The flutter of a uniform cantilever wing. Journal of Applied Mechanics. Transactions ASME 12 : 197-208,1945.

[5] A.V. Balakrishnan. Subsonic Flutter Suppression using self-straining actuators. Journal of the Franklin Institute 338: 149-170,2001.

[6] A.V. Balakrishnan , M.A. Shubov: Asymptotic and spectral properties of operator valued functions generated by aircraft wing model. Math. Meth. Appl.Sci. 2004, 27:329-362.

[7] A.V. Balakrishnan,æ K.W. Iliff: A Continuum AeroElastic Model for Inviscid Subsonic Bending-Torsion Wing Flutter. In Proceedings of the International Forum on AeroElasticity and Structural Dynamics, Amsterdam, June 4-6, 2003, Amsterdam.

[8] A.V. Balakrishnan. On the Transonic Small Disturbance Potential Equation. AIAA Journal 42: 1081-1088, 2003.

[9] A.V. Balakrishnan. On the NonNumeric Mathematical Foundations of Linear AeroElasticity. 4th Int. Conf. on NonLinear Problems in Aviation & Aerospace, European Conference Publications, U.K., 2003.

[10] C.E. Watkins, H.L. Runyon, D.S. Woolston. . The kernel function of the Integral Equation relating the lift to the downwash distribution in oscillating finite wings in subsonic flow. NACA TN 1234, 1955.

[11] E. Reissner: On the theory of oscillating air foils of finite span in subsonic compressible flow. NACA TR 1002,1950.

[12] S.G. Mikhlin: MultiDimensional singular integrals and integral equations. Pergamon 1965.

[13] R.L. Bisplinghof, H. Ashley and R.L Halfman: AeroElasticty, Addison-Wesley, 1955.

[14] A.V. Balakrishnan: Possio Integral of AeroElasticity Theory. Journal of AeroSpace Engineering. 16: 139 -154.

[15] E. Dowell et al. A Modern Course in AeroElasticity. Kluwer 2004.

BEYOND POSSIO EQUATION: THE LEGACY OF CAMILLO POSSIO TO FLIGHT DYNAMICS AND HYDRODYNAMICS

G. Avanzini[1] and A. De Santis,[2]
[1] *Politecnico di Torino, Dipartimento di Ingegneria Aeronautica e Spaziale, Turin, Italy, giulio.avanzini@polito.it* [2] *Università degli Studi di Roma "La Sapienza", Dipartimento di Informatica e Sistemistica, desantis@dis.uniroma1.it*

Abstract A review of the papers written by Camillo Possio on Flight Dynamics and Hydrodynamics is presented. The scope of the note is to underline how the versatile young researcher succeeded in delivering interesting contributions to the engineering sciences that go beyond the renown equation that bears his name.

keywords: aircraft stability derivatives, free surface, marine propeller

1. Introduction

The works of Camillo Possio on unsteady aerodynamics [1]–[7] have deeply influenced the studies in the field of aeroelasticity up to the present days, as demonstrated by some recent works that exploited his derivations [8]–[10]. For this reason the related papers are considered as his most important contributions to the engineering disciplines. As a matter of fact his scientific production was extremely prolific and diverse in spite of his untimely death under the last bombing that hit his native city, Turin, at the end of the Second World War, on April 5th, 1945.

Beyond his seminal contribution to unsteady aerodynamics, Possio's works ranged from the analysis of fluid motion [11, 12] to studies on physical properties of fluids [13], from flight mechanics [14]–[16] and experimental fluid dynamics [17] to free surface effects on the flow field generated by distributions of singularities [18]–[20]. These works deserve consideration not only because they reveal the versatility of his mind in applying a rigorous mathematical approach for describing physical phenomena, always preserving a deep practical understanding of the underlying physical system, but also because they are undoubtedly as interesting as the most renown ones. The present note will focus on five papers that represent the legacy of Possio to the fields of flight dynamics ([14] and [15]) and hydrodynamics ([18]–[20]), discussing

Please use the following format when citing this chapter:
Author(s) [insert Last name, First-name initial(s)], 2006, in IFIP International Federation for Information Processing, Volume 199, System Modeling and Optimization, eds. Ceragioli F., Dontchev A., Furuta H., Marti K., Pandolfi L., (Boston: Springer), pp. [insert page numbers].

their significance in the framework of the current studies of the time on the two subjects.

In Refs. [14] and [15], both written in 1938, Possio exploited his experience in managing basic models of oscillating lifting surfaces in order to derive some analytical results aimed at evaluating aircraft stability derivatives, that is the expressions of the perturbation terms of aerodynamic forces and moments that are required in the linearization of rigid aircraft equations of motion. The last part of his scientific activity was focused on the analysis of free–surface effects. In particular, between 1941 and 1943 he wrote a sequence of three papers that were devoted to the analysis of steady motion of heavy fluids under the influence of a distribution of singularities [18], the interaction of a vortex in steady motion with a free–surface [19], and to the estimate of marine propeller efficiency as a function of its depth [20], pushing further the analysis of the problem that had been given approximate solutions by other scientists of the time.

Without neglecting the originality and potential importance of the works that will not be discussed in the present note, the legacy of Camillo Possio to flight dynamics and hydrodynamics represents the major contribution (beyond Possio equation) that the young scientist could develop into a complete research, obtaining results that represented an advance in engineering knowledge at the time the papers were written, while preserving also nowadays a significant interest for the aerospace community.

2. Possio and aircraft stability derivatives

On January 1938 Camillo Possio published on the *Commentationes Pontificia Academia Scientiarum* (Proceedings of the Pontifical Academy of Sciencies) a work titled “On the Determination of Aerodynamic Coefficients for Aircraft Stability Analysis” [14]. The importance of this work can be easily understood when one considers that the formulation of rigid aircraft equations of motion had already reached full maturity for a long time in 1938, the seminal book by Bryan [21] dating back to 1911, but the evaluation of the so called dynamic stability derivatives was still an open problem.

The approach proposed by Bryan for the analysis of aircraft stability is based on the derivation of the equations of motion from first principles, the determination of aircraft equilibria (trimmed flight) and the linearization of the equations of motion in the neighborhood of the considered trim condition. In Bryan's theory, aerodynamic forces and moments are expanded in linear form as a function of aircraft state and control variables, under some reasonable simplifying assumptions.

After writing the linear model, the main problem for a meaningful aircraft stability analysis is to provide reasonable estimates of aircraft stability derivatives, that is, the partial derivatives of aerodynamic forces and moments with respect to state and control variables, divided by a mass parameter (the mass for force equations or the relevant moment of inertia for each one of the three attitude equations). Stability derivatives may change significantly, depending on the considered trim condition and aircraft configuration. The determination of the so called static stability derivatives (that is, stability derivatives made with respect to aerodynamic angles or velocity components) can be performed with different degrees of accuracy, but a reasonable estimate is not too a difficult task. On the converse, the evaluation of rotary and dynamic derivatives, that is stability derivatives with respect to angular velocity components and time derivatives of aerodynamic angles, is never trivial. In his earliest work on aircraft stability (that was written with W.E. Williams in 1903 before Wright brothers' first powered flight, and published on January 1904 only three weeks after the *Flyer* took–off!), Bryan simply neglected the terms depending on the rate of change of incidence and sideslip angles [22]. Flight experience and more accurate mathematical derivations rapidly demonstrated that unsteady downwash effects on the horizontal tail were sizable and cannot be neglected.

Unfortunately, dynamic derivatives escape also direct experimental evaluation, not only because of scale and wind–tunnel effects (the latter analyzed by Possio in [17]) but also because forced oscillation experiments cannot determine separately the contributions of angular velocity and rate of variation of aerodynamic angles to force and moment components.

As for evaluation of dynamic stability derivatives from theoretical aerodynamic results, Possio himself underlined how the effects of small amplitude oscillations on lift distribution cannot be described by a two–dimensional representation, that, together with other limitations, would make it impossible to evaluate stability derivatives with respect to roll angular velocity, nor by a simple strip theory, even under the assumption of a low frequency parameter $\omega = \Omega b/(2V_0)$, Ω being the frequency of the oscillation, b the wing span and V_0 the trim velocity [14]. Exploiting his analysis of the behavior of oscillating lifting surfaces, Possio derived a rigorous description of the vorticity and pressure distributions over a lifting surface in a harmonic oscillation, under the assumptions of small incidence and oscillation amplitude and large wing aspect ratio λ.

His derivation for a wing like that shown in Fig. 1 was based on Prandtl's acceleration potential approach [23]. The total derivative of the velocity $\mathbf{V}$ can be expressed as

$$\frac{d\mathbf{V}}{dt} = \mathrm{grad}\varphi \tag{1}$$

where the acceleration potential φ satisfies Laplace equation $\nabla^2\varphi = 0$. Writing Eq. (1) in linear, nondimensional form (that is assuming small perturbation of a uniform flow $\mathbf{V}_0$ of unity modulus), the vertical component v of the velocity

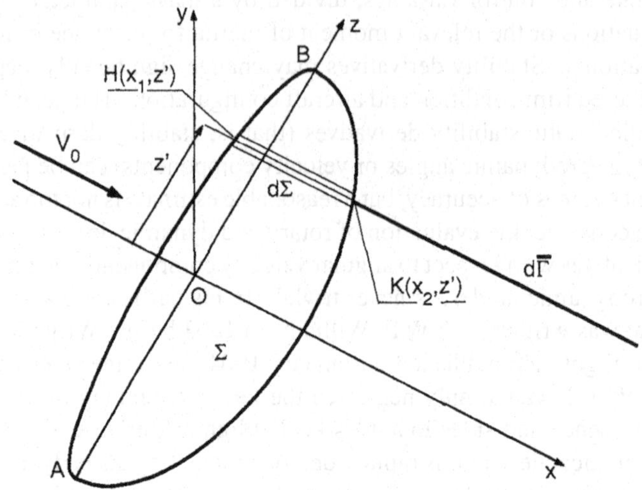

Figure 1. Elliptic wing in a uniform current [14]

field must satisfy the equation

$$\frac{\partial \varphi}{\partial y} = \frac{\partial v}{\partial x} + \frac{\partial v}{\partial t} \tag{2}$$

Taking into account that $v \to 0$ for $x \to -\infty$, it is possible to integrate Eq. (2), so that

$$v(x,y,z,t) = \int_{-\infty}^{x} \frac{\partial \varphi}{\partial y}(\xi, y, z, t + \xi - x)d\xi$$

The restriction of v over the surface Σ of the oscillating wing must satisfy the flow tangency condition on Σ, that is $v(x,z) = \partial\eta/\partial t - \alpha$, where η is the vertical displacement of the point $(x,z) \in \Sigma$ and α the wing incidence. For a harmonic oscillation all the terms can be expressed in complex form as $\varphi = \bar{\varphi}e^{i\omega t}$ and $v = \bar{v}e^{i\omega t}$, where i is the imaginary unit and $\bar{\varphi}$ and $\bar{v}$ are complex functions of x, y, and z. The restriction of $\bar{v}$ over Σ can thus be written as

$$\bar{v}(x,z) = \int_{-\infty}^{x} \frac{\partial \bar{\varphi}}{\partial y}(\xi, z)e^{i\omega(\xi - x)}d\xi \tag{3}$$

Noting that for incompressible fluids no time derivatives of φ are present in the Laplace equation governing acceleration potential, it is possible to state that $\bar{\varphi}$ is the acceleration potential of a steady flow field C_p with asymptotic velocity $\mathbf{V}_0$ around a lifting surface with the same shape of Σ. Indicating the horizontal and vertical velocity components of the steady flow field C_p with $1 + u'$ and w, respectively, and applying Bernouilli's theorem for small perturbations $u', w \ll 1$, it is $\bar{\varphi} = u'$. It is also $\partial\bar{\varphi}/\partial y = \partial w/\partial x$, because C_p is irrotational. By substituting w in Eq. (3) and integrating by parts, one gets

$$\bar{v}(x,z) = w(x,z) - i\omega \int_{-\infty}^{x} w(\xi,z)e^{i\omega(\xi-x)}d\xi \qquad (4)$$

The flow field C_p can be generated by a (steady) complex vorticity distribution $\bar{\gamma}(x,z)$ over Σ, that Possio determined following Prandtl theory for steady motion of finite wings, expressing $\bar{\gamma}$ in terms of Glaurt's trigonometric series. The details of the procedure are here omitted for the sake of brevity, but can be found on Possio's paper [14]. It must be observed that, as explicitly underlined by Possio himself, the flow field C_p and its vertical component w bear no physical meaning. In particular, although $w(x,z)$ can be decomposed into the sum of two contributions w_1 and w_2 related to the circulation distribution in C_p on the wing and downstream of it, respectively, the two corresponding terms $\bar{v}_1$ and $\bar{v}_2$ obtained by substituting w_1 and w_2 in Eq. (4) are not related to the unsteady distribution of vorticity over and downstream of the wing in the actual unsteady flow field.

In the final part of the paper Possio derived some simplified relations for the particular case of an elliptic wing. He also demonstrated that the approximation of his prediction for the unsteady case has the same accuracy of Prandtl's model for the steady case and developed a simple example for an isolated wing which undergoes a vertical harmonic oscillation. He anticipated that a complete set of results on aerodynamic forces and moments generated by roll, pitch, yaw and heave oscillations of a complete aircraft configuration would have been discussed in a subsequent note. On December 1938 this note was published by the italian journal *L'Aerotecnica* under the title "Determination of Aerodynamic Actions Due to Small Aircraft Oscillations" [15]. Figure 2 shows some of Possio's original plots, namely Figs. 3 through 9 taken from Ref. [15].

It should be noted that Possio presented his results adopting the italian notation in use at his time, where C_p represents the lift coefficient, from the word *portanza*, and C_{mr} is the roll moment coefficient, where mr stands for *momento di rollio*. In the paper, C_r indicates the drag coefficient (*resistenza*), C_d the side–force coefficient (*deriva*) while C_{mb} and C_{mi} indicate the pitch and yaw moments (*beccheggio* and *imbardata*), respectively.

In the last paragraph of the paper Possio carried out a numerical example for evaluating the differences with respect to what he called "the usual approximate methods" that neglect part of the unsteady aerodynamic effects, rigorously taken into account by his approach. While the terms in phase with the variation of the angle of attack where shown to be predicted well even by the approximate methods, the terms in quadrature with α were demonstrated to be significantly different, with variations that ranged from 28% for the pitch moment stability

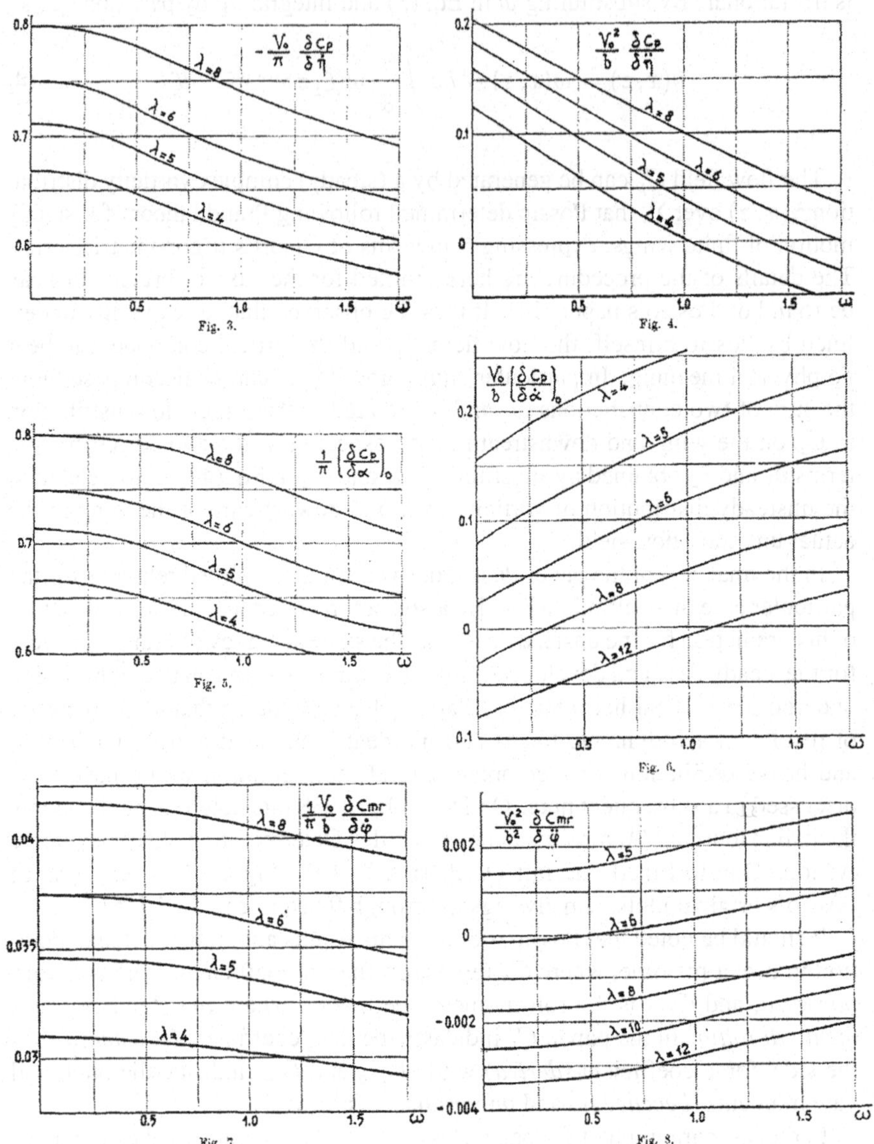

Figure 2. Lift coefficient C_p in plunge (Figs. 3 and 4) and pitch motion (Figs. 5 and 6), and roll coefficient C_{mr} for roll oscillations (Figs. 7 and 8) from Ref. [15]

derivative with respect to $\dot{\alpha}$ up to 200% for the lift derivative with respect to α [15].

3. Possio's analysis of free-surface effects

In 1941 Possio published his first work on free surface effects on the *Annali di Matematica Pura e Applicata* (Annals of Pure and Applied Mathematics) [18], analyzing the steady velocity field created by a distribution of singularities on the free surface of a heavy fluid.

Lamb and Havelock demonstrated that under the usual hypotheses of perfect fluid and infinitesimal perturbation an infinite set of solutions is analytically plausible for this problem [24],[25]. Indicating with V and V_0 the local and the asymptotic velocity, respectively, Rayleigh proposed a method to define a well-posed problem by assuming that an elementary viscous–like force per unit mass in the form $-\mu(V - V_0)$ acts on the fluid, where μ is a positive parameter. The flow field for a perfect fluid was thus obtained in the limit as $\mu \to 0$.

Possio provided an alternative formulation based on a more rigorous model. While retaining the assumption of perfect fluid, he considered the stationary motion as the limit of an unsteady one, started with the fluid at rest when a perturbation created by the singularities is suddenly applied at $t = 0$. The general solution can be expressed in terms of velocity potential as $\Phi(x, y, z, t) = \Phi_i(x, y, z) + \Phi_a(x, y, z, t)$, where Φ_i is the potential determined by the singularities distribution and its value is suddenly assumed at $t = 0$, while Φ_a is an additive potential satisfying the Laplace equation $\nabla^2\Phi_a = 0$, with continuous first derivatives in the half–space $z \geq 0$ and vanishing at infinity except in the downstream direction. The formulation of the unsteady problem is well posed and it admits a unique solution. If the perturbation is created by a distribution of sources or doublets, the solution is given by

$$\Phi_i(x, y, z) = \int_{-\pi}^{\pi} d\theta \int_0^{\infty} F(m, \theta) e^{mkz + imk\omega}\, dm \qquad (5)$$

$$\Phi_a(x, y, z, t) = \int_{-\pi}^{\pi} d\theta \int_0^{\infty} f(m, \theta, t) e^{mkz + imk\omega}\, dm \qquad (6)$$

with $k = g/V_0^2$ and $\omega = x\cos(\theta) + y\sin(\theta)$. $F(m, \theta)$ is a continuous bounded function, with continuous bounded first derivatives in the set $m \geq 0$ and $-\pi \leq \theta \leq \pi$, such that $F \to 0$ as e^{-mkl} if $m \to \infty$. Finally, f is an arbitrary function such that the integral is defined in $z \geq 0$ and differentiation under the integral sign is also defined.

Possio demonstrated that the stationary additional potential Φ_a^* obtained as $\lim_{t\to\infty} \Phi_a(x, y, z, t)$ coincides with the expression obtained by Rayleigh, and satisfies the limit condition imposed on the potential and its first derivatives at infinity. He also extended his approach to a more general situation where the perturbation is given by a pressure discontinuity, so as to include the flow field generated by a propeller or an airfoil.

This investigation was naturally developed into the study of the potential that describe the velocity field generated by a vortex with circuitation Γ in uniform motion at a depth h from a free surface [19]. In this case the potential can be expressed as

$$\Phi(x,y) = V_0 x + \frac{\Gamma}{2\pi}\tan^{-1}\left(\frac{y-h}{x}\right) + \Phi_a(x,y) \tag{7}$$

where Φ_a is the additional potential required to satisfy the boundary condition at the free surface, that is, pressure p in the fluid at the surface is equal to the external pressure p_a in the space over the surface. As usual, the potential function Φ_a must satisfy the Laplace equation $\nabla^2\Phi_a = 0$ in the half-space occupied by the fluid, the first derivatives vanishing upstream of the vortex, so that a well-posed problem is obtained.

Possio found the expression of Φ_a and investigated its numerical computation. The theory was then applied to the problem of the uniform motion of an airfoil of chord ℓ for small values of the ratio $\varepsilon = \ell/h$. The resulting lift and drag coefficients were expressed as

$$c_L = \frac{\pi\alpha}{1 + \pi\beta e^{-2\beta}\varepsilon}, \qquad c_D = \beta e^{-2\beta}\varepsilon c_L^2 \tag{8}$$

with $\beta = gh/V_0^2$. The aerodynamic coefficients were similar to those obtained by Prandtl for the small ε case, expressed in terms of a virtual aspect ratio $\lambda = 2e^{2\beta}/\pi\beta\varepsilon$. Possio determined λ for arbitrary values of ε, although his analysis was limited to the flat plate. In this case the virtual aspect ratio is given by

$$\lambda = \frac{2F^2 e^{2/(\varepsilon F^2)}}{\pi\left\{J_0^2\left[1/(2F^2)\right] + J_1^2\left[1/(2F^2)\right]\right\}} \tag{9}$$

where $J_0(\ )$ and $J_1(\ )$ are Bessel functions and $F = V_0/\sqrt{g\ell}$ is the Froud number. The diagram of Fig. 3 represent the behaviour of $1/\lambda$ as function of F for different values of ε.

In Ref. [20] Possio investigated free surface effects on propeller efficiency. This problem had been already addressed by Dickmann [26], under the assumption that the field external to the wake created by the propeller can be represented by a distribution of sinks over the propeller disc, thus exploiting the results of Havelock on wave resistance generated by a given singularities distribution [25].

Possio provided a more accurate computation of wave resistance and extended the investigation to the case of more than one propeller. Letting h be the depth of the propeller disc center, measured along the z axis oriented downward, and $\varepsilon = r/h$ the ratio between the disc radius and the depth, Possio

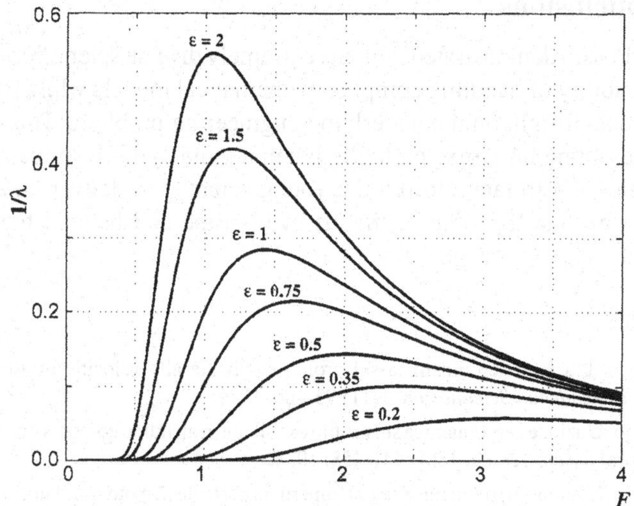

Figure 3. Reciprocal of virtual aspect ratio as a function of Froud number.

represented the wave resistance generated by a uniform distribution of sinks on the propeller disc σ as

$$R_0 = 8\pi\rho\left(\frac{\pi m r^2}{h}\right)^2 f, \qquad f = 2\sum_0^\infty a_n \varepsilon^{2n} I_n(\beta) \tag{10}$$

where $\beta = gh/V_0^2$ and

$$I_n(x) = x^{2n+2}e^{-x}\int_0^\infty e^{-x\cosh(2u)}\cosh^{2n+2}(u)\,du \tag{11}$$

Dickmann results refers to the case of a sink distribution concentrated in the disc center with $\varepsilon = 0$, that is just the first term in the series expansion of Eq. 10. On the converse, Possio computed f for different values of the ratio ε and of the Froud number.

He also evaluated propeller efficiency loss due to wave motion induced on the free surface. Denoting with u_m^* the average velocity increment at the propeller, with P and T the propeller power consumption and thrust, respectively, the actual propeller efficiency is $\eta_e = TV_0/P$, while the efficiency in an infinite fluid is $\bar{\eta} = T(V_0 + u_m^*)/P$. By taking into account that $u_m^*/V_0 \ll 1$, it is possible to write

$$\frac{\bar{\eta} - \eta_e}{\bar{\eta}} = \varepsilon^2 f \frac{u_d}{V_0} \tag{12}$$

where u_d denotes the velocity increment at the propeller disc.

4. Conclusions

Camillo Possio demonstrated with his extremely diverse scientific activity an unusual capability of handling complex mathematical models while preserving a deep physical insight on the underlying engineering problem. This note was aimed at presenting his works on flight dynamics and hydrodynamics subjects, the only fields where fate allowed the young scientist to deliver an articulate contribution beyond the state of the art of his time and beyond the seminal equation that bears his name.

References

[1] C. Possio. L'azione aerodinamica sul profilo oscillante alle velocità ultrasonore. *Acta Pontificia Academia Scientiarum*. 1(11):93-106, 1937.

[2] C. Possio. L'azione aerodinamica sul profilo oscillante in un fluido compressibile a velocità iposonora. *L'Aerotecnica*. 18(4):441-458, 1938.

[3] C. Possio. L'azione aerodinamica su una superficie portante in moto oscillatorio. *Atti Reale Acc. Lincei*. 26:194-200, 1938.

[4] C. Possio. Sul moto non stazionario di una superficie portante. *Atti Reale Acc. delle Scienze - Parte Fisica*. 74:285-299, 1939.

[5] C. Possio. L'azione aerodinamica su di una superficie portante in moto vario. *Atti Reale Acc. delle Scienze - Parte Fisica*. 74:537-557, 1939.

[6] C. Possio. Sul problema del moto discontinuo di un'ala - Nota 1a (Moto Piano). *L'Aerotecnica*. XX(9):655-681, 1940.

[7] C. Possio. Sul problema del moto discontinuo di un'ala - Nota 2a (Ala finita). *L'Aerotecnica*. XXI(3):205-230, 1941.

[8] G. Chiocchia, S. Prößdorf, D. Tordella. The lifting line equation for a curved wing in oscillatory motion. ZAMM, 77(4):295–315, 1997.

[9] J. Lin, K.W. Iliff. Aerodynamic Lift and Moment Calculations Using a Closed Form Solution of the Possio Equation. Nasa/TM-2000-209019, 2000.

[10] A.V. Balakrishnan. Possio Integral Equation of Aeroelasticity Theory. *J. Aerosp. Engrg*. 16(4):139-154, 2003.

[11] C. Possio. Sul moto rotazionale dei gas. *Atti Reale Acc. Lincei*. 25:455-461, 1937.

[12] C. Possio. Sul moto non stazionario di un fluido compressibile. *Rendiconti Reale Acc. Lincei*:481-487, 1939.

[13] C. Possio. L'influenza della viscosità e della conducibilità termica sulla propagazione del suono. *Atti Reale Acc. delle Scienze - Parte Fisica*. 77:274-293, 1942.

[14] C. Possio. Sulla determinazione dei coefficienti aerodinamici che interessano la stabilità del velivolo. *Commentationes Pontificia Academia Scientiarum*. III(6):141-169, 1938.

[15] C. Possio. Determinazione dell'azione aerodinamica corrispondente alle piccole oscillazioni del velivolo. *L'Aerotecnica*. XVIII(12):1323-1351, 1938.

[16] C. Possio. Sullo sparo di fianco da bordo di un aereo. *Atti Reale Acc. delle Scienze - Parte Fisica*. 74:276-285, 1939.

[17] C. Possio. L'interferenza della galleria aerodinamica nel caso di moto non stazionario. *VI Convegno Nazionale di Aerotecnica*,306-320, Rome, June 1940.

[18] C. Possio. Sulla teoria del moto stazionario di un fluido pesante con superficie libera. *Annali di Mat. Pura e Appl.*. 20:313-329, 1941.

[19] C. Possio. Campo di velocità creato da un vortice in un fluido pesante a superficie libera in moto uniforme. *Atti Reale Acc. Delle Scienze - Parte Fisica*. 76:365-388, 1941.

[20] C. Possio. L'influenza della superficie libera sul funzionamento dell'elica marina. *Sistemi di Propulsione e Navigazione aerea e marittima*, 334-352, Torino, Acc. Delle Scienze, 1943.

[21] G.H. Bryan. *Stability in Aviation*. MacMillan, London, 1911.

[22] G.H. Bryan, W.E. Williams. The Longitudinal Stability of Aerial Gliders. *Proc. Royal Soc. London* Ser. A, 73(489):100–116, 1904.

[23] L. Prandtl. Beitrag zur Theorie der tragenden Flache. *ZAMM*. 16(6):360-361, 1936.

[24] H. Lamb. *Hydrodynamics*, 6th ed. Cambridge University Press, Cambridge, 1932. Chapt. IX.

[25] T.H. Havelock. Wave resistance. *Proc. Royal Soc. London*. Ser. A, 118:24–33, 1928.

[26] H.E. Dickmann. Schiffskörpersog, Wellenwiderstand des Propellers und Wechselwirkung mit Schiffswellen. *Ing.-Arch.* 9, 1938.

ONE HUNDRED YEARS SINCE THE INTRODUCTION OF THE SET DISTANCE BY DIMITRIE POMPEIU

T. Birsan[1] and D. Tiba[2]
[1]*"Gh. Asachi" University, Department of Mathematics, Iasi, Romania, tbi@math.tuiasi.ro*
[2]*Institute of Mathematics, Romanian Academy, Bucharest, Romania, dan.tiba@imar.ro*

Abstract This paper recalls the work of D. Pompeiu who introduced the notion of set distance in his thesis published one century ago. The notion was further studied by F. Hausdorff, C. Kuratowski who acknowledged in their books the contribution of Pompeiu and it is frequently called the Hausdorff distance.

keywords: Hausdorff distance, Hausdorff-Pompeiu distance, Pompeiu functions, Pompeiu conjecture, Schiffer conjecture

1. The thesis of Pompeiu

On March 31-st, 1905, Dimitrie Pompeiu (1873-1954), a distinguished Romanian mathematician, defended at the Faculté de Sciences de Paris his Ph.D. work *Sur la continuité des fonctions de variables complexes* . The comission was chaired by H. Poincaré and included G. Koenigs and E. Goursat and the work was published the same year in [14]. P. Montel, a wellknown French mathematician, has appreciated the remarkable contribution of Pompeiu's thesis by the words: *Pour un coup d'essai, c'est un coup de maître*, [1]. Ideas, notions and results discussed in the thesis of Pompeiu have attracted the interest of important contemporary mathematicians: P. Painlevé, A. Denjoy, F. Hausdorff and is now a part of the universal mathematical heritage. Pompeiu studied a problem formulated by Painlevé [13] already in 1897, concerning the singularities of uniform analytic functions. At that time, it was generally admitted that a uniform analytic function cannot be continuously extended on the set of its singularities. However Pompeiu could construct examples of such functions which are continuous on the set of their singularities and this set has positive measure. The controversy that followed was solved in 1909 by Denjoy who confirmed that the arguments of Pompeiu are correct. This was a turning point in the theory of uniform analytic functions and Pompeiu was considered as the best specialist in this field, at the beginning of the last century [1]. Pom-

Please use the following format when citing this chapter:
Author(s) [insert Last name, First-name initial(s)], 2006, in IFIP International Federation for Information Processing, Volume 199, System Modeling and Optimization, eds. Ceragioli F., Dontchev A., Furuta H., Marti K., Pandolfi L., (Boston: Springer), pp. [insert page numbers].

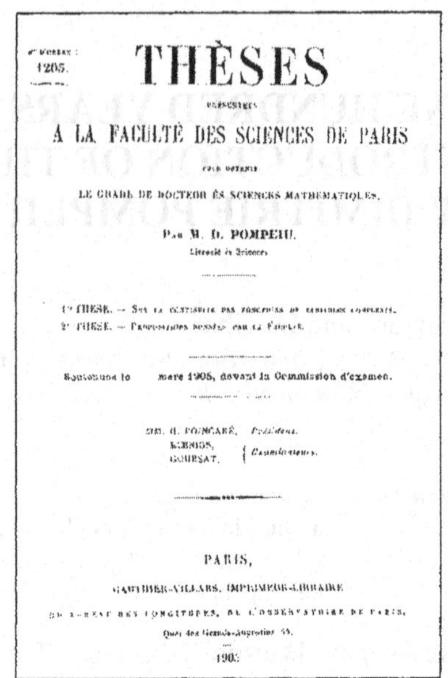

1205.

THÈSES

À LA FACULTÉ DES SCIENCES DE PARIS

LE GRADE DE DOCTEUR ÈS SCIENCES MATHÉMATIQUES,

Par M. D. POMPEIU.

Soutenues le mars 1905, devant la Commission d'examen.

MM. H. POINCARÉ, Président.
KOENIGS,
GOURSAT, Examinateurs.

PARIS,

GAUTHIER-VILLARS, IMPRIMEUR-LIBRAIRE

1905

peiu used an example due to A. Koepcke (pag.45 in [14]) which was rather complicated. Some years later, in [16], he constructed much simpler examples of the same type, that is real functions with bounded derivatives that have roots in any interval and are not identically zero in some interval. Such functions are now called *Pompeiu functions* and are included in many monographs on real functions.

Very well documented presentations of this moment from the history of mathematics may be found in [8], [9]. In his arguments in the complex plane, Pompeiu needed a notion of distance between closed curves. It is to be remarked that quite in the same period, M. Fréchet introduced in his Ph.D. thesis (published a little bit later in [3]) the distance between two elements and the axioms of metric spaces. This is the context in which D. Pompeiu defines in 1905 the notions of *écart* and *écart mutuel* between two sets ([14], p.17 and 18). Preserving his notations, let E_h and E_k be two compact subsets in the plane. The *écart* of E_h with respect to E_k, denoted by Δ_{hk} was defined as the maximum of the distances of an arbitrary point $P_h \in E_h$ to the set E_k. The sum $\Delta_{hk} + \Delta_{kh}$ was called the reciprocal distance (*écart mutuel*) between the sets E_h and E_k. This allows Pompeiu to see the compact subsets in the plane as the elements of another set and to define in a natural way limits, closure, etc. for this "set of sets". Consequently, Pompeiu is also considered as one of the founders of the theory of *hyperspaces*, see McAllister [10]. The thesis of D. Pompeiu is the birth certificate of the notion of distance between sets and its importance was immediately remarked by the contemporary mathematicians. Already in 1905, in his report for the Jahrbuch fuer Mathematik, on p.455 A. Gutzmer calls the Pompeiu distance *Einfuehrung eines neuen Begriffes der Mengenlehre* (the introduction of a new concept in set theory). In 1914,in his famous book [5], Hausdorff studies the notion of set distance, in the natural setting of metric spaces and with a small modification (the sum si replaced by the maximum) :

$$d(A, B) = \max\left\{\sup_{x\in A} d(x, B), \sup_{y\in B} d(y, A)\right\}.$$

He devotes a consistent paragraph to the study of the properties of the distance and he quotes correctly Pompeiu [5] on p. 463, as the author of this notion. Moreover he also establishes that the distances defined with the "sum" or with the "max" give the same toplogy, that is they are equivalent. The same considerations may be found in his book from 1927, [6], where Pompeiu is quoted at page 280. It is to be noted that at that time, the main text of a book did not include any reference to the origin of a subject and all these references were concentrated in a short appendix at the end (*Quellenangabe*). Probably, due to this reason, many mathematicians didn't notice the reference to Pompeiu's work [14] and the set distance is frequently termed as the *Hausdorff distance*. Other names used in the scientific literature are *Pompeiu distance* or *Pompeiu-Hausdorff distance, Hausdorff-Pompeiu distance*. The great Polish mathematician C. Kuratowski, in his treatise [7], on p. 106 mentions both the thesis of Pompeiu from 1905 and the book of Hausdorff from 1914. This is done in a footnote, on the page where the notion is introduced and is completely accessible to the reader. In 1978, McAllister [10] published a remarkable historical study devoted to the first 50 years in the theory of hyperspaces. The role of Pompeiu's thesis is underlined clearly: *"who [Pompeiu] may with some justice be said to have invented hyperspaces, and Hausdorff's use of them in 1914 in his treatise* Grundzuege der Mengenlehre *had made them very well known"* (p. 310) and similarly on p. 311 *"I have found no evidence of the Hausdorff metric itself before Pompeiu's thesis"*. There are many notions of set distance which are not equivalent with the Hausdorff-Pompeiu distance. For instance, one can define the distance between two Lebesgue measurable subsets in an Euclidean space as the measure of their symmetric difference. However, the Hausdorff-Pompeiu metric seems to be the only one with a remarkable compactness property: if (A_n) is a sequence of compacts, bounded with respect to n, then there is a compact subset A and a subsequence again indexed by n, such that :

$$\lim_{n\to\infty} d(A_n, A) = 0.$$

This property makes the Hausdorff-Pompeiu distance a fundamental notion in the study of the topologies on families of subsets and in the modern theory of shape optimization (*optimal design*). The interested reader may consult the recent monograph [11] and its references for a survey of the mathematical literature on geometric optimization.

2. The Pompeiu conjecture

The whole mathematical work of Pompeiu is characterized by a profound originality, by the introduction of many fruitful ideas and methods. We briefly recall the definition in 1912 of the *areolar derivative* (see [15]), further developed by M. Nicolescu, Gh. Calugareanu, N. Cioranescu, N. Teodorescu, Gr. Moisil and other. This is in fact the fundamental $\bar{\partial} - operator$ from complex analysis. In this context, Pompeiu also proved the $Cauchy-Pompeiu formula$ which appeared here for the first time, [9]. Probably the best known paper of Pompeiu is his Note [17] from 1929 (recent estimates indicate almost one thousand articles quoting it). Here one can find the famous *Pompeiu conjecture*, still unsolved completely :

Let f be a continuous function in the plane and D a compact subset such that

$$\int_{\sigma(D)} f(x,y)\,dx\,dy = 0$$

where $\sigma(D)$ denotes any compact subset in the plane obtained by rigid motions of D. Then, is it true that f is null in the plane ?

For the case when D is a disk, there are counterexamples [4] of the form $f(x,y) = sin(ax + by)$ with a, b appropriately chosen real numbers. For any other domains in the plane, the answer seems to be positive although just some special cases are solved [19]. We underline the fundamental character in the mathematical analysis of this property (if proved) and its relationship with the Schiffer conjecture (formulated later) concerning the eigenvalues of the Laplace operator with Cauchy conditions on the boundary, [2], [18].

O. Onicescu, a student of Pompeiu and a wellknown Romanian mathematician, said that the creations of Dimitrie Pompeiu are "*simple, plastic, global and full of significance in the world of science*", [12].

References

[1] G. Andonie. *Istoria Matematicii in Romania.* Ed. Ştiinţifică, Bucharest, 1965.

[2] C.A. Bernstein. An inverse spectral theorem and its relation to the Pompeiu problem. *J. d'Analyse Math.* 37:128-144, 1980.

[3] M. Fréchet. Sur quelques points du calcul fonctionnel (Thèse). *Rend. Circ. Mat. Palermo.* 22:1-74, 1906.

[4] N. Garofalo, F. Segala. Univalent functions and the Pompeiu problem. *Trans. A.M.S.* 346:137-146, 1994.

[5] F. Hausdorff. *Grundzuege der Mengenlehre.* Viet, Leipzig, 1914.

[6] F. Hausdorff. *Mengenlehre.* Walter de Gruyter, Berlin, 1927.

[7] C. Kuratowski. *Topologie I.* Polish Math. Soc., Warsaw, 1952.

[8] S. Marcus. Funcţiile lui Pompeiu. *Studii şi Cerc. Mat.* 5:413-419, 1954.

[9] M. Mitrea, F. Şabac. Pompeiu's integral representation formula. History and mathematics. *Rev.Roum.Math.Pures Appl.* 43:211-226, 1998.

[10] B.L. McAllister. Hyperspaces and multifunctions, the first halfcentury (1900-1950). *Nieuw Arch. Wisk.* 26:309-329, 1978.

[11] P. Neittaanmaki, J. Sprekels, D. Tiba. *Optimization of elliptic systems.Theory and applications.* Springer Verlag, New York, 2005.

[12] O. Onicescu. *Pe drumurile vieţii.* Ed. Şt. şi Enciclopedică, Bucharest, 1981.

[13] P. Painlevé. *Leçons sur la théorie analytique des équations differentielles, professées a Stockholm.* Hermann, Paris, 1897.

[14] D. Pompeiu. *Sur la continuité des fonctions de variables complexes (Thèse).* Gauthier-Villars, Paris, 1905; *Ann.Fac.Sci.de Toulouse* 7:264-315, 1905.

[15] D. Pompeiu. *Opera Matematică.* Ed.Acad.Române, Bucharest, 1959.

[16] D. Pompeiu. Sur les fonctions derivées. *Math.Ann.* 63:326-332, 1907.

[17] D. Pompeiu. Sur certains systemes d'équations linéires et sur une propriété intégrale des fonctions de plusieurs variables. *C.R.Acad.Sc.Paris.* 188:1138-1139, 1929.

[18] M. Vogelius. An inverse problem for the equation $\Delta u = -cu - d$. *Annales de l'Institut Fourier.* 44:1181-1209, 1994.

[19] S.A. Williams. A partial solution of the Pompeiu problem. *Math. Ann.* 223;183-190, 1976.

ANALYSIS OF A PDE MODEL FOR SANDPILE GROWTH

P. Cannarsa[1]
[1] *Department of Mathematics, University of Rome "Tor Vergata", Rome, Italy, cannarsa@axp.mat.uniroma2,it*

Abstract In the dynamical theory of granular matter, the so-called table problem consists in studying the evolution of a heap of matter poured continuously onto a bounded domain $\Omega \subset \mathbf{R}^2$. The mathematical description of the table problem, at an equilibrium configuration, can be reduced to a boundary value problem for a system of partial differential equations. The analysis of such a system, also connected with other mathematical models such as the Monge-Kantorovich problem, is the object of this paper. Our main result is an integral representation formula for the solution, in terms of the boundary curvature and of the normal distance to the cut locus of Ω.

keywords: granular matter, eikonal equation, singularities, viscosity solutions, optimal mass tranfer

1. A PDE model for sandpile growth

In recent years, the attention of many authors has been focussed on the system of partial differential equations

$$-\text{div}(vDu) = f \quad \text{in}\ \Omega\,, \qquad |Du| - 1 = 0 \quad \text{in}\ \{v > 0\} \tag{1}$$

in a given domain $\Omega \subset \mathbf{R}^n$.

For $n = 2$, a typical context of application for (1) is granular matter theory. The so-called 'table problem', for instance, consists of describing the evolution of a sandpile created by pouring dry matter onto a table. Different approaches to this problem have been proposed in the literature, such as: the **variational model** developed by Prigozhin [20]; the **ODE/PDE Model** introduced in [2] and [14] by Evans and co-authors; the **BCRE Model** initiated by Boutroux and de Gennes[3] and elaborated by Hadeler and Kuttler[17]. In our analysis, we shall be concerned with the BCRE model, where the table is represented by a bounded domain $\Omega \subset \mathbf{R}^2$, and the matter source by a function $f(t, x) \geq 0$. The physical description of the growing heap is based on the introduction of the so-called *standing* and *rolling layers*. The former collects the amount of

Please use the following format when citing this chapter:
Author(s) [insert Last name, First-name initial(s)], 2006, in IFIP International Federation for Information Processing, Volume 199, System Modeling and Optimization, eds. Ceragioli F., Dontchev A., Furuta H., Marti K., Pandolfi L., (Boston: Springer), pp. [insert page numbers].

matter that remains at rest, the latter represents matter moving down along the surface of the standing layer—eventually falling down when the base of the heap touches the boundary of the table.

As pointed out in [17], system (1) is related to equilibrium configurations that may occur in presence of a constant source. To explain this connection, let us denote by $u(x)$ and $v(x)$, respectively, the heights of the standing and rolling layers at a point $x \in \Omega$, for an equilibrium configuration. For physical reasons, the slope of the standing layer cannot exceed a given constant (the *angle of repose*)—typical of the matter under consideration—that we normalize to 1. Consequently, the standing layer must vanish on the boundary of the table. So, $|Du| \leq 1$ in Ω and $u = 0$ on $\partial\Omega$. Also, in the region where v is positive, the standing layer has to be 'maximal', for otherwise more matter would roll down there to rest. On the other hand, the rolling layer results from transporting matter, poured by the source, along the surface of the standing layer at a speed that is assumed proportional to the slope Du, with constant equal to 1. The above considerations lead to the boundary value problem

$$\begin{cases} -\mathrm{div}(vDu) = f & \text{in } \Omega \\ |Du| - 1 = 0 & \text{in } \{v > 0\} \\ |Du| \leq 1 \quad u, v \geq 0 & \text{in } \Omega \\ u = 0 & \text{on } \partial\Omega \end{cases} \tag{2}$$

Notice that (2) is the same equilibrium system one would obtain from Prigozhin's variational model.

1.1 Connection with optimal mass transfer

It is worth noting that system (1) arises in Monge-Kantorovich theory, as explained in the monograph [15], and futher analyzed in [1] and [16]. In the present context, we will just observe that the connection of the above system with optimal mass transfer can be obtained by looking at the so-called 'dual problem', which consists in maximizing

$$\int_{\Omega} u(x) f(x) dx \tag{3}$$

among all Lipschitz continuous functions $u : \Omega \to \mathbf{R}$, with $\mathrm{Lip}(u) \leq 1$, vanishing on $\partial\Omega$. Indeed, as proved in [4], the boundary value problem (2) turns out to be the system of necessary conditions satisfied by any maximizer u of (3), taking v equal to the associated Lagrange multiplier. Such a framework is also related to the optimization problem studied in [12].

1.2 Solution of the equilibrium system

The main purpose of the present work is to provide a full analysis of problem (2), including *existence*, *uniqueness*, and *regularity* of the solution. For existence and uniqueness, we shall follow the approach of [5] for the case of $n = 2$, and of [6] for the general case $n \geq 2$. As for regularity, we shall rely on the results of [7]. It is well-known that the eikonal equation

$$|Du| = 1$$

does not possess global smooth solutions in general, neither does the conservation law

$$-\text{div}(vDu) = f\,.$$

Therefore, we ought to explain what we mean by a solution of (2).

We say that a pair (u, v) of *continuous* functions in Ω is a solution of problem (2) if

- $u = 0$ on $\partial\Omega$, $\|Du\|_{\infty,\Omega} \leq 1$, and u is a viscosity solution of

$$|Du| = 1 \quad \text{in} \quad \{x \in \Omega \,:\, v(x) > 0\}$$

- $v \geq 0$ in Ω and, for every test function $\phi \in C_c^\infty(\Omega)$,

$$\int_\Omega v(x)\langle Du(x), D\phi(x)\rangle dx = \int_\Omega f(x)\phi(x)dx\,. \tag{4}$$

For the reader's sake, we now recall the definition of viscosity solution. The superdifferential of a function $u : \Omega \to \mathbf{R}$ at a point $x \in A$ is the set

$$D^+u(x) = \left\{p \in \mathbf{R}^n \;\middle|\; \limsup_{h\to 0} \frac{u(x+h) - u(x) - \langle p, h\rangle}{|h|} \leq 0\right\},$$

while the subdifferential D^-u is given by the formula $D^-u(x) = -D^+(-u)(x)$. We say that u is a viscosity solution of the eikonal equation $|Du| = 1$ in Ω if, for any $x \in \Omega$, we have

$$p \in D^-u(x) \quad \Rightarrow \quad |p| \geq 1\,, \qquad p \in D^+u(x) \quad \Rightarrow \quad |p| \leq 1\,.$$

Before describing our main results, we need to introduce some useful geometric properties of bounded domains with smooth boundary. This is the purpose of the next section.

2. Distance function

Let Ω be a bounded domain with C^2 boundary $\partial\Omega$. In what follows we denote by $d : \overline{\Omega} \to \mathbf{R}$ the distance function from the boundary of Ω, that is,

$$d(x) = \min_{y\in\partial\Omega} |y - x|\,,$$

and by Σ the singular set of d, that is, the set of points $x \in \Omega$ at which d is not differentiable. Since d is Lipschitz continuous, Σ has Lebesgue measure zero. Introducing the projection $\Pi(x)$ of x onto $\partial\Omega$ in the usual way, Σ is also the set of points x at which $\Pi(x)$ is not a singleton. So, if $\Pi(x) = \{\overline{x}\}$ for some $x \in \Omega$, then d is differentiable at x and

$$Dd(x) = \frac{x - \overline{x}}{|x - \overline{x}|}.$$

REMARK 1 We recall that the distance function d is the unique viscosity solution of the eikonal equation $|Du| = 1$ in Ω, with boundary condition $u = 0$ in $\partial\Omega$. Equivalently, d is the largest function such that $\|Du\|_{\infty,\Omega} \leq 1$ and $u = 0$ on $\partial\Omega$.

For any $x \in \partial\Omega$ and $i = 1, \ldots, n-1$, the number $\kappa_i(x)$ denotes the $i-th$ principal curvature of $\partial\Omega$ at the point x, corresponding to a principal direction $e_i(x)$ orthogonal to $Dd(x)$, with the sign convention $\kappa_i \geq 0$ if the normal section of Ω along the direction e_i is convex. Also, we will label in the same way the extension of κ_i to $\overline{\Omega} \setminus \Sigma$ given by

$$\kappa_i(x) = \kappa_i(\Pi(x)) \qquad \forall x \in \overline{\Omega} \setminus \Sigma. \tag{5}$$

Denoting by $p \otimes q$ the tensor product of two vectors $p, q \in \mathbf{R}^n$, defined as

$$(p \otimes q)(x) = p\,\langle q, x\rangle\,, \ \forall x \in \mathbf{R}^n\,,$$

for any $x \in \overline{\Omega}$ and any $y \in \Pi(x)$ we have

$$\kappa_i(y)d(x) \leq 1 \qquad \forall i = 1, \ldots, n-1.$$

If, in addition, $x \in \overline{\Omega}\setminus\overline{\Sigma}$, then

$$\kappa_i(x)d(x) < 1 \qquad \text{and} \qquad D^2 d(x) = -\sum_{i=1}^{n-1} \frac{\kappa_i(x)}{1 - \kappa_i(x)d(x)} e_i(x) \otimes e_i(x)$$

where $e_i(x)$ is the unit eigenvector corresponding to $\frac{\kappa_i(x)}{1-\kappa_i(x)d(x)}$.

We now turn our attention to the closure of Σ, a set that is also called the *cut locus* of $\partial\Omega$ in Ω, and to the function

$$\tau(x) = \begin{cases} \min\left\{t \geq 0 \ : \ x + tDd(x) \in \overline{\Sigma}\right\} & \forall x \in \overline{\Omega}\setminus\overline{\Sigma}, \\ 0 & \forall x \in \overline{\Sigma}. \end{cases} \tag{6}$$

Since the map $x \mapsto x + \tau(x)Dd(x)$ is a natural retraction of $\overline{\Omega}$ onto $\overline{\Sigma}$, we will refer to $\tau(\cdot)$ as the *maximal retraction length* of Ω onto $\overline{\Sigma}$ or *normal distance to* $\overline{\Sigma}$. The regularity properties of τ are described by the following theorem due to Li and Nirenberg [19] (see also [18]).

THEOREM 2 *Let Ω be a bounded domain in $\mathbf{R}^n$ with boundary of class $C^{2,1}$. Then the map τ defined in* (6) *is Lipschitz continuous on $\partial\Omega$.*

3. A representation formula for the solution

Before stating a precise result for our problem, let us show a formal derivation of the representation formula for the solution in the case of $n = 2$. Suppose (u, v) is a smooth solution of (2). In view of Remark 1, we can take $u = d$. Moreover, suppose that v vanishes on $\overline{\Sigma}$—this is reasonable from the point of view of the physical model, and can also be justified by a rigorous argument. Let us proceed to compute, for a given point $x \in \Omega\setminus\overline{\Sigma}$ and for any $t \in (0, \tau(x))$, the derivative

$$\begin{aligned}\frac{d}{dt}v(x + tDd(x)) &= \langle Dv(x + tDd(x), Dd(x)\rangle \\ &= -v(x + tDd(x))\Delta d(x + tDd(x)) - f(x + tDd(x))\end{aligned}$$

(recall that $Dd(x + tDd(x)) = Dd(x)$). Now, observe that

$$\Delta d(x + tDd(x)) = -\frac{\kappa(x)}{1 - (d(x) + t)\kappa(x)}$$

since $\kappa(x + tDd(x)) = \kappa(x)$ and $d(x + tDd(x)) = d(x) + t$. Hence, $V(t) := v(x + tDd(x))$ satisfies the Cauchy problem

$$\begin{cases} V'(t) - \frac{\kappa(x)}{1-(d(x)+t)\kappa(x)}V(t) + f(x + tDd(x)) = 0 \\ V(\tau(x)) = 0\,. \end{cases}$$

Thus, solving the above problem and noting that $v(x) = V(0)$, we conclude that, in $\Omega \setminus \overline{\Sigma}$, v must be given by the formula

$$v(x) = \int_0^{\tau(x)} f(x + tDd(x))\frac{1 - (d(x) + t)\kappa(x)}{1 - d(x)\kappa(x)}dt \qquad \forall x \in \Omega \setminus \overline{\Sigma}$$

REMARK 3 We note that the above formula entails that v vanishes at all points $x \in \Omega \setminus \overline{\Sigma}$ at which the half-line spanned by $Dd(x)$ fails to intersect the support of f before hitting $\overline{\Sigma}$. This description, which agrees with physical evidence, extends to dimension 2 the analogous result obtained in [17] for the one-dimensional case.

4. Existence, uniqueness, regularity

The following result, proved in [5] for $n = 2$ and in [6] for $n \geq 2$, ensures the existence and uniqueness of the solution of (2), as well as a representation formula for such solution.

THEOREM 4 *Let $\Omega \subset \mathbf{R}^n$ be a bounded domain with boundary of class C^2 and $f \geq 0$ be a continuous function in Ω. Then, a solution of system* (2) *is* *given by the pair* (d, v_f)*, where*

$$v_f(x) = \begin{cases} \int_0^{\tau(x)} f(x + tDd(x)) \prod_{i=1}^{n-1} \frac{1-(d(x)+t)\kappa_i(x)}{1-d(x)\kappa_i(x)}\, dt & \forall x \in \Omega \setminus \overline{\Sigma}, \\ 0 & \forall x \in \overline{\Sigma}, \end{cases} \tag{7}$$

where, $\kappa_i(x)$ *denotes the* $i-th$ *principal curvature of* $\partial\Omega$ *at the point* $\Pi(x)$*.*

Moreover, the above solution is unique in the following sense: if (u, v) *is another solution of* (2)*, then* $v = v_f$ *in* Ω*, and* $u = d$ *in* $\{x \in \Omega \mid v_f > 0\}$*.*

A noteworthy aspect of the above result is that we do construct a continuous solution v_f, instead of just a measure or a function in $L^1(\Omega)$. So, Theorem 4 could also be viewed as a regularity result. Moreover, formula (7) can be used to derive further regularity properties. This will be the object of our next section.

4.1 Regularity

A natural question to ask is what kind of regularity one can expect for the solution (d, v_f) of problem (2). For the first component, this is well understood: while d is of class C^2 on a neighborhood of $\partial\Omega$, the maximal regularity of d in the whole domain $\overline{\Omega}$ is *semiconcavity*, a generalization of concavity preserving most of the local properties of concave functions (see [8] for a detailed description of such a class of functions).

On the other hand, the situation is different for the second component. Indeed, formula (7) suggests that the regularity of v_f should depend on the regularity of f and τ. Therefore, our original problem leads to the question of the global regularity of the normal distance.

While it is easy to prove that τ is continuous on $\overline{\Omega}$, and locally Lipschitz in $\overline{\Omega} \setminus \overline{\Sigma}$ when $\partial\Omega \in C^{2,1}$, the following example shows that τ may fail to be globally Lipschitz continous on $\overline{\Omega}$.

EXAMPLE 5 (THE PARABOLA CASE) In the cartesian plane consider the set $\Omega := \{(x, y) \in \mathbf{R}^2 \mid y > x^2\}$, whose boundary is a parabola with vertex $(0, 0)$. By the symmetry of $\partial\Omega$ with respect to the vertical axis we deduce that $\overline{\Sigma}$ must be contained in such an axis. Moreover, $\overline{\Sigma} = \{(0, y) \mid y \geq 1/2\}$ and $\tau((s, s^2)) = \frac{1}{2}\sqrt{1 + 4s^2}$. Then, a straightforward computation shows that, for $a > 0$ sufficiently small,

$$|\tau((a, 1/2)) - \tau((0, 1/2))| \geq M|(a, 1/2) - (0, 1/2)|^{2/3},$$

for some $M > 0$. So, τ cannot be Lipschitz continuous in the whole set Ω.

On the positive side, we present two Hölder regularity results in $\mathbf{R}^2$ recently obtained in [7]—the former for τ the latter for v_f.

THEOREM 6 *Let Ω be a bounded simply connected domain in $\mathbf{R}^2$ with analytic boundary, different from a disk. Then, there exists an integer $m \geq 2$ such that τ is Hölder continuous in Ω with exponent $\frac{2m-2}{2m-1}$.*

In particular, τ is at least $2/3$-Hölder continuous, and Example 5 describes the 'worst' possible case.

THEOREM 7 *Assume that f is a Lipschitz continuous function and that Ω is a simply connected bounded domain in $\mathbf{R}^2$ with analytic boundary, different from a disk. Then v_f is a Hölder continuous function with exponent $\frac{1}{2m-1}$ for some integer $m \geq 2$.*

5. Application to a variational problem

We conclude this paper with an application to a problem in the calculus of variations which may seem quite unrelated to the present context at first glance. Let us consider an integral functional of the form

$$J(u) = \int_\Omega [h(|Du|) - f(x)u]\, dx\,, \qquad u \in W_0^{1,1}(\Omega), \tag{8}$$

where $f \in L^\infty(\Omega)$ is a nonnegative function and $h\colon [0, +\infty) \to [0, +\infty]$ is a lower semicontinuous function (possibly with non–convex values) satisfying

$$h(R) = 0, \quad h(s) \geq \max\{0, \Lambda(s - R)\} \text{ for some constants } R, \Lambda > 0. \tag{9}$$

In a pioneering work, Cellina [11] proved that, if Ω is a convex domain (that is, an open bounded convex set) in $\mathbf{R}^2$ with piecewise smooth (C^2) boundary and $f \equiv 1$, then J does attain its minimum in $W_0^{1,1}(\Omega)$, and a minimizer is explicitly given by the function

$$u_\Omega(x) = R\, d(x), \qquad x \in \Omega, \tag{10}$$

provided that the inradius r_Ω of Ω is small enough. (We recall that r_Ω is the supremum of the radii of all balls contained in Ω.) This result has been extended to convex domains in $\mathbf{R}^n$ and to more general functionals in subsequent works (see [9, 10, 13, 21, 22]). One common point of all these results is that the set Ω is always a *convex* subset of $\mathbf{R}^n$. In this paper we will see that, using the representation formula (7), the function u_Ω defined in (10) is a minimizer of J in $W_0^{1,1}(\Omega)$, even on possibly nonconvex domains.

We say that a set Ω is a *smooth K-admissible domain*, $K \in \mathbf{R}$, if it is a connected open bounded subset of $\mathbf{R}^n$ with C^2 boundary, such that the mean curvature of $\partial\Omega$ is bounded below by K, that is

$$H(y) := \frac{1}{n-1} \sum_{i=1}^{n-1} \kappa_i(y) \geq K \qquad \forall y \in \partial\Omega.$$

We note that every connected bounded open set $\Omega \subset \mathbf{R}^n$ with C^2 boundary is a K-admissible smooth domain for every K satisfying

$$K \leq \min_{y \in \partial\Omega} H(y).$$

The following is a special case of a more general result obtained in [6].

THEOREM 8 *Let h: $[0, \infty) \to [0, \infty]$ be a lower semicontinuous function satisfying (9), let $\Omega \subset \mathbf{R}^n$ be a smooth K-admissible domain, and let f be a nonnegative Lipschitz continuous function in Ω. If*

$$\|f\|_{\infty,\Omega}\, c(K, r_\Omega) \leq \Lambda, \tag{11}$$

where

$$c(K, r_\Omega) := \frac{1 - (1 - K r_\Omega)^n}{nK} \quad \text{if } K \neq 0, \quad c(K, r_\Omega) := r_\Omega \ \text{ if } K = 0, \tag{12}$$

then the function $u_\Omega(x) = R\, d(x)$ is a minimizer in $W_0^{1,1}(\Omega)$ of the functional J defined in (8).

We will now sketch the proof of Theorem 8 in order to point out the connection of this problem with (2). Given f as above denote once again by v_f the continuous function defined in (7). The first step of the proof, we will comment no further on in this context, consists of estimating the integrand in the representation formula for v_f as in [6], to show that

$$0 \leq v_f(x) \leq \|f\|_{\infty,\Omega}\, c(K, r_\Omega), \qquad \forall x \in \Omega. \tag{13}$$

Therefore, in view of assumption (11), we have

$$0 \leq v_f(x) \leq \Lambda, \quad \forall x \in \Omega. \tag{14}$$

Let $u \in W_0^{1,1}(\Omega)$. Since h satisfies (9) and v_f satisfies (14), we have that

$$\begin{aligned} h(|Du(x)|) &\geq v_f(x)(|Du(x)| - R) \\ &\geq h(|Du_\Omega(x)|) + v_f(x)\langle Dd(x), Du(x) - Du_\Omega(x)\rangle, \end{aligned}$$

hence

$$J(u) \geq J(u_\Omega) + \Delta,$$

where

$$\Delta = \int_\Omega [v_f(x)\langle Dd(x), Du(x) - Du_\Omega(x)\rangle - f(x)(u(x) - u_\Omega(x))]\, dx.$$

Since v_f is bounded, by a density argument equation (4) holds for every $\phi \in W_0^{1,1}(\Omega)$. Choosing $\phi = u - u_\Omega$, we obtain that Δ vanishes, so that $J(u) \geq$

$J(u_\Omega)$. Since u was an arbitrary function in $W_0^{1,1}(\Omega)$, we have proven that u_Ω is a minimizer of J in $W_0^{1,1}(\Omega)$.

REMARK 9 We note:

1) The result of Theorem 8 can be extended to nonsmooth domains, such as domains satisfying a uniform exterior sphere condition. See [6] for details.

2) If Ω is a convex domain with smooth boundary, then condition

$$L\, c(K, r_\Omega) \leq \Lambda$$

is certainly satisfied provided that

$$L r_\Omega \leq \Lambda. \tag{15}$$

Namely, it is enough to observe that a convex domain is a 0-admissible domain, and that $c(0, r_\Omega) = r_\Omega$. Condition (15) was first introduced in [11] in the case of $f \equiv 1$. In [9] it was proven that, if (15) does not hold, then J needs not have minimizers in $W_0^{1,1}(\Omega)$.

3) Assumption (11) of Theorem 8 for the existence of a minimizer of J is optimal in the following sense. Let $h(s) = \max\{0, \Lambda(s-R)\}$ for some positive constants Λ and R, let $f(x) = 1$ and let $\Omega = B_r(0) \subset \mathbf{R}^n$. Then $r_\Omega = r$, and Ω is a $(1/r)$-admissible domain. Since $c(1/r, r) = r/n$, Theorem 8 states that the function $u_\Omega(x) = Rd(x)$ is a minimizer of J provided that $r \leq n\Lambda$. This condition is optimal: indeed, functional J is not even bounded from below if $r > n\Lambda$. Let us define the sequence of functions in $W_0^{1,1}(\Omega)$

$$u_k(x) = \begin{cases} k(r - |x|) & \text{if } n\Lambda < |x| < r \\ R(r - |x|) & \text{if } |x| \leq n\Lambda \end{cases}$$

A straightforward computation shows that, for $k \geq R$,

$$J(u_k) = \frac{\omega_n}{n+1}\left[\psi(n\Lambda) - \psi(r)\right] k + A,$$

where ω_n is the n-dimensional Lebesgue measure of the unit ball of $\mathbf{R}^n$, A is a constant independent of k, and $\psi(\rho) = \rho^{n+1} - (n+1)\Lambda\rho^n$. Since ψ is strictly increasing for $\rho \geq n\Lambda$, and $r > n\Lambda$, we have that $\psi(n\Lambda) - \psi(r) < 0$, hence $\lim_{k\to\infty} J(u_k) = -\infty$.

References

[1] L. Ambrosio. Optimal transport maps in Monge-Kantorovich problem. In *Proceedings of the International Congress of Mathematicians* Higher Ed. Press, 2002.

[2] G. Aronsson, L. C. Evans, Y. Wu. Fast/slow diffusion and growing sandpiles. *J. Differential Equations* 131:304-335, 1996.

[3] T. Boutreux, P.-G. de Gennes. Surface flows of granular mixtures. I. General principles and minimal model. *J. Phys. I France* 6: 1295-1304, 1996.

[4] G. Bouchitté, G. Buttazzo, P. Seppechere. Shape optimization solutions via Monge-Kantorovich equation. *C. R. Acad. Sci. Paris Ser. I Math. 324* 10: 1185-1191, 1997.

[5] P. Cannarsa, P. Cardaliaguet. Representation of equilibrium solutions to the table problem for growing sandpile. *J. Eur. Math. Soc.* 6: 1-30, 2004.

[6] P. Cannarsa, P. Cardaliaguet, G. Crasta, E. Giorgieri. A boundary value problem for a PDE model in mass transfer theory: representation of solutions and applications. *Calc. Var.* DOI 10.1007/s00526-005-0328-7, 2005.

[7] P. Cannarsa, P. Cardaliaguet, E. Giorgieri. Hölder regularity of the normal distance with an application to a PDE model for growing sandpiles. Pre-print.

[8] P. Cannarsa, C. Sinestrari, *Semiconcave functions, Hamilton–Jacobi equations, and optimal control.* Birkhäuser progress in nonlinear differential equations and their applications. Boston, 2004.

[9] P. Celada, A. Cellina. Existence and non existence of solutions to a variational problem on a square. *Houston J. Math.* 24: 345-375, 1998.

[10] P. Celada, S. Perrotta, G. Treu. Existence of solutions for a class of non convex minimum problems. *Math. Z.* 228: 177-199, 1997.

[11] A. Cellina. Minimizing a functional depending on ∇u and on u. *Ann. Inst. H. Poincaré, Anal. Non Linéaire* 14: 339-352, 1997.

[12] A. Cellina, S. Perrotta. On the validity of the maximum principle and of the Euler-Lagrange equation for a minimum problem depending on the gradient. *SIAM J. Control Optim.* 36: 1987-1998, 1997.

[13] G. Crasta, A. Malusa. Geometric constraints on the domain for a class of minimum problems. *ESAIM Control Optim. Calc. Var.* 9:125-133, 2003.

[14] L.C. Evans, M. Feldman, R. Gariepy. Fast/slow diffusion and growing sandpiles. *J. Differential Equations* 137:166-209, 1997.

[15] L.C. Evans, W. Gangbo. *Differential equations methods for the Monge-Kantorovich mass transfer problem.* Mem. Amer. Math. Soc. 137, no. 653, 1999.

[16] M. Feldman, R. J. McCann. Uniqueness and transport density in Monge's mass transportation problem. *Calc. Var.* 15:81-113, 2002.

[17] K. P. Hadeler, C. Kuttler. Dynamical models for granular matter. *Granular Matter* 2:9-18, 1999.

[18] J. Itoh, M. Tanaka. The Lipschitz continuity of the distance function to the cut locus. *Trans. Am. Math. Soc.* 353:21-40, 2001.

[19] Y. Y. Li, L. Nirenberg. The distance function to the boundary, Finsler geometry and the singular set of viscosity solutions of some Hamilton–Jacobi equations. *Comm. Pure Appl. Math.* 58:85-146, 2005.

[20] L. Prigozhin. Variational model of sandpile growth. *European J. Appl. Math.* 7:225-235, 1996.

[21] G. Treu. An existence result for a class of non convex problems of the Calculus of Variations. *J. Convex Anal.* 5:31-44, 1998.

[22] M. Vornicescu. A variational problem on subsets of $\mathbf{R}^n$. *Proc. Roy. Soc. Edinburgh Sect. A* 127:1089-1101, 1997.

ON WARM STARTS FOR INTERIOR METHODS

A. Forsgren[1]
[1]*Optimization and Systems Theory, Department of Mathematics, Royal Institute of Technology, SE-100 44 Stockholm, Sweden, andersf@kth.se**

Abstract An appealing feature of interior methods for linear programming is that the number of iterations required to solve a problem tends to be relatively insensitive to the choice of initial point. This feature has the drawback that it is difficult to design interior methods that efficiently utilize information from an optimal solution to a "nearby" problem. We discuss this feature in the context of general nonlinear programming and specialize to linear programming. We demonstrate that warm start for a particular nonlinear programming problem, given a near-optimal solution for a "nearby" problem, is closely related to an SQP method applied to an equality-constrained problem. These results are further refined for the case of linear programming.

keywords: nonlinear programming, linear programming, interior method, warm start.

1. Introduction

This paper concerns the solution of a nonlinear program in the form

$$\begin{array}{ll} \underset{x \in \Re^n}{\text{minimize}} & f(x) \\ \text{subject to} & c(x) \geq 0, \end{array} \tag{1}$$

where $f : \Re^n \to \Re$ and $c : \Re^n \to \Re^m$ are twice-continuously differentiable. Our interest is the situation where we want to solve (1) given the solution to a "nearby" problem. This situation is commonly referred to as *warm start*. It may for example be the case that one is interested in resolving the problem when only some constraints have been changed. Our discussion specifically concerns *interior methods*. We will study properties of the search directions generated for the nonlinear programming case, and then specialize further to linear programming. For related discussions concerning linear programming, see, e.g. Jansen et al. [11], Kim, Park and Park [12], Gondzio and Grothey [9],

*Research supported by the Swedish Research Council (VR).

Please use the following format when citing this chapter:
Author(s) [insert Last name, First-name initial(s)], 2006, in IFIP International Federation for Information Processing, Volume 199, System Modeling and Optimization, eds. Ceragioli F., Dontchev A., Furuta H., Marti K., Pandolfi L., (Boston: Springer), pp. [insert page numbers].

Yildirim and Todd [18, 19], Yildirim and Wright [20], Gonzalez-Lima, Wei and Wolkowicz [10]. For extensions to linear semidefinite programming, see Yildirim [17].

2. Background

Methods for solving (1) all have to solve a combinatorial problem of identifying the constraints that are active at the solution. This can roughly be done in two ways: (i) by a "hard" estimate of the active constraints at each iteration or by (ii) a "soft" estimate. Our focus is interior methods, which belong to the latter class. However, in the discussion, sequential quadratic programming methods, which belong to the former type of methods, arise too. In this section, we review basic properties of these methods, and also give a brief review of optimality conditions.

2.1 Optimality conditions

Given a suitable *constraint qualification*, an optimal solution x to (1), together with a *Lagrange multiplier vector* $\lambda \in \Re^m$, has to satisfy the *first-order necessary optimality conditions* associated with (1). These conditions may be written in the form

$$g(x) - A(x)^T y = 0, \quad (2)$$

$$c(x) - s = 0, \quad (3)$$

$$YSe = 0, \quad y \geq 0, \quad s \geq 0, \quad (4)$$

where $g(x) = \nabla f(x)$, $A(x) = c'(x)$ and e is an m-dimensional vector of ones. Here, and throughout the paper, we denote by upper-case letters Y and S, the diagonal matrices formed by y and s respectively. In (2)–(4) we have introduced the slack variables s associated with the constraints $c(x) \geq 0$. They need not be present, since s may be eliminated from (3). They do not affect the discussion, but simplify the notation. The analogous discussion could be made without introducing s. A constraint qualification ensures that a linearization of the constraints around a point of interest gives a suitable approximation to the constraints. We will throughout the paper assume that a constraint qualification holds at all points which we consider. For a more detailed discussion of constraint qualifications in the context of interior methods, see, e.g., Forsgren, Gill and Wright [6, Section 2.2].

Second-order optimality conditions typically involve the Hessian of the *Lagrangian* $L(x, y)$ with respect to x, where $L(x, y) = f(x) - y^T c(x)$. We denote this Hessian by $H(x, y)$, i.e.,

$$H(x, y) = \nabla^2_{xx} L(x, y) = \nabla^2 f(x) - \sum_{i=1}^{m} y_i \nabla^2 c_i(x).$$

Given a suitable constraint qualification, the curvature of the objective function on the surface of the active constraints is captured by the curvature of $H(x, y)$ on the tangent surface of the active constraints. For further discussion, see, e.g., Forsgren, Gill and Wright [6, Section 2.2].

2.2 Interior methods

The interior methods of interest to this paper are based on approximately following the *barrier trajectory*. This trajectory is defined as the set of solutions to the perturbed optimality conditions

$$\begin{aligned} g(x) - A(x)^T y &= 0, && (5)\\ c(x) - s &= 0, && (6)\\ YSe &= \mu e, && (7) \end{aligned}$$

where $y > 0$ and $s > 0$ are held implicitly. Here, μ is a positive parameter, referred to as the *barrier parameter.* As in (2)–(4) , we have introduced slack variables s. This slack reformulation is not of importance for the discussion, but convenient for the notation.

A primal-dual interior method computes approximate solution to (5)–(7) for decreasing values of μ by Newton's method. This means that the steps Δx, Δs and Δy are computed from the linear equation

$$\begin{pmatrix} H(x,y) & & -A(x)^T \\ A(x) & -I & \\ & Y & S \end{pmatrix} \begin{pmatrix} \Delta x \\ \Delta s \\ \Delta y \end{pmatrix} = - \begin{pmatrix} g(x) - A(x)^T y \\ c(x) - s \\ YSe - \mu e \end{pmatrix}. \tag{8}$$

Equivalently, we may eliminate Δs and solve

$$\begin{pmatrix} H(x,y) & A(x)^T \\ A(x) & -Y^{-1}S \end{pmatrix} \begin{pmatrix} \Delta x \\ -\Delta y \end{pmatrix} = - \begin{pmatrix} g(x) - A(x)^T y \\ c(x) - \mu S^{-1} e \end{pmatrix}. \tag{9}$$

Note that there is no loss in sparsity when forming (9) from (8), since Y is diagonal. Local convexity is typically deduced by the inertia of the matrix of (9). If the inertia is such that the matrix has n positive eigenvalues and m negative eigenvalues, the equations are solved. Otherwise, some modification is made. The solution of this equation can either be done by factorization methods or by iterative methods. See, e.g., Forsgren [4] and Forsgren, Gill and Griffin [5] for a discussion of these issues.

In order to enforce convergence of the method, typically a linesearch strategy, a trust-region strategy or a filter strategy may be used. We shall not be concerned with the precise method, but focus on the linear equations to be solved. For more detailed descriptions on interior methods, see e.g., Forsgren, Gill and Wright [6] or Wright [15]. Note that for a solution of (5)–(7) , no constraints are active, since $\mu > 0$. Hence, the active constraints at the solution are determined implicitly as μ tends to zero.

2.3 Sequential quadratic programming methods

In contrast to an interior method, where one system of linear equations is solved at each iteration, a sequential programming method has a subproblem which is an inequality-constrained quadratic program on the form

$$\begin{array}{ll} \underset{p\in\Re^n}{\text{minimize}} & \frac{1}{2}p^T H(x,y)p + g(x)^T p \\ \text{subject to} & A(x)p \geq -c(x). \end{array} \tag{10}$$

If the problem is locally convex, this subproblem is well defined. Otherwise, some modification is made. We will assume that local convexity holds in our discussion, and not consider the modifications. If p^* denotes the optimal solution of (10) and y^* denotes the corresponding Lagrange multiplier vector, then the next iterate for solving (1) is given by $x + p^*$, and the next Lagrange multiplier estimate is given by y^*. Again, some strategy is required to ensure convergence, but the basis of the subproblem is the solution of a quadratic program on the form (10). Note that the prediction of the constraints that are active at the solution of (1) are given by the constraints active at the solution of (10). For a thorough discussion on sequential quadratic programming methods, see, e.g., Nocedal and Wright [13, Chapter 18].

3. Warm starts of interior methods for nonlinear programming

We now return to the issue of solving (1) by a primal-dual interior method. Specifically, we consider the warm-start situation. Assume that the initial point is given as a near-optimal solution on the trajectory to a different problem

$$\begin{array}{ll} \underset{x\in\Re^n}{\text{minimize}} & \tilde{f}(x) \\ \text{subject to} & \tilde{c}(x) \geq 0, \end{array} \tag{11}$$

where $\tilde{f}$ and $\tilde{c}$ have the same properties as f and c of (1). We denote this point by $\tilde{x}$. This means that for a small barrier parameter $\tilde{\mu}$, we assume that $\tilde{x}$, $\tilde{s}$ and $\tilde{y}$ solve

$$\tilde{g}(\tilde{x}) - \tilde{A}(\tilde{x})^T\tilde{y} = 0, \tag{12}$$

$$\tilde{c}(\tilde{x}) - \tilde{s} = 0, \tag{13}$$

$$\tilde{Y}\tilde{S}e = \tilde{\mu}e, \quad \tilde{y} > 0, \quad \tilde{s} > 0, \tag{14}$$

where $\tilde{g}(\tilde{x}) = \nabla\tilde{f}(\tilde{x})$ and $\tilde{A}(\tilde{x}) = \tilde{c}'(\tilde{x})$. Throughout the paper, we will consider quantities related to matrices that are implicitly dependent on $\tilde{\mu}$. The notation $O(\tilde{\mu})$ will be used to denote a quantity that converges to zero at least as fast as $\tilde{\mu}$. Analogously, $\Theta(\tilde{\mu})$ denotes a quantity that converges to zero at the same rate as $\tilde{\mu}$, $\Theta(1)$ denotes a bounded quantity that is bounded away from zero as $\tilde{\mu} \to 0$ and $\Omega(1/\tilde{\mu})$ denotes a quantity whose inverse converges to zero at least as fast as $\tilde{\mu}$.

From the point given by (11), we want to take a primal-dual interior step towards solving (1) for a given barrier parameter μ. By (8) and (12)–(14), the Newton equations take the form

$$\begin{pmatrix} H(\tilde{x},\tilde{y}) & & -A(\tilde{x})^T \\ A(\tilde{x}) & -I & \\ & \tilde{Y} & \tilde{S} \end{pmatrix} \begin{pmatrix} \Delta x \\ \Delta s \\ \Delta y \end{pmatrix} = \begin{pmatrix} -g(\tilde{x}) + A(\tilde{x})^T\tilde{y} \\ \tilde{c}(\tilde{x}) - c(\tilde{x}) \\ (\mu - \tilde{\mu})e \end{pmatrix}. \tag{15}$$

In order to make a more detailed analysis, we assume that the set $\{1, 2, \ldots, m\}$ can be partitioned into two sets, A and I, according to which constraints of (11) that are "almost active" at $\tilde{x}$, and which constraints that are "not almost active" at $\tilde{x}$. This means that $A \cup I = \{1, \ldots, m\}$, where $A = \{i \in \{1, 2, \ldots, m\} : \tilde{c}_i(\tilde{x}) = \Theta(\tilde{\mu})\}$ and $I = \{i \in \{1, 2, \ldots, m\} : \tilde{c}_i(\tilde{x}) = \Theta(1)\}$. This would typically be the case in the neighborhood of a local minimizer of (1) where strict complementarity holds. We will throughout this section let subscript "A" denote quantities associated with A, and similarly for I. For example, the matrix $A(\tilde{x})$ is partitioned into $A_A(\tilde{x})$ and $A_I(\tilde{x})$. By the above assumption, it follows that $\tilde{y}_I = \Theta(\tilde{\mu})$, and that $\tilde{s}_A = \Theta(\tilde{\mu})$. We will also make the assumption that $A_A(\tilde{x})$ has full row rank, and that $H(\tilde{x}, \tilde{y})$ is positive definite on the nullspace of $A_A(\tilde{x})$. With this partition, (15) may be written as

$$\begin{pmatrix} H(\tilde{x},\tilde{y}) & & & -A_A(\tilde{x})^T & -A_I(\tilde{x})^T \\ A_A(\tilde{x}) & -I & & & \\ A_I(\tilde{x}) & & -I & & \\ & \tilde{Y}_A & & \tilde{S}_A & \\ & & \tilde{Y}_I & & \tilde{S}_I \end{pmatrix} \begin{pmatrix} \Delta x \\ \Delta s_A \\ \Delta s_I \\ \Delta y_A \\ \Delta y_I \end{pmatrix}$$

$$= \begin{pmatrix} -g(\tilde{x}) + A(\tilde{x})^T\tilde{y} \\ -c_A(\tilde{x}) \\ \tilde{c}_I(\tilde{x}) - c_I(\tilde{x}) \\ (\mu - \tilde{\mu})e \\ (\mu - \tilde{\mu})e \end{pmatrix}.$$

We may now approximate these equations by ignoring the $\Theta(\tilde{\mu})$ terms in the matrix, which gives a related system of equations

$$\begin{pmatrix} H(\tilde{x},\tilde{y}) & & & -A_A(\tilde{x})^T & -A_I(\tilde{x})^T \\ A_A(\tilde{x}) & -I & & & \\ A_I(\tilde{x}) & & -I & & \\ & \tilde{Y}_A & & & \\ & & & & \tilde{S}_I \end{pmatrix} \begin{pmatrix} u \\ v_A \\ v_I \\ z_A \\ z_I \end{pmatrix}$$

$$= \begin{pmatrix} -g(\tilde{x}) + A(\tilde{x})^T\tilde{y} \\ -c_A(\tilde{x}) \\ \tilde{c}_I(\tilde{x}) - c_I(\tilde{x}) \\ (\mu - \tilde{\mu})e \\ (\mu - \tilde{\mu})e \end{pmatrix}.$$

where v_A and Δy_I may be eliminated so as to give the equivalent equations

$$v_A = (\mu - \tilde{\mu})\tilde{Y}_A^{-1}e, \tag{16}$$

$$z_I = (\mu - \tilde{\mu})\tilde{S}_I^{-1}e, \tag{17}$$

$$\begin{pmatrix} H(\tilde{x},\tilde{y}) & A_A(\tilde{x})^T \\ A_A(\tilde{x}) & \end{pmatrix} \begin{pmatrix} u \\ -z_A \end{pmatrix} = \begin{pmatrix} -g(\tilde{x}) + A(\tilde{x})^T\tilde{y} + A_I(\tilde{x})^T z_I \\ -c_A(\tilde{x}) + v_A \end{pmatrix}, \tag{18}$$

$$v_I = c_I(\tilde{x}) - \tilde{c}_I(\tilde{x}) + A_I(\tilde{x})u. \tag{19}$$

From (16)–(19), we may identify u and $\tilde{y}_A + z_A$ as the solution and Lagrange multipliers of an equality-constrained quadratic program. By our assumptions, the difference between Δx, Δs, Δy, and u, v and z, respectively, is $O(\tilde{\mu})$, as the following lemma states.

LEMMA 1 *Let $\tilde{x}$, $\tilde{y}$, and $\tilde{s}$ satisfy* (12)–(14). *Assume that (i) $A_A(\tilde{x})$ has full row rank, (ii) that $H(\tilde{x},\tilde{y})$ is positive definite on the nullspace of $A_A(\tilde{x})$, (iii) that $\tilde{y}_I = \Theta(\tilde{\mu})$, and (iv) that $\tilde{s}_A = \Theta(\tilde{\mu})$. Further, let Δx, Δs and Δy satisfy* (15), *and let u, v and z satisfy* (16)–(19). *Then, $\Delta x = u + O(\tilde{\mu})$, $\Delta s = v + O(\tilde{\mu})$ and $\Delta y = z + O(\tilde{\mu})$.*

Proof The quantities u, v and z are solutions of a system of linear equations whose matrix is bounded and nonsingular as $\tilde{\mu} \to 0$, by our assumptions. Hence, since Δx, Δy and Δs satisfy (12)–(14), where the only difference is that some $O(\tilde{\mu})$ elements have been added to the matrix, the result follows. ∎

This means that we may use properties of u, v and w to deduce properties of our desired quantities Δx, Δs and Δy, as stated in the following proposition.

THEOREM 2 *Let $\tilde{x}$, $\tilde{y}$, and $\tilde{s}$ satisfy* (12)–(14). *Assume that (i) $A_A(\tilde{x})$ has full row rank, (ii) that $H(\tilde{x},\tilde{y})$ is positive definite on the nullspace of $A_A(\tilde{x})$, (iii) that $\tilde{y}_I = \Theta(\tilde{\mu})$, and (iv) that $\tilde{s}_A = \Theta(\tilde{\mu})$. Further, let Δx, Δs and Δy satisfy* (15). *Then, Δx differs by $O(\tilde{\mu})$ from the optimal solution to the equality-constrained quadratic program*

$$\begin{array}{ll} \underset{p\in\Re^n}{\text{minimize}} & \frac{1}{2}p^T H(\tilde{x},\tilde{y})p + (g(\tilde{x}) - (\mu - \tilde{\mu})A_I(\tilde{x})^T\tilde{S}_I^{-1}e)^T p \\ \text{subject to} & A_A(\tilde{x})p = -c_A(\tilde{x}) + (\mu - \tilde{\mu})\tilde{Y}_A^{-1}e, \end{array} \tag{20}$$

and $\tilde{y}_A + \Delta y_A$ differs by $O(\tilde{\mu})$ from the associated Lagrange multipliers.

Proof The optimality conditions for (20) are given by

$$H(\tilde{x},\tilde{y})p^* + g(\tilde{x}) - (\mu - \tilde{\mu})A_I(\tilde{x})^T\tilde{S}_I^{-1}e \;=\; A_A(\tilde{x})^T\lambda_A^*, \tag{21}$$

$$A_A(\tilde{x})p^* = -c_A(\tilde{x}) + (\mu - \tilde{\mu})\tilde{Y}_A^{-1}e, \tag{22}$$

for an optimal solution p^* together with a Lagrange multiplier vector λ_A^*. Rearrangement of (21)–(22), taking into account that $\tilde{x}$, $\tilde{s}$ and $\tilde{y}$ satisfy (12)–(14), gives

$$H(\tilde{x},\tilde{y})p^* - A_A(\tilde{x})^T(\lambda_A^* - \tilde{y}_A) = -g(\tilde{x}) + A_A(\tilde{x})^T\tilde{y}_A$$
$$+(\mu - \tilde{\mu})A_I(\tilde{x})^T\tilde{S}_I^{-1}e, \qquad A_A(\tilde{x})p^* = -c_A(\tilde{x}) + (\mu - \tilde{\mu})\tilde{Y}_A^{-1}e.$$

By comparing these conditions with (18), it follows that $p^* = u$ and $\lambda_A^* - \tilde{y}_A = z_A$. Lemma 1 now gives $\Delta x = p^* + O(\tilde{\mu})$ and $\lambda_A^* - \tilde{y}_A = \Delta y_A + O(\tilde{\mu})$, as required. ∎

A consequence of Theorem 2 is that if $\mu = \tilde{\mu}$, then the step is near-optimal to the equality-constrained problem where the active constraint are set as equalities, as summarized in the following corollary.

COROLLARY 3 *Let $\tilde{x}$, $\tilde{y}$, and $\tilde{s}$ satisfy* (12)–(14). *Assume that (i) $A_A(\tilde{x})$ has full row rank, (ii) that $H(\tilde{x},\tilde{y})$ is positive definite on the nullspace of $A_A(\tilde{x})$, (iii) that $\tilde{y}_I = \Theta(\tilde{\mu})$, and (iv) that $\tilde{s}_A = \Theta(\tilde{\mu})$. Further, let Δx, Δs and Δy satisfy* (15) *for $\mu = \tilde{\mu}$. Then, Δx differs by $O(\tilde{\mu})$ from the optimal solution to the equality-constrained quadratic program*

$$\begin{array}{ll} \underset{p\in\Re^n}{\text{minimize}} & \frac{1}{2}p^T H(\tilde{x},\tilde{y})p + g(\tilde{x})^T p \\ \textit{subject to} & A_A(\tilde{x})p = -c_A(\tilde{x}), \end{array}$$

and $\tilde{y}_A + \Delta y_A$ differs by $O(\tilde{\mu})$ from the associated Lagrange multipliers.

Another consequence is that for $\mu = \tilde{\mu}$, the step is near-optimal to the "appropriate" inequality-constrained quadratic programming problem.

COROLLARY 4 *Let $\tilde{x}$, $\tilde{y}$, and $\tilde{s}$ satisfy* (12)–(14). *Assume that (i) $A_A(\tilde{x})$ has full row rank, (ii) that $H(\tilde{x},\tilde{y})$ is positive definite on the nullspace of $A_A(\tilde{x})$, (iii) that $\tilde{y}_I = \Theta(\tilde{\mu})$, and (iv) that $\tilde{s}_A = \Theta(\tilde{\mu})$. Further, let Δx, Δs and Δy satisfy* (15) *for $\mu = \tilde{\mu}$. Then, Δx differs by $O(\tilde{\mu})$ from the optimal solution to the equality-constrained quadratic program*

$$\begin{array}{lll} \underset{p\in\Re^n}{\text{minimize}} & \frac{1}{2}p^T H(\tilde{x},\tilde{y})p + g(\tilde{x})^T p & \\ \textit{subject to} & a_i(\tilde{x})^T p \ge -c_i(\tilde{x}), & i \in I_A^+, \\ & a_i(\tilde{x})^T p \le -c_i(\tilde{x}), & i \in I_A \backslash I_A^+, \end{array} \tag{23}$$

and $\tilde{y} + \Delta y$ *differ by* $O(\tilde{\mu})$ *from the associated Lagrange multipliers, where* $I_A^+ = \{i \in A : \tilde{y}_i + z_i \geq 0\}$, *where* z *is given by* (16)–(19).

The conclusion is that if μ is small, of the order of $\tilde{\mu}$, the primal-dual step behaves like the sequential quadratic programming step, i.e., for small perturbations we may expect the step to give a near-optimal solution, but for larger perturbations, the step is likely to violate both inactive constraints (that are ignored) and nonnegativity of the multipliers. In addition, we have not taken into account the implicit requirement that y and s have to be maintained positive.

4. Warm starts of interior methods for linear programming

The above analysis applies also to linear programming. However, linear programming is special in the sense that there always exists a strictly complementary optimal solution, if an optimal solution exists, and the analysis may be specialized further. To be consistent with the discussion in Section 3, we will consider the linear program

$$\begin{array}{ll} \underset{x \in \Re^n}{\text{minimize}} & d^T x \\ \text{subject to} & Ax \geq b, \end{array} \tag{24}$$

and a near-optimal solution to the related linear program

$$\begin{array}{ll} \underset{x \in \Re^n}{\text{minimize}} & \tilde{d}^T x \\ \text{subject to} & Ax \geq \tilde{b}. \end{array} \tag{25}$$

An underlying assumption is that the constraint matrix A has full column rank. Our analysis is "classical" sensitivity analysis in the sense that the constraint matrix A is assumed fixed, whereas the cost coefficients and the right hand sides may differ.

The analysis of the previous section applies. However, here we need not make any assumption about nonsingularity of the resulting limiting Newton system. Again, we assume that a "small" barrier $\tilde{\mu}$ is given, and that $\tilde{x}$, $\tilde{s}$ and $\tilde{y}$ solve

$$\tilde{d} - A^T \tilde{y} = 0, \tag{26}$$

$$A\tilde{x} - \tilde{b} - \tilde{s} = 0, \tag{27}$$

$$\tilde{Y}\tilde{s}e = \tilde{\mu}, \quad \tilde{y} > 0, \quad \tilde{s} > 0. \tag{28}$$

In addition to the assumption on full column rank of A, we make the assumption that the barrier trajectory for the perturbed problem (25) is well defined, i.e., $\{(x, s) : Ax - s = \tilde{b}, s > 0\} \neq \emptyset$ and $\{y : A^T y = \tilde{c}, y > 0\} \neq \emptyset$.

From the initial point given by $\tilde{x}$, $\tilde{s}$ and $\tilde{y}$, we want to take a primal-dual interior step towards solving (1) for a given barrier parameter μ. The Newton equations may be written as

$$\begin{pmatrix} & & A^T \\ A & -I & \\ & \tilde{Y} & \tilde{S} \end{pmatrix} \begin{pmatrix} \Delta x \\ \Delta s \\ \Delta y \end{pmatrix} = \begin{pmatrix} d - \tilde{d} \\ b - \tilde{b} \\ 0 \end{pmatrix} + (\mu - \tilde{\mu}) \begin{pmatrix} 0 \\ 0 \\ e \end{pmatrix}. \tag{29}$$

If Δs is eliminated from (29), the resulting equivalent system of equations becomes

$$\begin{aligned} \begin{pmatrix} & A^T \\ A & \tilde{W}^{-1} \end{pmatrix} \begin{pmatrix} \Delta x \\ \Delta y \end{pmatrix} &= \begin{pmatrix} d - \tilde{d} \\ b - \tilde{b} \end{pmatrix} + (\mu - \tilde{\mu}) \begin{pmatrix} 0 \\ \tilde{Y}^{-1} e \end{pmatrix}, && (30) \\ \Delta s &= A\Delta x + \tilde{b} - b, && (31) \end{aligned}$$

where $\tilde{W} = \tilde{S}^{-1}\tilde{Y}$. By further eliminating Δy, we obtain

$$\begin{aligned} A^T \tilde{W} A \Delta x &= \tilde{d} - d + A^T \tilde{W}(b - \tilde{b}) + (\mu - \tilde{\mu}) A^T \tilde{S}^{-1} e, && (32) \\ \tilde{W}^{-1} \Delta y &= b - \tilde{b} - A\Delta x + (\mu - \tilde{\mu}) \tilde{Y}^{-1} e, && (33) \\ \Delta s &= A\Delta x + \tilde{b} - b. && (34) \end{aligned}$$

From this latter system, we obtain explicit expressions for Δx, Δy and Δs as

$$\begin{aligned} \Delta x &= (A^T \tilde{W} A)^{-1} (\tilde{d} - d) \\ &+ (A^T \tilde{W} A)^{-1} A^T \tilde{W} (b - \tilde{b} + (\mu - \tilde{\mu}) \tilde{Y}^{-1} e), && (35) \\ \Delta y &= \tilde{W} A (A^T \tilde{W} A)^{-1} (d - \tilde{d}) \\ &+ \tilde{W}(I - A(A^T \tilde{W} A)^{-1} A^T \tilde{W})(b - \tilde{b} + (\mu - \tilde{\mu}) \tilde{Y}^{-1} e) && (36) \\ &= \tilde{W} A (A^T \tilde{W} A)^{-1} (d - \tilde{d}) \\ &+ Z(Z^T \tilde{W}^{-1} Z)^{-1} Z^T (b - \tilde{b} + (\mu - \tilde{\mu}) \tilde{Y}^{-1} e), && (37) \\ \Delta s &= A\Delta x + \tilde{b} - b, && (38) \end{aligned}$$

where Z is a matrix whose columns form a basis for $\text{null}(A^T)$. Note that (37) is obtained from (36) from the relations

$$\begin{aligned} &\tilde{W}(I - A(A^T \tilde{W} A)^{-1} A^T \tilde{W}) \\ &= \tilde{W}^{1/2}(I - \tilde{W}^{1/2} A (A^T \tilde{W} A)^{-1} A^T \tilde{W}^{1/2}) \tilde{W}^{1/2} \\ &= \tilde{W}^{1/2}(\tilde{W}^{-1/2} Z (Z^T \tilde{W}^{-1} Z)^{-1} Z^T \tilde{W}^{-1/2}) \tilde{W}^{1/2} \\ &= Z(Z^T \tilde{W}^{-1} Z)^{-1} Z^T. && (39) \end{aligned}$$

A special property of linear programming is that the elements of $\tilde{w}$ can be split into "large" and "small" elements, denoted by $\mathcal{A}$ and $\mathcal{I}$ in consistency with the analysis of Section 3, as stated in the following lemma.

LEMMA 5 *Assume that* $\{(x,s) : Ax - s = \tilde{b}, s > 0\} \neq \emptyset$ *and* $\{y : A^T y = \tilde{c}, y > 0\} \neq \emptyset$. *Let* $\tilde{x}$, $\tilde{y}$ *and* $\tilde{s}$ *satisfy* (26)–(28). *Then, we may partition* $\{1,\ldots,m\} = A \cup I$, *with* $A \cap I = \emptyset$, *such that*

$$\begin{aligned}\tilde{s}_i &= \Theta(1),\ \tilde{y}_i = \Theta(\tilde{\mu}),\ \tilde{w}_i = \Theta(\tilde{\mu}), \quad i \in I,\\ \tilde{s}_i &= \Theta(\tilde{\mu}),\ \tilde{y}_i = \Theta(1),\ \tilde{w}_i = \Theta(1/\tilde{\mu}), \quad i \in A,\end{aligned}$$

where $\tilde{w}_i = \tilde{y}_i/\tilde{s}_i$, $i = 1,\ldots,m$.

Proof See, e.g., Wright [16, Lemma 5.13]. ∎

We may get a further description of Δx, Δs and Δy by utilizing the following result, which essentially states that we may obtain $(A^T\tilde{W}A)^{-1}A^T\tilde{W}$ as a convex combination of solutions obtained from nonsingular $n \times n$ submatrices of A.

THEOREM 6 (DIKIN [2]) *Let A be an $m \times n$ matrix of full column rank, let g be a vector of dimension m, and let D be a positive definite diagonal $m \times m$ matrix. Then,*

$$(A^T DA)^{-1}A^T Dg = \sum_{J \in \mathcal{J}(A)} \left(\frac{\det(D_J)\det(A_J)^2}{\sum_{K\in\mathcal{J}(A)} \det(D_K)\det(A_K)^2} \right) A_J^{-1} g_J,$$

where $\mathcal{J}(A)$ is the collection of sets of row indices associated with nonsingular $n \times n$ submatrices of A.

Proof See, e.g., Ben-Tal and Teboulle [1, Corollary 2.1]. ∎

Note that an implication of Theorem 6 is that $\|(A^T\tilde{W}A)^{-1}A^T\tilde{W}\|$ is bounded when $\tilde{W}$ varies over the set of positive definite and diagonal matrices. For further discussions on this issue, see, e.g., Forsgren [3] and Forsgren and Sporre [7].

In order to give the results on the search directions, we first need a result on the behavior of $(A^T\tilde{W}A)^{-1}(\tilde{d} - d)$. As in the previous section, we will let subscript "A" denote quantities associated with A, and similarly for I.

LEMMA 7 *It holds that* $\|(A^T\tilde{W}A)^{-1}(\tilde{d} - d)\| = O(\tilde{\mu})$ *if* $\tilde{d} - d \in \text{range}(A_A^T)$ *and* $\|(A^T\tilde{W}A)^{-1}(\tilde{d} - d)\| = \Omega(1/\tilde{\mu})$ *otherwise.*

Proof First assume that $\tilde{d} - d \in \text{range}(A_A^T)$. Then, $\tilde{d} - d = A_A^T u$ for some u. We then get

$$(A^T\tilde{W}A)^{-1}(\tilde{d} - d) = (A^T\tilde{W}A)^{-1}A_A^T u = (A^T\tilde{W}A)^{-1}A^T\tilde{W}\tilde{W}^{-1}I_A u, \tag{40}$$

where I_A is the diagonal matrix with ones in diagonals corresponding to A and zeros in diagonals corresponding to I. Taking norms in (40), taking into account that $\|\tilde{W}^{-1}I_A\| = \|\tilde{W}_A^{-1}\|$, gives

$$\|(A^T\tilde{W}A)^{-1}(\tilde{d}-d)\| \leq \|(A^T\tilde{W}A)^{-1}A^T\tilde{W}\|\|\tilde{W}_A^{-1}\|\|u\|. \tag{41}$$

Theorem 6 gives $\|(A^T\tilde{W}A)^{-1}A^T\tilde{W}\| = O(1)$, Lemma 5 shows that $\|\tilde{W}_A^{-1}\| = O(\tilde{\mu})$ and u is a fixed vector. Consequently, (41) gives $\|(A^T\tilde{W}A)^{-1}(\tilde{d}-d)\| = O(\tilde{\mu})$, as required.

Now assume that $\tilde{d} - d \notin \text{range}(A_A^T)$. Then, $Z_A^T(\tilde{d} - d) \neq 0$, where Z_A is a matrix whose columns form a basis for the null space of A_A. Let $\tilde{v} = (A^T\tilde{W}A)^{-1}(\tilde{d}-d)$, i.e., $\tilde{v}$ solves

$$(A_A^T\tilde{W}_AA_A + A_I^T\tilde{W}_IA_I)\tilde{v} = \tilde{d} - d. \tag{42}$$

Premultiplication of (42) by Z_A^T gives

$$Z_A^TA_I^T\tilde{W}_IA_I\tilde{v} = Z_A^T(\tilde{d}-d),$$

and consequently

$$\|\tilde{v}\| \geq \frac{\|Z_A^T(\tilde{d}-d)\|}{\|Z_A^TA_I^T\tilde{W}_IA_I\|}. \tag{43}$$

Lemma 5 shows that $\|Z_A^TA_I^T\tilde{W}_IA_I\| = O(\tilde{\mu})$. Since $Z_A^T(\tilde{d}-d) \neq 0$, it follows from (43) that $\|\tilde{v}\| = \Omega(1/\tilde{\mu})$, as required. ▮

Analogously, a result on the behavior of $\tilde{W}(I - A(A^T\tilde{W}A)^{-1}A^T\tilde{W})(b-\tilde{b})$ is needed.

LEMMA 8 *It holds that* $\|\tilde{W}(I - A(A^T\tilde{W}A)^{-1}A^T\tilde{W})(b-\tilde{b})\| = O(\tilde{\mu})$ *if* $\tilde{b}_A - b_A \in \text{range}(A_A)$ *and* $\|\tilde{W}(I - A(A^T\tilde{W}A)^{-1}A^T\tilde{W})(b-\tilde{b})\| = \Omega(1/\tilde{\mu})$ *otherwise.*

Proof First assume that $b_A - \tilde{b}_A \in \text{range}(A_A)$. Then, $b - \tilde{b} = Au + r$ for some u and r with $r_A = 0$. We then get

$$\begin{aligned}&\tilde{W}(I - A(A^T\tilde{W}A)^{-1}A^T\tilde{W})(b-\tilde{b})\\&= (I - \tilde{W}A(A^T\tilde{W}A)^{-1}A^T)\tilde{W}(Au+r)\\&= (I - \tilde{W}A(A^T\tilde{W}A)^{-1}A^T)\tilde{W}r.\end{aligned} \tag{44}$$

Taking norms in (44) gives

$$\|\tilde{W}(I - A(A^T\tilde{W}A)^{-1}A^T\tilde{W})(b-\tilde{b})\| \leq \|I - \tilde{W}A(A^T\tilde{W}A)^{-1}A^T\|\|\tilde{W}r\|. \tag{45}$$

Theorem 6 gives $\|I - \tilde{W}A(A^T\tilde{W}A)^{-1}A^T\| = O(1)$. Note that since $r_A = 0$, we obtain $\|\tilde{W}r\| = \|\tilde{W}_I r_I\|$. Lemma 5 shows that $\|\tilde{W}_I\| = O(\tilde{\mu})$ and r is a fixed vector. Consequently, (45) gives $\|\tilde{W}(I - A(A^T\tilde{W}A)^{-1}A^T\tilde{W})(b - \tilde{b})\| = O(\tilde{\mu})$, as required.

Now assume that $b_A - \tilde{b}_A \notin \text{range}(A_A)$. Then, $\text{null}(A_A^T) \neq \emptyset$, and there is an orthonormal matrix Z_A whose columns form a basis for $\text{null}(A_A^T)$. Since $b_A - \tilde{b}_A \notin \text{range}(A_A)$, it holds that $Z_A^T(b_A - \tilde{b}_A) \neq 0$. Moreover, there is an orthonormal matrix Z whose columns form a basis for $\text{null}(A^T)$ of the form

$$Z = \begin{pmatrix} Z_A & Z_1 \\ 0 & Z_2 \end{pmatrix},$$

where Z_1 and Z_2 are suitably dimensioned, possibly empty. It follows from (39) that

$$\tilde{W}(I - A(A^T\tilde{W}A)^{-1}A^T\tilde{W})(b - \tilde{b}) = Z(Z^T\tilde{W}^{-1}Z)^{-1}Z^T(b - \tilde{b}). \quad (46)$$

Consequently,

$$\|\tilde{W}(I - A(A^T\tilde{W}A)^{-1}A^T\tilde{W})(b - \tilde{b})\| = \|(Z^T\tilde{W}^{-1}Z)^{-1}Z^T(b - \tilde{b})\|.$$

Let $y = (Z^T\tilde{W}^{-1}Z)^{-1}Z^T(b - \tilde{b})$. Then $Z^T\tilde{W}^{-1}Zy = Z^T(b - \tilde{b})$, or equivalently

$$\begin{pmatrix} Z_A^T & 0 \\ Z_1^T & Z_2^T \end{pmatrix} \begin{pmatrix} \tilde{W}_A^{-1} & 0 \\ 0 & \tilde{W}_I^{-1} \end{pmatrix} \begin{pmatrix} Z_A & Z_1 \\ 0 & Z_2 \end{pmatrix} \begin{pmatrix} y_1 \\ y_2 \end{pmatrix}$$
$$= \begin{pmatrix} Z_A^T & 0 \\ Z_1^T & Z_2^T \end{pmatrix} \begin{pmatrix} b_A - \tilde{b}_A \\ b_I - \tilde{b}_I \end{pmatrix}.$$

The first block of equations implies that

$$\begin{pmatrix} Z_A^T\tilde{W}_A^{-1} & 0 \end{pmatrix} \begin{pmatrix} Z_A & Z_1 \\ 0 & Z_2 \end{pmatrix} \begin{pmatrix} y_1 \\ y_2 \end{pmatrix} = Z_A^T(b_A - \tilde{b}_A).$$

Consequently,

$$\left\| \begin{pmatrix} Z_A^T\tilde{W}_A^{-1} & 0 \end{pmatrix} \right\| \|Z\| \|y\| \geq \|Z_A^T(b_A - \tilde{b}_A)\|.$$

By assumption, $\|Z_A^T(b_A - \tilde{b}_A)\| \neq 0$. Lemma 5 shows that $\|\tilde{W}_A^{-1}\| = O(\tilde{\mu})$, and Z is a fixed matrix. Consequently, (46) gives

$$\|\tilde{W}(I - A(A^T\tilde{W}A)^{-1}A^T\tilde{W})(b - \tilde{b})\| = \Omega(1/\tilde{\mu}),$$

as required. ∎

The following proposition now gives a characterization of Δx. Note that Δx is unbounded if $\tilde{d} - d \notin \text{range}(A_A^T)$.

THEOREM 9 *If* $\tilde{d} - d \in \text{range}(A_A^T)$, *then*

$$\Delta x = (A^T \tilde{W} A)^{-1} A^T \tilde{W} (b - \tilde{b} + (\mu - \tilde{\mu}) \tilde{Y}^{-1} e) + O(\tilde{\mu}).$$

Otherwise, $\Delta x = \Omega(1/\tilde{\mu})$.

Proof This is a consequence of Lemma 7 in conjunction with (35). ∎

Analogously, the following proposition now gives a characterization of Δy. Note that Δy is unbounded if $\tilde{b}_A - b_A \notin \text{range}(A_A)$.

THEOREM 10 *If* $\tilde{b}_A - b_A \in \text{range}(A_A)$, *then*

$$\begin{aligned}
\Delta y &= \tilde{W} A (A^T \tilde{W} A)^{-1} (d - \tilde{d}) \\
&+ (\mu - \tilde{\mu}) \tilde{W} (I - A (A^T \tilde{W} A)^{-1} A^T \tilde{W}) \tilde{Y}^{-1} e + O(\tilde{\mu}) \\
&= \tilde{W} A (A^T \tilde{W} A)^{-1} (d - \tilde{d}) \\
&+ (\mu - \tilde{\mu}) Z (Z^T \tilde{W}^{-1} Z)^{-1} Z^T \tilde{Y}^{-1} e + O(\tilde{\mu}).
\end{aligned}$$

Otherwise, $\Delta y = \Omega(1/\tilde{\mu})$.

Proof This is a consequence of Lemma 8 in conjunction with (37). ∎

Finally, for the case of primal and dual nondegeneracy, it follows that both primal and dual steps are bounded.

COROLLARY 11 *If* A_A *is square and nonsingular, then*

$$\begin{aligned}
\Delta x &= A_A^{-1} (b_A - \tilde{b}_A + (\mu - \tilde{\mu}) \tilde{Y}_A^{-1} e) + O(\tilde{\mu}), \\
\Delta s_A &= (\mu - \tilde{\mu}) \tilde{Y}_I^{-1} e + O(\tilde{\mu}), \\
\Delta s_I &= A_I A_A^{-1} (b_A - \tilde{b}_A + (\mu - \tilde{\mu}) \tilde{Y}_A^{-1} e) - b_I + \tilde{b}_I + O(\tilde{\mu}), \\
\Delta y_A &= A_A^{-T} (d - \tilde{d} - (\mu - \tilde{\mu}) A_I^T \tilde{S}_I^{-1} e) + O(\tilde{\mu}), \\
\Delta y_I &= (\mu - \tilde{\mu}) \tilde{S}_I^{-1} e + O(\tilde{\mu}).
\end{aligned}$$

Proof If A_A is square and nonsingular, Theorem 6 in conjunction with Lemma 5 gives

$$(A^T \tilde{W} A)^{-1} A_A^T \tilde{W}_A = A_A^{-1} + O(\tilde{\mu}) \quad \text{and} \quad \|(A^T \tilde{W} A)^{-1}\| = O(\tilde{\mu}). \tag{47}$$

Consequently, $\|(A^T \tilde{W} A)^{-1} A_I^T \tilde{W}_I Y_I^{-1}\| = O(\tilde{\mu})$. The result for Δx now follows by using (47) in Theorem 9. The result for Δs follows from Δx.

Analogously, we may let

$$Z = \begin{pmatrix} Z_A \\ Z_I \end{pmatrix} = \begin{pmatrix} -A_A^{-T} A_I \\ I \end{pmatrix},$$

for which

$$(Z^T \tilde{W}^{-1} Z)^{-1} Z_I^T \tilde{W}_I^{-1} = I + O(\tilde{\mu}) \quad \text{and} \quad \|(Z^T \tilde{W}^{-1} Z)^{-1}\| = O(\tilde{\mu}). \tag{48}$$

Consequently, $\|(Z^T \tilde{W}^{-1} Z)^{-1} Z_A^T \tilde{Y}_A^{-1}\| = O(\tilde{\mu})$. The result for Δy now follows by using (47) and (48) in Theorem 10. ∎

Corollary 11 gives the result from the nonlinear programming case, specialized to linear programming, since the limiting Newton equations are nonsingular in this situation.

4.1 Example linear programming problem

Consider the example linear programming problem where A, $\tilde{b}$ and $\tilde{d}$ are given by

$$A = \begin{pmatrix} 1 & 0 \\ -1 & 0 \\ 0 & 1 \end{pmatrix}, \quad \tilde{b} = \begin{pmatrix} -1 \\ -1 \\ 0 \end{pmatrix}, \quad \text{and} \quad \tilde{d} = \begin{pmatrix} 0 \\ 1 \end{pmatrix}.$$

Then, $\tilde{x} = (0\ \tilde{\mu})^T$, $\tilde{y} = (\tilde{\mu}\ \tilde{\mu}\ 1)^T$ and $\tilde{s} = (1\ 1)^T$. Hence, $\tilde{w} = \tilde{S}^{-1}\tilde{Y}e = (\tilde{\mu}\ \tilde{\mu}\ 1/\tilde{\mu})^T$. Accordingly,

$$\tilde{W}A = \begin{pmatrix} \tilde{\mu} & 0 \\ -\tilde{\mu} & 0 \\ 0 & \frac{1}{\tilde{\mu}} \end{pmatrix}, \quad A^T \tilde{W} A = \begin{pmatrix} 2\tilde{\mu} & 0 \\ 0 & \frac{1}{\tilde{\mu}} \end{pmatrix}.$$

$$Z = \begin{pmatrix} 1 \\ 1 \\ 0 \end{pmatrix}, \quad \tilde{W}^{-1} Z = \begin{pmatrix} \frac{1}{\tilde{\mu}} \\ \frac{1}{\tilde{\mu}} \\ 0 \end{pmatrix}, \quad (Z^T \tilde{W}^{-1} Z)^{-1} = \frac{\tilde{\mu}}{2}.$$

We may use Theorem 6 to express

$$\begin{aligned} (A^T \tilde{W} A)^{-1} A^T \tilde{W} &= \frac{1}{2}\begin{pmatrix} 1 & 0 & 0 \\ 0 & 0 & 1 \end{pmatrix} + \frac{1}{2}\begin{pmatrix} 0 & -1 & 0 \\ 0 & 0 & 1 \end{pmatrix} \\ &= \begin{pmatrix} \frac{1}{2} & -\frac{1}{2} & 0 \\ 0 & 0 & 1 \end{pmatrix}. \end{aligned} \tag{49}$$

In this example $A_A = (0\ 1)$. Since A_A has full row rank, a combination of Theorem 10 and (49) gives

$$\Delta y = \begin{pmatrix} \frac{1}{2} & 0 \\ -\frac{1}{2} & 0 \\ 0 & 1 \end{pmatrix} \begin{pmatrix} d_1 \\ d_2 - 1 \end{pmatrix} + (\mu - \tilde{\mu}) \begin{pmatrix} 1 \\ 1 \\ 0 \end{pmatrix} + O(\tilde{\mu}).$$

However, since A_A does not have full column rank, Theorem 9 shows that Δx is $\Omega(1/\tilde{\mu})$ unless $d_1 = \tilde{d}_1$. For $d_1 = \tilde{d}_1$, a combination of Theorem 9 and (49) gives

$$\Delta x = \begin{pmatrix} \frac{1}{2} & -\frac{1}{2} & 0 \\ 0 & 0 & 1 \end{pmatrix} \begin{pmatrix} b_1 + 1 \\ b_2 + 1 \\ b_3 \end{pmatrix} + (\mu - \tilde{\mu}) \begin{pmatrix} 0 \\ 1 \end{pmatrix} + O(\tilde{\mu}).$$

For large-scale problems, the explicit representations from this small problem are naturally not available, but we have included the example to give a feeling of what Theorems 9 and 10 say.

5. Summary

We have characterized search directions that would arise in warm starts for interior methods, first for the general nonlinear programming case, and then more specialized results for the linear programming case. The difficulties in warm starts for interior methods are emphasized by these characterizations, since the directions are similar to directions what would arise in a sequential-quadratic-programming method applied to the active constraints only.

The results are related to "false convergence" of interior methods on non-convex problems in that the iterates are close to the boundary of the inequality constraints, but not well centered with respect to the trajectory. See e.g., Wächter and Biegler [14] or Forsgren and Sporre [8], for further discussions on false convergence.

References

[1] A. Ben-Tal and M. Teboulle. A geometric property of the least squares solution of linear equations. *Linear Algebra Appl.*, 139:165–170, 1990.

[2] I. I. Dikin. On the speed of an iterative process. *Upravlyaemye Sistemi*, 12:54–60, 1974.

[3] A. Forsgren. On linear least-squares problems with diagonally dominant weight matrices. *SIAM J. Matrix Anal. Appl.*, 17:763–788, 1996.

[4] A. Forsgren. Inertia-controlling factorizations for optimization algorithms. *Appl. Num. Math.*, 43:91–107, 2002.

[5] A. Forsgren, P. E. Gill, and J. D. Griffin. Iterative solution of augmented systems arising in interior methods. Report TRITA-MAT-2005-OS3, Department of Mathematics, Royal Institute of Technology, Stockholm, Sweden, 2005.

[6] A. Forsgren, P. E. Gill, and M. H. Wright. Interior methods for nonlinear optimization. *SIAM Rev.*, 44(4):525–597 (electronic) (2003), 2002.

[7] A. Forsgren and G. Sporre. On weighted linear least-squares problems related to interior methods for convex quadratic programming. *SIAM J. Matrix Anal. Appl.*, 23:42–56, 2001.

[8] A. Forsgren and G. Sporre. Relations between divergence of multipliers and convergence to infeasible points in primal-dual interior methods for nonconvex nonlinear programming. Report TRITA-MAT-2002-OS7, Department of Mathematics, Royal Institute of Technology, Stockholm, Sweden, 2002.

[9] J. Gondzio and A. Grothey. Reoptimization with the primal-dual interior point method. *SIAM J. Optim.*, 13(3):842–864 (electronic) (2003), 2002.

[10] M. Gonzalez-Lima, H. Wei, and H. Wolkowicz. A stable iterative method for linear programming. Report CORR 2004-26, Department of Combinatorics and Optimization, University of Waterloo, 2004.

[11] B. Jansen, J. J. de Jong, C. Roos, and T. Terlaky. Sensitivity analysis in linear programming:just be careful! *European J. Oper. Res.*, 101:15–28, 1997.

[12] W.-J. Kim, C.-K. Park, and S. Park. An ϵ-sensitivity analysis in the primal-dual interior point method. *European J. Oper. Res.*, 116:629–639, 1999.

[13] J. Nocedal and S. J. Wright. *Numerical Optimization*. Springer, New York, 1999. ISBN 0-387-98793-2.

[14] A. Wächter and L. T. Biegler. Failure of global convergence for a class of interior point methods for nonlinear programming. *Math. Program.*, 88:565–574, 2000.

[15] M. H. Wright. The interior-point revolution in optimization: history, recent developments, and lasting consequences. *Bull. Amer. Math. Soc. (N.S.)*, 42(1):39–56 (electronic), 2005.

[16] S. J. Wright. *Primal-Dual Interior-Point Methods*. SIAM, Society for Industrial and Applied Mathematics, Philadelphia, 1997. ISBN 0-89871-382-X.

[17] E. A. Yıldırım. An interior-point perspective on sensitivity analysis in semidefinite programming. *Math. Oper. Res.*, 28(4):649–676, 2003.

[18] E. A. Yıldırım and M. J. Todd. Sensitivity analysis in linear programming and semidefinite programming using interior-point methods. *Math. Program.*, 90(2, Ser. A):229–261, 2001.

[19] E. A. Yildirim and M. J. Todd. An interior-point approach to sensitivity analysis in degenerate linear programs. *SIAM J. Optim.*, 12(3):692–714 (electronic), 2002.

[20] E. A. Yildirim and S. J. Wright. Warm-start strategies in interior-point methods for linear programming. *SIAM J. Optim.*, 12(3):782–810 (electronic), 2002.

RECENT ADVANCES IN BOUND CONSTRAINED OPTIMIZATION

W. W. Hager,[1] and H. Zhang[1]

[1] *PO Box 118105, Department of Mathematics, University of Florida, Gainesville, FL 32611-8105,{hager, hzhang}@math.ufl.edu, http://www.math.ufl.edu/~hager, http://www.math.ufl.edu/~hzhang**

Abstract A new active set algorithm (ASA) for large-scale box constrained optimization is introduced. The algorithm consists of a nonmonotone gradient projection step, an unconstrained optimization step, and a set of rules for switching between the two steps. Numerical experiments and comparisons are presented using box constrained problems in the CUTEr and MINPACK test problem libraries.

keywords: Nonmonotone gradient projection, box constrained optimization, active set algorithm, ASA, cyclic BB method, CBB, conjugate gradient method, CG_DESCENT, degenerate optimization

1. Introduction

We present a new active set algorithm for solving the box constrained optimization problem

$$\min \{f(\mathbf{x}) : \mathbf{x} \in \mathcal{B}\}, \tag{1}$$

where f is a real-valued, continuously differentiable function defined on the box

$$\mathcal{B} = \{\mathbf{x} \in \Re^n : \mathbf{l} \le \mathbf{x} \le \mathbf{u}\}. \tag{2}$$

Here $\mathbf{l} < \mathbf{u}$; possibly, $l_i = -\infty$ or $u_i = \infty$. The following notation is used throughout the paper: $\|\cdot\|$ is the Euclidean norm of a vector, the subscript k is used for the iteration number, while x_{ki} stands for the i-th component of the iterate $\mathbf{x}_k$. The gradient $\nabla f(\mathbf{x})$ is a row vector while $\mathbf{g}(\mathbf{x}) = \nabla f(\mathbf{x})^{\mathsf{T}}$ is a column vector and ^{T} denotes transpose. The gradient at the iterate $\mathbf{x}_k$ is $\mathbf{g}_k = \mathbf{g}(\mathbf{x}_k)$, the Hessian of f at $\mathbf{x}$ is $\nabla^2 f(\mathbf{x})$, and the ball with center $\mathbf{x}$ and radius ρ is $B_\rho(\mathbf{x})$.

*This paper is based upon work supported by the National Science Foundation under Grant No. 0203270.

Please use the following format when citing this chapter:
Author(s) [insert Last name, First-name initial(s)], 2006, in IFIP International Federation for Information Processing, Volume 199, System Modeling and Optimization, eds. Ceragioli F., Dontchev A., Furuta H., Marti K., Pandolfi L., (Boston: Springer), pp. [insert page numbers].

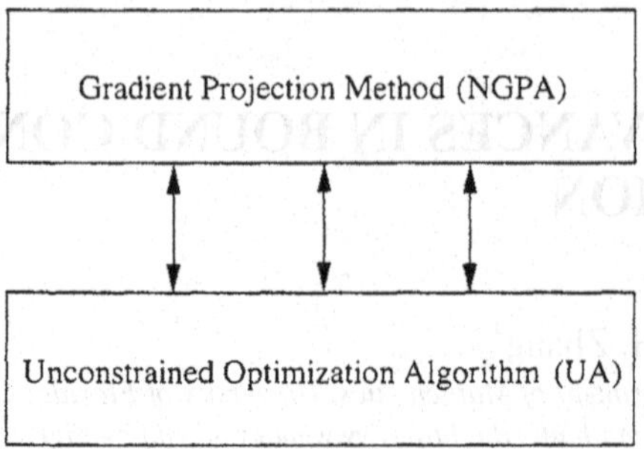

Figure 1. Structure of ASA.

The problem (1) may result from the discretization of a variational inequality such as the obstacle problem [42, 49]:

$$\min \int_{\Omega} \|\nabla u(x)\|^2 + 2f(x)u(x)dx$$
$$\text{subject to } u(x) \geq \psi(x) \ \ a.e.$$

It may come from the discretization of a control problem such as

$$\min \ f(\mathbf{x}, \mathbf{u})$$
$$\text{subject to } \dot{\mathbf{x}} = \mathbf{A}\mathbf{x} + \mathbf{B}\mathbf{u}, \ \ \mathbf{x}(0) = \mathbf{x}_0, \ \ \mathbf{u} \geq \mathbf{0} \ \ a.e.,$$

where $\mathbf{A}$ and $\mathbf{B}$ are operators and the dot denotes time derivative. It also appears as the subproblem in augmented Lagrangian or penalty methods [18, 4, 26, 27, 30, 32, 47]. For example, in an augmented Lagrangian approach to the nonlinear optimization

$$\min f(\mathbf{x}) \text{ subject to } \mathbf{h}(\mathbf{x}) = \mathbf{0}, \ \ \mathbf{x} \geq \mathbf{0},$$

we might solve the box constrained subproblem

$$\min f(\mathbf{x}) + \boldsymbol{\lambda}^{\mathsf{T}}\mathbf{h}(\mathbf{x}) + p\|\mathbf{h}(\mathbf{x})\|^2 \text{ subject to } \mathbf{x} \geq \mathbf{0},$$

where p is the penalty parameter and $\boldsymbol{\lambda}$ is an approximation to a Lagrange multiplier for the equality constraint. Thus efficient algorithms for large-scale box constrained optimization problems are important, both in theory and practice.

2. Gradient projection methods

Our active set algorithm (ASA) has two phases as indicated in Figure 1 a gradient projection phase and an unconstrained optimization phase. For the unconstrained optimization phase, we exploit the box structure of the constraints

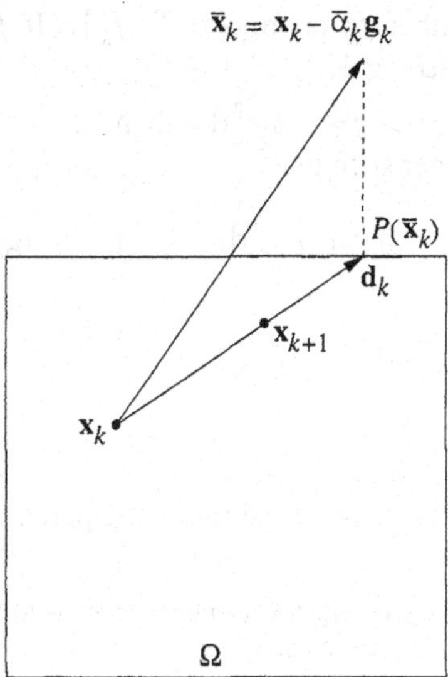

Figure 2. The gradient projection step.

in (1), while the gradient projection phase can be applied to any problem with a closed, convex feasible set. Hence, in this section, we consider a more general problem in which the box $\mathcal{B}$ is replaced by a nonempty, closed convex set Ω:

$$\min \{f(\mathbf{x}) : \mathbf{x} \in \Omega\}. \tag{3}$$

Let P denote the projection onto Ω. The gradient projection step at iteration k is depicted in Figure 2. Starting from the current iterate $\mathbf{x}_k$, we take a positive step $\bar{\alpha}_k$ along the negative gradient arriving at $\bar{\mathbf{x}}_k = \mathbf{x}_k - \bar{\alpha}_k \mathbf{g}_k$. If $\bar{\mathbf{x}}_k$ is outside Ω, then we apply the projection P to obtain a point $P(\bar{\mathbf{x}}_k)$ on the boundary of Ω. The search direction $\mathbf{d}_k$ is along the line segment $[\mathbf{x}_k, P(\bar{\mathbf{x}}_k)]$. The new iterate $\mathbf{x}_{k+1}$ is obtained by a line search along the search direction.

A more precise statement of the gradient projection algorithm follows:

Nonmonotone Gradient Projection Algorithm (NGPA)

Initialize $k = 0$, $\mathbf{x}_0 =$ starting guess, and $f_{-1}^r = f(\mathbf{x}_0)$.

While $\|P(\mathbf{x}_k - \mathbf{g}_k) - \mathbf{x}_k\| > \epsilon$

1. Choose $\bar{\alpha}_k \in [\alpha_{\min}, \alpha_{\max}]$ and set $\mathbf{d}_k = P(\mathbf{x}_k - \bar{\alpha}_k \mathbf{g}_k) - \mathbf{x}_k$.
2. Choose f_k^r so that $f(\mathbf{x}_k) \le f_k^r \le \max\{f_{k-1}^r, f_k^{\max}\}$ and $f_k^r \le f_k^{\max}$ infinitely often.

3. Let f_R be either f_k^r or $\min\{f_k^{\max}, f_k^r\}$. If $f(\mathbf{x}_k + \mathbf{d}_k) \le f_R + \delta \mathbf{g}_k^\mathsf{T} \mathbf{d}_k$, then $\alpha_k = 1$.
4. If $f(\mathbf{x}_k + \mathbf{d}_k) > f_R + \delta \mathbf{g}_k^\mathsf{T} \mathbf{d}_k$, then $\alpha_k = \eta^j$ where $j > 0$ is the smallest integer such that

$$f(\mathbf{x}_k + \eta^j \mathbf{d}_k) \le f_R + \eta^j \delta \mathbf{g}_k^\mathsf{T} \mathbf{d}_k. \tag{4}$$

5. Set $\mathbf{x}_{k+1} = \mathbf{x}_k + \alpha_k \mathbf{d}_k$ and $k = k + 1$.

End

The statement of NGPA involves the following parameters:

$\epsilon \in [0, \infty)$	–	convergence tolerance ($P(\mathbf{x}_k - \mathbf{g}_k) = \mathbf{x}_k$ if and only if $\mathbf{x}_k$ is a stationary point)
$[\alpha_{\min}, \alpha_{\max}] \subset (0, \infty)$	–	bound on the stepsize in Step 1
$f_k^{\max}$	–	$\max\{f(\mathbf{x}_{k-i}) : 0 \le i \le \min(k, M-1)\}$ (local maximum of function values near $\mathbf{x}_k$, $M > 0$)
$\delta, \eta \in (0, 1)$	–	parameters entering the Armijo line search in Step 4

In Step 2, the requirement that "$f_k^r \le f_k^{\max}$ infinitely often" is needed for global convergence. This is a rather weak condition which can be satisfied by many strategies. For example, every L iteration, we could simply set $f_k^r = f_k^{\max}$. Another strategy, closer in spirit to the one used in the numerical experiments, is to choose a decrease parameter $\Delta > 0$ and an integer $L > 0$ and set $f_k^r = f_k^{\max}$ if

$$f(\mathbf{x}_{k-L}) - f(\mathbf{x}_k) \le \Delta.$$

Thus we set $f_k^r = f_k^{\max}$ when the function values decrease "too slowly."

For our numerical experiments, the initial step $\bar{\alpha}_k$ in Step 1 is generated by a cyclic Barzilai-Borwein method developed in [20]. The traditional Barzilai and Borwein stepsize [3] is

$$\overline{\alpha}_{k+1}^{BB} = \frac{\mathbf{s}_k^\mathsf{T} \mathbf{s}_k}{\mathbf{s}_k^\mathsf{T} \mathbf{y}_k}, \tag{5}$$

where $\mathbf{s}_k = \mathbf{x}_{k+1} - \mathbf{x}_k$ and $\mathbf{y}_k = \mathbf{g}_{k+1} - \mathbf{g}_k$. If the same BB stepsize is repeated for several iterations, then even faster convergence can be achieved (see [20]). These schemes in which the same BB stepsize are repeated for several iterations are called cyclic BB schemes (CBB). The CBB update formula is

$$\overline{\alpha}_{k+j} = \overline{\alpha}_k^{BB} \quad \text{for } j = 0, \ldots, m-1,$$

where m is the cycle length, and k is a multiple of m. The cycle length can be chosen in a adaptive way, as explained in [20].

Our line search along the search direction $\mathbf{d}_k$ is an Armijo type line search [1], which may be viewed as a relaxed version of the Grippo, Lampariello, and Lucidi nonmonotone line search [34] (denoted GLL). For NGPA, the GLL scheme corresponds to $f_k^r = f_k^{\max}$ for each k. In practice, we have obtained faster convergence results by allowing the reference function value f_k^r to decay more slowly on average than $f_k^{\max}$.

Our statement of the gradient projection algorithm employs a direction operator $\mathbf{d}^\alpha(\mathbf{x})$ given by

$$\mathbf{d}^\alpha(\mathbf{x}) = P(\mathbf{x} - \alpha\mathbf{g}(\mathbf{x})) - \mathbf{x},$$

where α is a scalar. Some properties of $\mathbf{d}^\alpha$ are summarized below (see [37] for further details concerning these properties and other results presented in this paper):

PROPOSITION 1 *P and $\mathbf{d}^\alpha$ have the following properties:*

P1. $\|P(\mathbf{x}) - P(\mathbf{y})\| \leq \|\mathbf{x} - \mathbf{y}\|$ *for all* $\mathbf{x}$ *and* $\mathbf{y} \in \Re^n$.

P2. *For any* $\mathbf{x} \in \Omega$ *and* $\alpha > 0$, $\mathbf{d}^\alpha(\mathbf{x}) = \mathbf{0}$ *if and only if* $\mathbf{x}$ *is a stationary point for* (3).

P3. *Suppose* $\mathbf{x}^*$ *is a stationary point for* (3). *If for some* $\mathbf{x} \in \Re^n$, *there exist positive scalars* λ *and* γ *such that*

$$(\mathbf{g}(\mathbf{x}) - \mathbf{g}(\mathbf{x}^*))^\mathsf{T}(\mathbf{x} - \mathbf{x}^*) \geq \gamma\|\mathbf{x} - \mathbf{x}^*\|^2 \tag{6}$$

and

$$\|\mathbf{g}(\mathbf{x}) - \mathbf{g}(\mathbf{x}^*)\| \leq \lambda\|\mathbf{x} - \mathbf{x}^*\|, \tag{7}$$

then we have

$$\|\mathbf{x} - \mathbf{x}^*\| \leq \left(\frac{1+\lambda}{\gamma}\right)\|\mathbf{d}^1(\mathbf{x})\|. \tag{8}$$

P4. *Suppose that* f *is twice-continuously differentiable near a stationary point* $\mathbf{x}^*$ *of* (1) *satisfying the strong second-order sufficient optimality condition; that is, there exists* $\gamma > 0$ *such that*

$$\mathbf{d}^\mathsf{T}\nabla^2 f(\mathbf{x}^*)\mathbf{d} \geq \gamma\|\mathbf{d}\|^2 \tag{9}$$

for all $\mathbf{d} \in \Re^n$ *with the property that* $d_i = 0$ *when* $x_i = 0$ *and* $g_i(\mathbf{x}^*) > 0$. *Then there exists* $\rho > 0$ *with the following property:*

$$\|\mathbf{x} - \mathbf{x}^*\| \leq \sqrt{1 + \left(\frac{(1+\lambda)^2}{.5\gamma}\right)^2}\,\|\mathbf{d}^1(\mathbf{x})\| \tag{10}$$

whenever $\mathbf{x} \in B_\rho(\mathbf{x}^*)$, *where* λ *is any Lipschitz constant for* ∇f *on* $B_\rho(\mathbf{x}^*)$.

In P3 we assume a convexity/monotonicity type condition at x; for any x which satisfies (6) and the Lipschitz condition (7), we can estimate the error in x in accordance with (8). In P4, we make a convexity type assumption at $\mathbf{x}^*$ (the strong second-order sufficient optimality condition), and we have the error estimate (10) in a neighborhood of $\mathbf{x}^*$. Based on P3 and P4, the Lipschitz continuity of $\mathbf{d}^1(\cdot)$ implied by P1, and the fact P2 that $\mathbf{d}^1(\mathbf{x}) = \mathbf{0}$ if and only if x is a stationary point, the function $\mathbf{d}^1(\mathbf{x})$ can be used to measure the error in any iterate $\mathbf{x}_k$. In particular the convergence condition $\|P(\mathbf{x}_k - \mathbf{g}_k) - \mathbf{x}_k\| \le \epsilon$ in NGPA is equivalent to $\|\mathbf{d}^1(\mathbf{x}_k)\| \le \epsilon$.

Sufficient conditions for the global convergence of NGPA are given below.

THEOREM 1 *Let* $\mathcal{L}$ *be the level set defined by*

$$\mathcal{L} = \{\mathbf{x} \in \Omega : f(\mathbf{x}) \le f(\mathbf{x}_0)\}. \tag{11}$$

Assume the following conditions hold:

G1. f *is bounded from below on* $\mathcal{L}$ *and* $d_{\max} = \sup_k \|\mathbf{d}_k\| < \infty$.

G2. *If* $\bar{\mathcal{L}}$ *is the collection of* $\mathbf{x} \in \Omega$ *whose distance to* $\mathcal{L}$ *is at most* $d_{\max}$, *then* ∇f *is Lipschitz continuous on* $\bar{\mathcal{L}}$.

Then NGPA with $\epsilon = 0$ *either terminates in a finite number of iterations at a stationary point, or we have*

$$\liminf_{k \to \infty} \|\mathbf{d}^1(\mathbf{x}_k)\| = 0.$$

When f is a strongly convex function, Theorem 1 can be strengthened as follows:

COROLLARY 2 *Suppose* f *is strongly convex and twice continuously differentiable on* Ω, *and there is a positive integer* L *with the property that for each* k, *there exists* $j \in [k, k+L)$ *such that* $f_j^r \le f_j^{\max}$. *Then the iterates* $\mathbf{x}_k$ *of NGPA with* $\epsilon = 0$ *converge to the global minimizer* $\mathbf{x}^*$.

3. Active Set Algorithm

In this section, we focus on the active set algorithm. Unlike the gradient projection algorithm where the feasible set can be any closed, convex set, we now restrict ourselves to box constraints. Moreover, to simplify the discussion, we consider (without loss of generality) the special case $\mathbf{l} = \mathbf{0}$ and $\mathbf{u} = \infty$. In other words, the constraint is $\mathbf{x} \ge \mathbf{0}$.

Although the gradient projection scheme NGPA has an attractive global convergence theory, the convergence rate can be slow in a neighborhood of a local minimizer. We accelerate the convergence by exploiting a superlinearly convergent algorithm for unconstrained minimization. For the numerical experiments, we utilize the conjugate gradient code CG_DESCENT [35, 38, 36, 39] for the unconstrained algorithm (UA). In general, any UA satisfying the following conditions can be employed:

Unconstrained Algorithm (UA) Requirements

U1. $\mathbf{x}_k \geq 0$ and $f(\mathbf{x}_{k+1}) \leq f(\mathbf{x}_k)$ for each k.

U2. $\mathcal{A}(\mathbf{x}_k) \subset \mathcal{A}(\mathbf{x}_{k+1})$ for each k where $\mathcal{A}(\mathbf{x}) = \{i \in [1,n] : x_i = 0\}$.

U3. If $x_{ji} > 0$ for $j \geq k$, then $\liminf_{j \geq k} |g_i(\mathbf{x}_j)| = 0$.

U4. Whenever the unconstrained algorithm is started, compute $\mathbf{x}_{k+1} = P(\mathbf{x}_k - \alpha_k \mathbf{g}_I(\mathbf{x}_k))$, where α_k is obtained from a Wolfe line search. That is, α_k is chosen to satisfy

$$\phi(\alpha_k) \leq \phi(0) + \delta\alpha_k\phi'(0) \quad \text{and} \quad \phi'(\alpha_k) \geq \sigma\phi'(0),$$

where $\phi(\alpha) = f(P(\mathbf{x}_k - \alpha\mathbf{g}_I(\mathbf{x}_k)))$, $0 < \delta < \sigma < 1$, and $\mathbf{g}_I(\mathbf{x})$ is the part of the gradient associated with inactive constraints:

$$g_{Ii}(\mathbf{x}) = \begin{cases} 0 & \text{if } x_i = 0, \\ g_i(\mathbf{x}) & \text{if } x_i > 0. \end{cases}$$

Conditions U1–U3 are sufficient for global convergence, while U1–U4 are sufficient for the local convergence results summarized below. U4 could be replaced by another descent condition for the initial line search, however, the local analysis in [37] has been carried out under U4.

The active set algorithm is based on a set of rules which determine when we switch between NGPA and UA. These rules correspond to the double arrows in Figure 1. Before presenting the switching rules, we give some motivation. A fundamental set embedded in our switching rules is the "undecided index set" $\mathcal{U}$:

$$\mathcal{U}(\mathbf{x}) = \{i \in [1,n] : |g_i(\mathbf{x})| \geq \|\mathbf{d}^1(\mathbf{x})\|^\alpha \text{ and } x_i \geq \|\mathbf{d}^1(\mathbf{x})\|^\beta\}, \tag{12}$$

where $\alpha \in (0,1)$ and $\beta \in (1,2)$ are fixed constants. In the numerical experiments, we take $\alpha = 1/2$ and $\beta = 3/2$. Observe that at a local minimizer $\mathbf{x}^*$, the only components of the gradient which do not vanish are associated with components of $\mathbf{x}^*$ at the boundary of the feasible set. The undecided index set consists of indices of large gradient components with large $\mathbf{x}$ components (in the sense of (12)).

We show [37] that if f is twice continuously differentiable, then for any algorithm converging to a stationary point where each iterate is generated by either NGPA or a UA satisfying U1–U4, the set $\mathcal{U}(\mathbf{x}_k)$ is empty for k sufficiently large. This result does not depend on the rules used to switch between NGPA and UA. When $\mathcal{U}(\mathbf{x}_k)$ becomes empty while performing NGPA, we feel that the strictly active constraints at a stationary point are almost identified and we may switch to UA to exploit its faster convergence.

Another quantity which enters into our switching rules is the ratio between the norm of the inactive gradient components $\|\mathbf{g}_I(\mathbf{x})\|$ and the error estimator $\|\mathbf{d}^1(\mathbf{x})\|$. By U3, $\mathbf{g}_I(\mathbf{x}_k)$ tends to zero as iterates are generated by the UA. By U2, UA does not free constraints; hence, any limit, say $\mathbf{y}^*$, of iterates typically does not solve the original problem (1). In other words, $\mathbf{d}^1(\mathbf{y}^*)$ may not be $\mathbf{0}$. We stop the UA and switch to the NGPA when $\|\mathbf{g}_I(\mathbf{x}_k)\|$ is sufficiently small relative to $\|\mathbf{d}^1(\mathbf{x}_k)\|$. More precisely, we introduce a parameter $\mu > 0$ and we branch from UA to NGPA when

$$\|\mathbf{g}_I(\mathbf{x}_k)\| \le \mu\|\mathbf{d}^1(\mathbf{x}_k)\|. \tag{13}$$

Unlike UA where bound components are fixed by U2, NGPA allows bound components of $\mathbf{x}_k$ to move into the interior of the feasible set. Hence, by switching from UA to NGPA, the iterates are able to move to a new face of the feasible set. In NGPA we may decrease μ, in which case the accuracy with which we solve subproblems in UA increases.

Assuming f is twice continuously differentiable, we show in [37] that for a local minimizer $\mathbf{x}^*$ satisfying the strong second-order sufficient optimality condition (9) and for any sequence of iterates generated by either NGPA or a UA satisfying U1–U4, there exists a scalar $\mu^* > 0$ such that

$$\|\mathbf{g}_I(\mathbf{x}_k)\| \ge \mu^*\|\mathbf{d}^1(\mathbf{x}_k)\| \tag{14}$$

for k sufficiently large. As a result, when μ becomes sufficiently small, condition (13) is never satisfied; hence, if the switch from UA to NGPA is dictated by (13), we conclude that the iterates will never leave the UA. In other words, we eventually solve (1) using the unconstrained optimization algorithm.

With these insights, we now state ASA, or equivalently, we give the switching rules:

Active Set Algorithm (ASA)

1. While $\|\mathbf{d}^1(\mathbf{x}_k)\| > \epsilon$ execute NGPA and check the following:

   a. If $\mathcal{U}(\mathbf{x}_k) = \emptyset$, then

If $\|\mathbf{g}_I(\mathbf{x}_k)\| < \mu\|\mathbf{d}^1(\mathbf{x}_k)\|$, then $\mu = \rho\mu$.
Otherwise, goto Step 2.

b. Else if $\mathcal{A}(\mathbf{x}_k) = \mathcal{A}(\mathbf{x}_{k-1}) = \ldots = \mathcal{A}(\mathbf{x}_{k-n_1})$, then
If $\|\mathbf{g}_I(\mathbf{x}_k)\| \geq \mu\|\mathbf{d}^1(\mathbf{x}_k)\|$, then goto Step 2.

End

2. While $\|\mathbf{d}^1(\mathbf{x}_k)\| > \epsilon$ execute UA and check the following:

a. If $\|\mathbf{g}_I(\mathbf{x}_k)\| < \mu\|\mathbf{d}^1(\mathbf{x}_k)\|$, then restart NGPA (Step 1).

b. If $|\mathcal{A}(\mathbf{x}_{k-1})| < |\mathcal{A}(\mathbf{x}_k)|$, then
If $\mathcal{U}(\mathbf{x}_k) = \emptyset$ or $|\mathcal{A}(\mathbf{x}_k)| > |\mathcal{A}(\mathbf{x}_{k-1})| + n_2$, restart UA at $\mathbf{x}_k$.
Else restart NGPA.

End

End

In addition to the convergence tolerance ϵ introduced previously, ASA utilizes the following four parameters:

$\mu \in (0, 1)$	-	$\|\mathbf{g}_I(\mathbf{x}_k)\| < \mu\|\mathbf{d}^1(\mathbf{x}_k)\|$ implies the UA subproblem solved with sufficient accuracy
$\rho \in (0, 1)$	-	decay factor used to decrease μ in NGPA
$n_1, n_2 \in [1, n)$	-	integers connected with active set repetitions or change

A strong convergence theory can be developed for this algorithm. The following global convergence property holds:

THEOREM 3 *Let $\mathcal{L}$ be the level set defined by*

$$\mathcal{L} = \{\mathbf{x} \in \mathcal{B} : f(\mathbf{x}) \leq f(\mathbf{x}_0)\}.$$

Assume the following conditions hold:

A1. *f is bounded from below on $\mathcal{L}$ and* $d_{\max} = \sup_k \|\mathbf{d}_k\| < \infty$.

A2. *If $\bar{\mathcal{L}}$ is the collection of $\mathbf{x} \in \mathcal{B}$ whose distance to $\mathcal{L}$ is at most $d_{\max}$, then ∇f is Lipschitz continuous on $\bar{\mathcal{L}}$.*

A3. *UA satisfies* U1–U3.

Then ASA with $\epsilon = 0$ either terminates in a finite number of iterations at a stationary point, or we have

$$\liminf_{k\to\infty} \|\mathbf{d}^1(\mathbf{x}_k)\| = 0. \tag{15}$$

For strongly convex objective functions, the global convergence result can be strengthened as follows.

THEOREM 4 *If f is strongly convex and twice continuously differentiable on $\mathcal{B}$, and assumptions* A2 *and* A3 *of Theorem* 3 *are satisfied, then the iterates* $\mathbf{x}_k$ *of ASA with* $\epsilon = 0$ *converge to the global minimum.*

Under the hypotheses of the following theorem, ASA eventually reduces to the unconstrained algorithm with a fixed active constraint set. In other words, the constrained problem is eventually solved by the unconstrained algorithm.

THEOREM 5 *If f is twice-continuously differentiable and the iterates* $\mathbf{x}_k$ *generated by ASA with* $\epsilon = 0$ *converge to a stationary point satisfying the strong second-order sufficient optimality condition, then after a finite number of iterations, ASA performs only the UA without restarts.*

When f is a strongly convex quadratic function, the iterates $\mathbf{x}_k$ converge to the global minimizer $\mathbf{x}^*$ by Theorem 4. Thus, if the UA is based on the conjugate gradient method, it follows from Theorem 5 that ASA converges in a finite number of iterations, since the conjugate gradient method has finite convergence when applied to a convex quadratic.

In our analysis, summarized above, we never claim that the active indices at a stationary point $\mathbf{x}^*$ can be identified in a finite number of iterations. In fact, there is a fundamental difference between the gradient projection algorithm presented in this paper, and algorithms based on a "piecewise projected gradient" [11–13]. For our gradient projection algorithm, we perform a single projection, and then we back track towards the starting point. We are unable to show that the active constraints are identified in a finite number of iterations. In the piecewise projected gradient approach, where a series of projections may be performed, the active constraints can be identified in a finite number of iterations. Even though we do not identify the active constraints, we show in [37] that the components of $\mathbf{x}_k$ corresponding to the strictly active constraints are on the order of $\|\mathbf{x}_k - \mathbf{x}^*\|^2$. Moreover, in our experience, the single-projection approach is more efficient in practice.

4. Numerical Experiments

In this section, we compare the CPU time performance of ASA to the performance of other algorithms for box constrained optimization. We begin with a brief overview of algorithm development for box constrained optimization.

One important line of research focused on the development of conjugate gradient methods for box constrained problems with a quadratic objective function. Polyak's 1969 seminal work [50] considers a convex, quadratic cost function. The conjugate gradient method is used to explore a face of the feasible set, and the negative gradient is used to leave a face. Since Polyak's algorithm only added or dropped one constraint in each iteration, Dembo and Tulowitzki proposed [21] an algorithm CGP which could add and drop many constraints in an iteration. Later, Yang and Tolle [55] further developed this algorithm so as to obtain finite termination, even when the problem was degenerate at a local minimizer $\mathbf{x}^*$. That is, for some i, $x_i^* = 0$ and $g_i(\mathbf{x}^*) = 0$. Another variation of the CGP algorithm, for which there is a rigorous convergence theory, is developed by Wright [53]. Moré and Toraldo [49] point out that when the CGP scheme starts far from the solution, many iterations may be required to identify a suitable working face. Hence, they propose using the gradient projection method to identify a working face, followed by the conjugate gradient method to explore the face. Their algorithm, called GPCG, has finite termination for nondegenerate quadratic problems. Recently, adaptive conjugate gradient algorithms have been developed by Dostál *et al.* [24, 25, 27] which have finite termination for a strictly convex quadratic cost function, even when the problem is degenerate.

For general nonlinear functions, some of the earlier research [4, 14, 33, 44, 48] focused on gradient projection methods. To accelerate the convergence, more recent research has developed Newton and trust region methods. In [1, 13, 18, 29] superlinear and quadratic convergence is established for nondegenerate problems, while [31, 32, 43, 46] establish analogous convergence results, even for degenerate problems. Although computing a Newton step can be expensive computationally, approximation techniques, such as a sparse, incomplete Cholesky factorization [45], could be used to reduce the computational expense. Nonetheless, for large-dimensional problems or for problems where the initial guess is far from the solution, the Newton/trust region approach can be inefficient. In cases where the Newton step is unacceptable, a gradient projection step is preferred.

The affine-scaling interior point method of Coleman and Li [10, 15–17] is a different approach to (1), related to the trust region algorithm. More recent research on this strategy includes [22, 40, 41, 52, 56]. These methods are based on a reformulation of the necessary optimality conditions obtained by multiplication with a scaling matrix. The resulting system is often solved by Newton-type methods. Without assuming strict complementarity (i. e. for degenerate problems), the affine-scaling interior-point method converges superlinearly or quadratically, for a suitable choice of the scaling matrix, when the strong second-order sufficient optimality condition [51] holds. When the dimension is large, forming and solving the system of equations at each iteration can be time consuming, unless the problem has special structure. Recently, Zhang [56] proposes an interior-point gradient approach for solving the system at each iteration. Convergence results for other interior-point methods applied to more general constrained optimization appear in [28, 54].

We compare the performance of ASA to the following four codes:

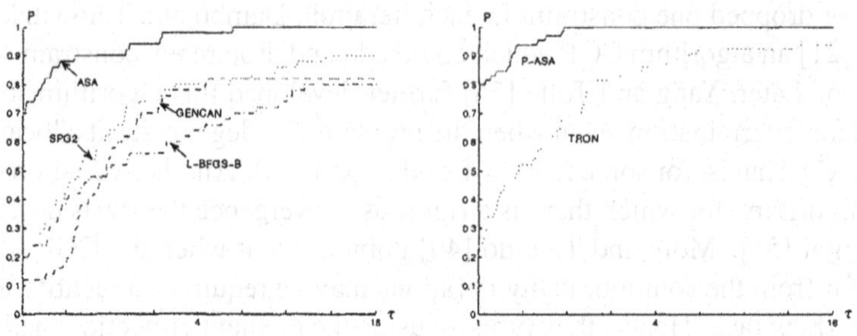

Figure 3. Performance profiles, 50 CUTEr test problems (left), 42 sparsest CUTEr problems, 23 MINPACK-2 problems (right)

- L-BFGS-B [57]: The limited memory quasi-Newton method of Zhu, Byrd, Nocedal (ACM Algorithm 778).
- SPG2 Version 2.1 [7, 8]: The nonmonotone spectral projected gradient method of Birgin, Martínez, and Raydan (ACM Algorithm 813).
- GENCAN [6]: The monotone active set method with spectral projected gradients developed by Birgin and Martínez.
- TRON Version 1.2 [46]: A Newton trust region method with incomplete Cholesky preconditioning developed by Lin and Moré.

These codes are all carefully written, high quality codes that reflect the different approaches to box constrained optimization summarized above. All codes are written in Fortran and compiled with f77 (default compiler settings) on a Sun workstation. The stopping condition was

$$\|P(\mathbf{x} - \mathbf{g}(\mathbf{x})) - \mathbf{x}\|_\infty \leq 10^{-6},$$

where $\|\cdot\|_\infty$ denotes the sup-norm of a vector. In running any of these codes, default values were used for all parameters. Our test problem set consisted of all 50 box constrained problems in the CUTEr library [9] with dimensions between 50 and 15,625, and all 23 box constrained problems in the MINPACK-2 library [2] with dimension 2500. The performance of the algorithms, relative to CPU time, was evaluated using the performance profiles of Dolan and Moré [23]. That is, for each method, we plot the fraction P of problems for which the method is within a factor τ of the best time.

TRON is somewhat different from the other codes since it employs Hessian information and an incomplete Cholesky preconditioner, while the other codes only utilize gradient information. In Figure 3, left, we compare the performance of the four gradient based codes ASA, L-BFGS-B, SPG2, and GENCAN using

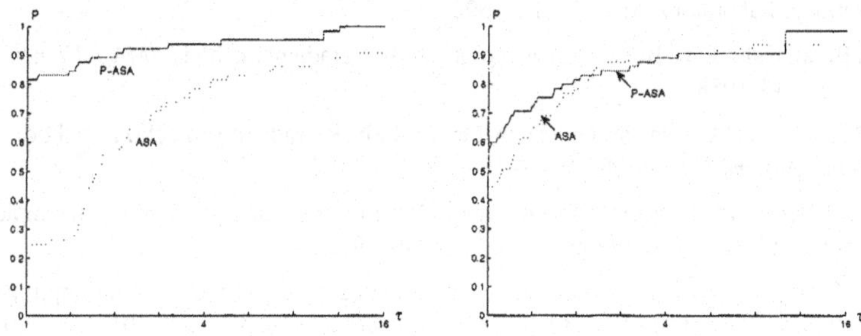

Figure 4. Performance comparison for P-ASA and ASA, $\epsilon = 10^{-6}$ (left), for $\epsilon = 10^{-2}\|\mathbf{d}^1(\mathbf{x}_0)\|_\infty$ (right)

the 50 CUTEr test problems. In a performance profile, the top curve corresponds to the method which solved the largest fraction of problems in a time within a factor τ of the best time. According to Figure 3, left, ASA achieves better CPU time performance than the other methods for this test set.

In order to compare ASA to the Hessian-based code TRON, we incorporated preconditioning in the conjugate gradient iteration. The preconditioner was the inverse of the incomplete Cholesky factorization of the Hessian at the current iterate. That is, we extracted the incomplete Cholesky factorization from TRON and used it in our code; hence, the two codes were using precisely the same approximation to the Hessian at each iterate. We let P-ASA denote this preconditioned version of ASA. Since TRON is targeted to large-sparse problems, such as the MINPACK problems, we compare P-ASA to TRON using the 23 MINPACK-2 problems and the 42 sparsest CUTEr problems (the number of nonzeros in the Hessian at most 1/5 the total number of entries in the Hessian). In Figure 3, right, we see that P-ASA has better CPU time performance than TRON in this test set.

In Figure 4, left, we compare the performance of P-ASA to that of ASA using the 42 sparsest CUTEr problems and the 23 MINPACK-2 problems. Clearly, the preconditioning was effective for this problem set and the convergence tolerance $\epsilon = 10^{-6}$. In Figure 4, right, the convergence tolerance is relaxed to $\epsilon = 10^{-2}\|\mathbf{d}^1(\mathbf{x}_0)\|_\infty$. With this relaxed convergence tolerance, there is not much difference between the preconditioned and the unconditioned codes.

References

[1] L. Armijo. Minimization of functions having Lipschitz continuous first partial derivatives. *Pacific J. Math.*, 16:1–3, 1966.

[2] B. M. Averick, R. G. Carter, J. J. Moré, and G. L. Xue. The MINPACK-2 test problem collection. Technical report, Mathematics and Computer Science Division, Argonne

National Laboratory, Argonne, IL, 1992.

[3] J. Barzilai and J. M. Borwein. Two point step size gradient methods. *IMA J. Numer. Anal.*, 8:141–148, 1988.

[4] D. P. Bertsekas. On the Goldstein-Levitin-Polyak gradient projection method. *IEEE Trans. Automatic Control*, 21:174–184, 1976.

[5] D. P. Bertsekas. Projected Newton methods for optimization problems with simple constraints. *SIAM J. Control Optim.*, 20:221–246, 1982.

[6] E. G. Birgin and J. M. Martínez. Large-scale active-set box-constrained optimization method with spectral projected gradients. *Comput. Optim. Appl.*, 23:101–125, 2002.

[7] E. G. Birgin, J. M. Martínez, and M. Raydan. Nonmonotone spectral projected gradient methods for convex sets. *SIAM J. Optim.*, 10:1196–1211, 2000.

[8] E. G. Birgin, J. M. Martínez, and M. Raydan. Algorithm 813: SPG - software for convex-constrained optimization. *ACM Trans. Math. Software*, 27:340–349, 2001.

[9] I. Bongartz, A. R. Conn, N. I. M. Gould, and P. L. Toint. CUTE: constrained and unconstrained testing environments. *ACM Trans. Math. Software*, 21:123–160, 1995.

[10] M.A. Branch, T.F. Coleman, and Y. Li. A subspace, interior, and conjugate gradient method for large-scale bound-constrained minimization problems. *SIAM J. Sci. Comput.*, 21:1–23, 1999.

[11] J. V. Burke and J. J. Moré. On the identification of active constraints. *SIAM J. Numer. Anal.*, 25:1197–1211, 1988.

[12] J. V. Burke and J. J. Moré. Exposing constraints. *SIAM J. Optim.*, 25:573–595, 1994.

[13] J. V. Burke, J. J. Moré, and G. Toraldo. Convergence properties of trust region methods for linear and convex constraints. *Math. Prog.*, 47:305–336, 1990.

[14] P. Calamai and J. Moré. Projected gradient for linearly constrained problems. *Math. Prog.*, 39:93–116, 1987.

[15] T. F. Coleman and Y. Li. On the convergence of interior-reflective Newton methods for nonlinear minimization subject to bounds. *Math. Prog.*, 67:189–224, 1994.

[16] T. F. Coleman and Y. Li. An interior trust region approach for nonlinear minimization subject to bounds. *SIAM J. Optim.*, 6:418–445, 1996.

[17] T. F. Coleman and Y. Li. A trust region and affine scaling interior point method for nonconvex minimization with linear inequality constraints. Technical report, Cornell University, Ithaca, NY, 1997.

[18] A. R. Conn, N. I. M. Gould, and Ph. L. Toint. Global convergence of a class of trust region algorithms for optimization with simple bounds. *SIAM J. Numer. Anal.*, 25:433–460, 1988.

[19] A. R. Conn, N. I. M. Gould, and Ph. L. Toint. A globally convergent augmented Lagrangian algorithm for optimization with general constraints and simple bounds. *SIAM J. Numer. Anal.*, 28:545–572, 1991.

[20] Y. H. Dai, W. W. Hager, K. Schittkowski, and H. Zhang. The cyclic Barzilai-Borwein method for unconstrained optimization. *IMA J. Numer. Anal.*, submitted, 2005.

[21] R. S. Dembo and U. Tulowitzki. On the minimization of quadratic functions subject to box constraints. Technical report, School of Organization and Management, Yale University, New Haven, CT, 1983.

[22] J. E. Dennis, M. Heinkenschloss, and L. N. Vicente. Trust-region interior-point algorithms for a class of nonlinear programming problems. *SIAM J. Control Optim.*, 36:1750–1794, 1998.

[23] E. D. Dolan and J. J. Moré. Benchmarking optimization software with performance profiles. *Math. Program.*, 91:201–213, 2002.

[24] Z. Dostál. Box constrained quadratic programming with proportioning and projections. *SIAM J. Optim.*, 7:871–887, 1997.

[25] Z. Dostál. A proportioning based algorithm for bound constrained quadratic programming with the rate of convergence. *Numer. Algorithms*, 34:293–302, 2003.

[26] Z. Dostál, A. Friedlander, and S. A. Santos. Solution of coercive and semicoercive contact problems by FETI domain decomposition. *Contemp. Math.*, 218:82–93, 1998.

[27] Z. Dostál, A. Friedlander, and S. A. Santos. Augmented Lagrangians with adaptive precision control for quadratic programming with simple bounds and equality constraints. *SIAM J. Optim.*, 13:1120–1140, 2003.

[28] A. S. El-Bakry, R. A. Tapia, T. Tsuchiya, and Y. Zhang. On the formulation and theory of the primal-dual Newton interior-point method for nonlinear programming. *J. Optim. Theory Appl.*, 89:507–541, 1996.

[29] F. Facchinei, J. Júdice, and J. Soares. An active set Newton's algorithm for large-scale nonlinear programs with box constraints. *SIAM J. Optim.*, 8:158–186, 1998.

[30] F. Facchinei and S. Lucidi. A class of penalty functions for optimization problems with bound constraints. *Optimization*, 26:239–259, 1992.

[31] F. Facchinei, S. Lucidi, and L. Palagi. A truncated Newton algorithm for large-scale box constrained optimization. *SIAM J. Optim.*, 4:1100–1125, 2002.

[32] A. Friedlander, J. M. Martínez, and S. A. Santos. A new trust region algorithm for bound constrained minimization. *Appl. Math. Optim.*, 30:235–266, 1994.

[33] A .A. Goldstein. Convex programming in Hilbert space. *Bull. Amer. Math. Soc.*, 70:709–710, 1964.

[34] L. Grippo, F. Lampariello, and S. Lucidi. A nonmonotone line search technique for Newton's method. *SIAM J. Numer. Anal.*, 23:707–716, 1986.

[35] W. W. Hager and H. Zhang. CG_DESCENT user's guide. Technical report, Dept. Math., Univ. Florida, 2004.

[36] W. W. Hager and H. Zhang. A new conjugate gradient method with guaranteed descent and an efficient line search. *SIAM J. Optim.*, 16:170–192, 2005.

[37] W. W. Hager and H. Zhang. A new active set algorithm for box constrained optimization. *SIAM J. Optim.*, submitted, 2005.

[38] W. W. Hager and H. Zhang. CG_DESCENT, a conjugate gradient method with guaranteed descent. *ACM Trans. Math. Software*, to appear 2006.

[39] W. W. Hager and H. Zhang. A survey of nonlinear conjugate gradient methods. *Pacific J. Optim.*, to appear 2006.

[40] M. Heinkenschloss, M. Ulbrich, and S. Ulbrich. Superlinear and quadratic convergence of affine-scaling interior-point Newton methods for problems with simple bounds without strict complementarity assumption. *Math. Prog.*, 86:615–635, 1999.

[41] C. Kanzow and A. Klug. On affine-scaling interior-point Newton methods for nonlinear minimization with bound constraints. *Comput. Optim. Appl.*, 2006, to appear.

[42] D. Kinderlehrer and G. Stampacchia. *An introduction to variational inequalities and their applications*, volume 31 of *Classics in Applied Mathematics*. SIAM, Philadelphia, PA, 2000.

[43] M. Lescrenier. Convergence of trust region algorithms for optimization with bounds when strict complementarity does not hold. *SIAM J. Numer. Anal.*, 28:476–495, 1991.

[44] E. S. Levitin and B. T. Polyak. Constrained minimization problems. *USSR Comput. Math. Math. Physics*, 6:1–50, 1966.

[45] C. J. Lin and J. J. Moré. Incomplete cholesky factorizations with limited memory. *SIAM J. Sci. Comput.*, 21:24–45, 1999.

[46] C. J. Lin and J. J. Moré. Newton's method for large bound-constrained optimization problems. *SIAM J. Optim.*, 9:1100–1127, 1999.

[47] J. M. Martínez. BOX-QUACAN and the implementation of augmented Lagrangian algorithms for minimization with inequality constraints. *J. Comput. Appl. Math.*, 19:31–56, 2000.

[48] G. P. McCormick and R. A. Tapia. The gradient projection method under mild differentiability conditions. *SIAM J. Control*, 10:93–98, 1972.

[49] J. J. Moré and G. Toraldo. On the solution of large quadratic programming problems with bound constraints. *SIAM J. Optim.*, 1:93–113, 1991.

[50] B. T. Polyak. The conjugate gradient method in extremal problems. *USSR Comp. Math. Math. Phys.*, 9:94–112, 1969.

[51] S. M. Robinson. Strongly regular generalized equations. *Math. Oper. Res.*, 5:43–62, 1980.

[52] M. Ulbrich, S. Ulbrich, and M. Heinkenschloss. Global convergence of affine-scaling interior-point Newton methods for infinite-dimensional nonlinear problems with pointwise bounds. *SIAM J. Control Optim.*, 37:731–764, 1999.

[53] S. J. Wright. Implementing proximal point methods for linear programming. *J. Optim. Theory Appl.*, 65:531–554, 1990.

[54] H. Yamashita and H. Yabe. Superlinear and quadratic convergence of some primal-dual interior-point methods for constrained optimization. *Math. Prog.*, 75:377–397, 1996.

[55] E. K. Yang and J. W. Tolle. A class of methods for solving large convex quadratic programs subject to box constraints. *Math. Prog.*, 51:223–228, 1991.

[56] Y. Zhang. Interior-point gradient methods with diagonal-scalings for simple-bound constrained optimization. Technical Report TR04-06, Department of Computational and Applied Mathematics, Rice University, Houston, Texas, 2004.

[57] C. Zhu, R. H. Byrd, and J. Nocedal. Algorithm 778: L-BFGS-B, Fortran subroutines for large-scale bound-constrained optimization. *ACM Trans. Math. Software*, 23:550–560, 1997.

P-FACTOR-APPROACH TO DEGENERATE OPTIMIZATION PROBLEMS

O. A. Brezhneva,[1] and A. A. Tret'yakov[2]

[1] *Department of Mathematics and Statistics, Miami University, Oxford, OH 45056, USA, brezhnoa@muohio.edu,* [2] *Center of the Russian Academy of Sciences, Vavilova 40, Moscow, GSP-1, Russia and University of Podlasie in Siedlce, 3 Maja, Siedlce, Poland, tret@ap.siedlce.pl*

Abstract The paper describes and analyzes an application of the p-regularity theory to *nonregular*, (*irregular, degenerate*) nonlinear optimization problems. The p-regularity theory, also known as the *factor-analysis of nonlinear mappings*, has been developing successfully for the last twenty years. The p-factor-approach is based on the construction of a p-factor-operator, which allows us to describe and analyze nonlinear problems in the degenerate case.

First, we illustrate how to use the p-factor-approach to solve degenerate optimization problems with equality constraints, in which the Lagrange multiplier associated with the objective function might be equal to zero. We then present necessary and sufficient optimality conditions for a degenerate optimization problem with inequality constraints. The p-factor-approach is also used for solving mathematical programs with equilibrium constraints (MPECs). We show that the constraints are 2-regular at the solution of the MPEC. This property allows us to localize the minimizer independently of the objective function. The same idea is applied to some other nonregular nonlinear programming problems and allows us to reduce these problems to a regular system of equations without an objective function.

keywords: Lagrange optimality conditions, degeneracy, p-regularity

1. Introduction

The main goal of this paper is to describe and analyze an application of the p-regularity theory to *nonregular*, (*irregular, degenerate*) nonlinear optimization problems. In the first part of the paper, we recall some definitions of the p-regularity theory [2, 3]. In the second part, we illustrate how to use the p-factor-approach to solve degenerate optimization problems with equality constraints, in which the Lagrange multiplier associated with the objective function might be equal to zero. In the third part, we present necessary and sufficient optimality conditions for a degenerate optimization problem with inequality constraints. In

Please use the following format when citing this chapter:
Author(s) [insert Last name, First-name initial(s)], 2006, in IFIP International Federation for Information Processing, Volume 199, System Modeling and Optimization, eds. Ceragioli F., Dontchev A., Furuta H., Marti K., Pandolfi L., (Boston: Springer), pp. [insert page numbers].

the last part of the paper we consider mathematical programs with equilibrium constraints (MPECs).

Notation. Let $\mathcal{L}(X,Y)$ be the space of all continuous linear operators from X to Y and for a given linear operator $\Lambda : X \to Y$, we denote its image by $\mathrm{Im}\Lambda = \{y \in Y \mid y = \Lambda x \text{ for some } x \in X\}$. Also, $\Lambda^* : Y^* \to X^*$ denotes the adjoint of Λ, where X^* and Y^* denote the dual spaces of X and Y, respectively. Let p be a natural number and let $B : X \times X \times \ldots \times X$ (with p copies of X) $\to Y$ be a continuous symmetric p-multilinear mapping. The p-form associated to B is the map $B[\cdot]^p : X \to Y$ defined by $B[x]^p = B(x, x, \ldots, x)$, for $x \in X$.

If $F : X \to Y$ is a differentiable mapping, its derivative at a point $x \in X$ will be denoted by $F'(x) : X \to Y$. If $F : X \to Y$ is of class C^p, we let $F^{(p)}(x)$ be the pth derivative of F at the point x (a symmetric multilinear map of p copies of X to Y) and the associated p-form, also called the pth–order mapping, is $F^{(p)}(x)[h]^p = F^{(p)}(x)(h, h, \ldots, h)$.

2. The p-factor-operator and the p-regular mappings

In this section, we recall some definitions of the p–regularity theory [2, 3].

Consider a sufficiently smooth mapping F from a Banach space X to a Banach space Y. The mapping F is called *regular* at some point $\bar{x} \in X$ if

$$\mathrm{Im}\, F'(\bar{x}) = Y.$$

We are interested in the case when the mapping F is *nonregular (irregular, degenerate)* at $\bar{x}$, i.e., when

$$\mathrm{Im} F'(\bar{x}) \neq Y. \tag{1}$$

We recall the definition of the p-regular mapping and of the p-factor-operator. We construct a p-factor-operator under an assumption that the space Y is decomposed into a direct sum

$$Y = Y_1 \oplus \ldots \oplus Y_p, \tag{2}$$

where $Y_1 = \mathrm{cl}\,\mathrm{Im}\, F'(\bar{x})$, $Y_i = \mathrm{cl}\,\mathrm{Sp}\ (\mathrm{Im} P_{Z_i} F^{(i)}(\bar{x})[\cdot]^i)$, $i = 2, \ldots, p-1$, $Y_p = Z_p$, Z_i is a closed complementary subspace for $(Y_1 \oplus \ldots \oplus Y_{i-1})$ with respect to Y, $i = 2, \ldots, p$, and $P_{Z_i} : Y \to Z_i$ is the projection operator onto Z_i along $(Y_1 \oplus \ldots \oplus Y_{i-1})$ with respect to Y, $i = 2, \ldots, p$.

Define the mappings [2]

$$f_i(x) : X \to Y_i, \quad f_i(x) = P_{Y_i} F(x), \quad i = 1, \ldots, p, \tag{3}$$

where $P_{Y_i} : Y \to Y_i$ is the projection operator onto Y_i along $(Y_1 \oplus \ldots \oplus Y_{i-1} \oplus Y_{i+1} \oplus \ldots \oplus Y_p)$ with respect to Y, $i = 1, \ldots, p$.

The p-factor-operator plays the central role in the p-regularity theory. The number p is chosen as the minimum number for which (2) holds. We give the following definition of the p-factor-operator.

DEFINITION 1 *The linear operator* $\Psi_p(h) \in \mathcal{L}(X, Y_1 \oplus \ldots \oplus Y_p)$, *defined by*

$$\Psi_p(h) = f_1'(\bar{x}) + f_2''(\bar{x})[h] + \ldots + f_p^{(p)}(\bar{x})[h]^{p-1}, \quad h \in X,$$

is called a p-factor-operator of the mapping $F(x)$ *at the point* $\bar{x}$.

Now we are ready to introduce another very important definition in the p-regularity theory.

DEFINITION 2 *We say that the mapping* F *is* p*–regular at* $\bar{x}$ *along an element* h *if* $\mathrm{Im}\Psi_p(h) = Y$.

The following definition is a specific form of Definition 1 for the case of $p = 2$ and $F : \mathcal{R}^n \to \mathcal{R}^n$.

DEFINITION 3 *A linear operator* $\Psi_2(h) : \mathcal{R}^n \to \mathcal{R}^n$,

$$\Psi_2(h) = F'(\bar{x}) + P^{\perp}F''(\bar{x})h, \quad h \in \mathcal{R}^n, \quad \|h\| = 1,$$

is said to be the 2–factor-operator, where $P^{\perp}$ *is a matrix of the orthoprojector onto* $(\mathrm{Im}\, F'(\bar{x}))^{\perp}$, *which is an orthogonal complementary subspace to the image of the first derivative of* F *evaluated at* $\bar{x}$.

The next definition is a specific form of Definition 2.

DEFINITION 4 *The mapping* F *is called 2-regular at* $\bar{x}$ *along an element* h *if*

$$\mathrm{Im}\, \Psi_2(h) = \mathcal{R}^n.$$

3. Degenerate optimization problems with equality constraints

In this section, we consider the nonlinear optimization problem with equality constraints:

$$\underset{x \in X}{\text{minimize}}\ f(x) \quad \text{subject to} \quad F(x) = 0, \tag{4}$$

where $f : X \to R$. We will denote a local solution of (4) by $\bar{x}$. We assume that F, f are C^{p+1} in some neighborhood of $\bar{x}$ and that the mapping $F : X \to Y$ is nonregular at $\bar{x}$, i.e., the condition (1) holds.

The Lagrangian for problem (4) is defined as

$$L(x, \lambda_0, \lambda) = \lambda_0 f(x) + \langle \lambda, F(x) \rangle, \tag{5}$$

where $(\lambda_0, \lambda) \in (\mathcal{R} \times Y^*) \backslash \{0\}$ is a generalized Lagrange multiplier.

The classical first-order Euler-Lagrange necessary optimality conditions for problem (4) state that there exists a generalized Lagrange multiplier $(\lambda_0, \bar{\lambda})$ such that

$$\lambda_0 f'(\bar{x}) + (F'(\bar{x}))^* \bar{\lambda} = 0,$$
$$F(\bar{x}) = 0,$$
$$\lambda_0^2 + \|\bar{\lambda}\|^2 = 1.$$

In other words, the point $(\bar{x}, \lambda_0, \bar{\lambda})$ is a solution of the following system of equations:

$$\mathcal{L}(x, \lambda_0, \lambda) = \begin{pmatrix} \lambda_0 f'(x) + (F'(x))^* \lambda \\ F(x) \\ \lambda_0^2 + \|\lambda\|^2 - 1 \end{pmatrix} = 0. \tag{6}$$

We are interested in the case, when the Lagrange multiplier λ_0 might be equal to zero. In this case, the mapping $\mathcal{L}$ and the system (6) are degenerate at $(\bar{x}, 0, \bar{\lambda})$. However, if the mapping $\mathcal{L}$ is p-regular at $(\bar{x}, 0, \bar{\lambda})$ with respect to some vector h, then solving system (6) can be reduced to solving a regular system of equations. This result is stated in the following Theorem 5.

Before we give the theorem, introduce the notation $z = (x, \lambda_0, \lambda)$ and the functions $l_i(z)$ associated with the mapping $\mathcal{L}(z)$ introduced in (6). We define functions $l_i(z)$ by (3) with $f_i(x) = l_i(z)$ and $F(x) = \mathcal{L}(z)$. We also define $\bar{\Psi}_p(h)$ to be a p-factor-operator of the mapping $\mathcal{L}(z)$ at the point $\bar{z} = (\bar{x}, 0, \bar{\lambda})$.

THEOREM 5 *Let $\bar{x}$ be a solution of* (4) *and let* $(0, \bar{\lambda})$ *be a generalized Lagrange multiplier such that* $(\bar{x}, 0, \bar{\lambda})$ *is a solution of* (6). *Assume that the mapping* $\mathcal{L}(z)$, *defined in* (6) *is p-regular at the point* $\bar{z} = (\bar{x}, 0, \bar{\lambda})$ *with respect to some vector* h. *Assume also that* $\operatorname{Ker}\bar{\Psi}_p(h) = \{0\}$. *Then the following system has a locally unique regular solution* $\bar{z}$:

$$\bar{\mathcal{L}}(z) = l_1(z) + l_2'(z)[h] + \ldots + l_p^{(p-1)}[h]^{p-1} = 0. \tag{7}$$

Example. Consider the problem

$$\underset{(x,y)}{\text{minimize}}\ y \quad \text{subject to} \quad F(x, y) = x^2 - y^3 = 0, \tag{8}$$

This problem has a solution $(\bar{x}, \bar{y})^T = (0, 0)^T$ and $F'(0, 0) = (0, 0)$. The Lagrangian for (8) is defined as

$$L(x, y, \lambda_0, \lambda) = \lambda_0 y + (x^2 - y^3)\lambda.$$

Then the system (6) for problem (8) has the form

$$\mathcal{L}(x, y, \lambda_0, \lambda) = \begin{pmatrix} 2x\lambda \\ \lambda_0 - 3y^2\lambda \\ x^2 - y^3 \\ \lambda_0^2 + \lambda^2 - 1 \end{pmatrix} = 0.$$

This system has a unique degenerate solution $(\bar{x}, \bar{y}, \lambda_0, \bar{\lambda}) = (0, 0, 0, 1)$.

Since $\mathcal{L}(x, y, \lambda_0, \lambda)$ is 3-regular at $(0, 0, 0, 1)^T$ with respect to the vector $h = (0, 1, 0, 0)^T$, we get by Theorem 5 that the system (7) is defined as

$$\bar{\mathcal{L}}(x, y, \lambda_0, \lambda) = \begin{pmatrix} 2x\lambda \\ \lambda_0 - 3y^2\lambda \\ -3y^2 - 6y \\ \lambda_0^2 + \lambda^2 - 1 \end{pmatrix} = 0.$$

This system has a locally unique regular solution $(0, 0, 0, 1)^T$.

4. Optimality conditions for degenerate optimization problems

In this section, we consider the nonlinear optimization problem with inequality constraints

$$\underset{x \in X}{\text{minimize}} \; f(x) \quad \text{subject to} \quad g(x) = (g_1(x), \ldots, g_m(x))^T \geq 0, \tag{9}$$

where $f, g_i : X \to \mathcal{R}^1$ and X is a Banach space. We will denote a local solution of (9) by $\bar{x}$.

For some $p \geq 2$, we say that we have the *completely degenerate case* if

$$g_i^{(r)}(\bar{x}) = 0, \quad r = 1, \ldots, p-1, \quad i \in I(\bar{x}) = \{i \,|\, g_i(\bar{x}) = 0\}. \tag{10}$$

Introduce the sets

$$H_p(\bar{x}) = \{h \in X \,|\, g_i{}^{(p)}(\bar{x})\,[h]^p \geq 0, \, i \in I(\bar{x})\,\}, \tag{11}$$

$$I_p(\bar{x}, h) = \{i \in I(\bar{x}) \,|\, g_i{}^{(p)}(\bar{x})\,[h]^p = 0, \, h \in H_p(\bar{x})\},$$

and

$$H_\alpha = \{h \in H_p(\bar{x}) \,|\, |g_i{}^{(p)}(\bar{x})\,[h]^p| \leq \alpha, \, i \in I(\bar{x}), \, \|h\| = 1\}.$$

Introduce the definition.

DEFINITION 6 *A mapping $g(x)$ is called strongly p-regular at $\bar{x}$ if there exists $\alpha > 0$ such that* $\sup_{h \in H_\alpha} \|\{\Psi_p(h)\}^{-1}\| < \infty$.

Let fix some element $h \in X$ and introduce the *p-factor-Lagrange function* [3]:

$$\mathcal{L}_p(x, h, \lambda(h)) = f(x) - \sum_{i=1}^{m} \lambda_i(h)\, g_i^{(p-1)}(x)[h]^{p-1}. \tag{12}$$

THEOREM 7 *Let $\bar{x}$ be a local solution to problem* (9). *Assume that there exists a vector h, $\|h\| = 1$, $h \in H_p(\bar{x})$, such that the vectors $\{g_i{}^{(p)}(\bar{x})[h]^{p-1}$, $i \in I_p(\bar{x}, h)\}$ are linearly independent. Then there exist Lagrange multipliers $\bar{\lambda}_i(h)$ such that*

$$\mathcal{L}_{px}(\bar{x}, h, \bar{\lambda}(h)) = f'(\bar{x}) - \sum_{i=1}^{m} \bar{\lambda}_i(h)(g_i^{(p)}(\bar{x})[h]^{p-1}) = 0,$$

$$g_i(\bar{x}) \geq 0, \quad \bar{\lambda}_i(h) \geq 0, \quad \bar{\lambda}_i(h) g_i(\bar{x}) = 0, \quad i = 1, \ldots, m.$$

Furthermore, suppose that the mapping $g(x)$ is strongly p-regular at $\bar{x}$. If there exist $w > 0$ and $\bar{\lambda}(h)$ such that

$$\mathcal{L}_{px}(\bar{x}, h, \bar{\lambda}(h)) = 0 \quad \textit{and} \quad \mathcal{L}_{pxx}\left(\bar{x}, h, \frac{2\bar{\lambda}(h)}{p(p+1)}\right)[h]^2 \geq w\|h\|^2 \tag{13}$$

for all $h \in H_p(\bar{x})$, then $\bar{x}$ is an isolated solution to problem (9).

5. Mathematical programs with equilibrium constraints (MPECs)

The MPEC considered in this section is a mathematical program with nonlinear complementary problem (NCP) constraints:

$$\underset{x}{\text{minimize}}\ f(x) \quad \text{subject to} \quad g(x) \geq 0, \quad x \geq 0, \quad \langle g(x), x \rangle = 0, \tag{14}$$

where $f : \mathcal{R}^n \to \mathcal{R}$, $g : \mathcal{R}^n \to \mathcal{R}^n$ are twice continuously differentiable functions. We are interested in the case when the strict complementarity condition does *not* hold at the solution $\bar{x}$, i.e., when there exists at least one index j such that $g_j(\bar{x}) = 0$ and $\bar{x}_j = 0$.

We show that under a certain condition, the problem (14) can be reduced to solving a system of nonlinear equations, which is independent of the objective function.

By introducing the slack variables s_j^2 and y_j^2, we reduce (14) to the problem with only equality constraints in the form:

$$F(x,s,y) = \begin{pmatrix} g_1(x) - s_1^2 \\ \dots \\ g_n(x) - s_n^2 \\ x_1 - y_1^2 \\ \dots \\ x_n - y_n^2 \\ s_1 y_1 \\ \dots \\ s_n y_n \end{pmatrix} = 0. \tag{15}$$

System (15) is a system of the $3n$ equations in the $3n$ unknowns x, s, and y. The corresponding Jacobian is given by

$$F'(x,s,y) = \begin{pmatrix} g_1'(x) & -2s_1(e^1)^T & 0 \\ \dots & \dots & \dots \\ g_n'(x) & -2s_n(e^n)^T & 0 \\ (e^1)^T & 0 & -2y_1(e^1)^T \\ \dots & \dots & \dots \\ (e^n)^T & 0 & -2y_n(e^n)^T \\ 0 & y_1(e^1)^T & s_1(e^1)^T \\ \dots & \dots & \dots \\ 0 & y_n(e^n)^T & s_n(e^n)^T \end{pmatrix}, \tag{16}$$

where $e^1, \dots, e^n$ denotes the standard basis in $\mathcal{R}^n$. If there exists an index j such that $y_j = 0$ and $s_j = 0$ (the strict complementarity condition does not hold), then the Jacobian matrix (16) is singular.

Assume that we can identify the set I_0 of the weakly active constraint indices, that is, $I_0 = \{i = 1, \dots, n \mid g_i(\bar{x}) = 0 \text{ and } \bar{x}_i = 0\}$. There are different techniques to identify the set I_0, for example, ones described in [1] or in [4].

Define the vector $h = (h_1, \dots, h_n)$ as

$$h_i = \begin{cases} 1, & i \in I_0 \\ 0, & i \notin I_0 \end{cases}$$

and a vector $\bar{h} \in \mathcal{R}^{3n}$ as $\bar{h}^T = (0_n^T, h^T, h^T)$.

From the explicit form of the Jacobian (16), the orthoprojector $P^\perp$ onto $(\operatorname{Im} F'(x,s,y))^\perp$ in $\mathcal{R}^{3n}$ is a diagonal matrix $P^\perp = \operatorname{diag}(p_j)_{j=1}^{3n}$ that is constant in some neighborhood of $(\bar{x}, \bar{s}, \bar{y})$ and that is given by

$$p_i = \begin{cases} 1, & i = 2n + j, \quad \text{and } j \in I_0 \\ 0, & \text{otherwise.} \end{cases}$$

Construct the mapping

$$\Psi(x,s,y) = F(x,s,y) + P^{\perp}F'(x,s,y)\bar{h}.$$

Without loss of generality, we assume that $I_0 = \{1,\dots,r\}$. Then

$$\Psi(x,s,y) = \begin{pmatrix} g_1(x) - s_1^2 \\ \dots \\ g_n(x) - s_n^2 \\ x_1 - y_1^2 \\ \dots \\ x_n - y_n^2 \\ s_1y_1 + s_1 + y_1 \\ \dots \\ s_ry_r + s_r + y_r \\ s_{r+1}y_{r+1} \\ \dots \\ s_ny_n \end{pmatrix} = 0. \qquad (17)$$

The Jacobian of (17) is nonsingular at $(\bar{x},\bar{s},\bar{y})$. Consequently the system (17) has a locally unique regular solution $(\bar{x},\bar{s},\bar{y})$. Thus, we have reduced the solution of the problem (14) to solving system (17) that is independent of the objective function $f(x)$.

References

[1] F. Facchinei, A. Fisher, and C. Kanzow, On the accurate identification of active constraints, *SIAM J. Optim.*, 9 (1998), pp. 14–32.

[2] A. F. Izmailov, and A. A. Tret'yakov, *Factor-analysis of nonlinear mappings*, Nauka, Moscow, 1994 (in Russian).

[3] A. A. Tret'yakov, and J. E. Marsden, Factor-analysis of nonlinear mappings: p–regularity theory, *Communications on Pure and Applied Analysis*, 2 (2003), pp. 425–445.

[4] S. J. Wright, An algorithm for degenerate nonlinear programming with rapid local convergence, *SIAM Journal on Optimization*, 15 (2005), pp. 673-696.

NON MONOTONE ALGORITHMS FOR UNCONSTRAINED MINIMIZATION: UPPER BOUNDS ON FUNCTION VALUES

U.M. Garcia-Palomares[1]
[1] *Universidad Simón Bolívar, Dep Procesos y Sistemas, Caracas, Venezuela, garciap@usb.ve*[*]

Abstract Non monotone algorithms allow a possible increase of function values at certain iterations. This paper gives a suitable control on this increase to preserve the convergence properties of its monotone counterpart. A new efficient MultiLineal Search is also proposed for minimization algorithms.

keywords: Non Monotone, Lineal Search, Trust Region.

1. Introduction

This paper is concerned with algorithms for solving the unconstrained minimization problem of finding a *local* minimizer $\bar{x}$ and the *local* minimum value $\bar{f} = f(\bar{x})$ of a scalar function $f(\cdot) \in C^1 : S \subset I\!R^n \to I\!R$. Armijo's inequality (1) has been frequently used by monotone algorithms: given $x_i, d_i \in I\!R^n$, the algorithm must determine a stepsize λ_i so that the new iterate x_{i+1} gives a sufficient decrease in the function value,

$$f(x_{i+1}) = f(x_i + \lambda_i d_i) \leq f(x_i) + 0.01\lambda_i \nabla f(x_i)^T d_i. \tag{1}$$

Under suitable assumptions (**A1-A4** below) a (*sub*)sequence $\{x_i\}_{i \in I}$ fulfilling (1) converges to a point $\bar{x}$ satisfying the first order necessary optimality condition; namely $\nabla f(\bar{x}) = 0$ [14]. Additional conditions, mainly in the choice of $\{d_i\}_1^\infty$, are obviously required to ensure a superlinear rate of convergence. Monotone algorithms force strict decrease of function values, i.e., $f(x_{i+1}) < f(x_i)$. This stringent condition may impair the convergence of the algorithm. Although the asymptotic rate of convergence is preserved, narrow valleys may demand an excessive number of function evaluations, which is normally considered a poor performance index when comparing optimization

*Paper written with support of USB and Universidade de Vigo, Spain.

Please use the following format when citing this chapter:
Author(s) [insert Last name, First-name initial(s)], 2006, in IFIP International Federation for Information Processing, Volume 199, System Modeling and Optimization, eds. Ceragioli F., Dontchev A., Furuta H., Marti K., Pandolfi L., (Boston: Springer), pp. [insert page numbers].

algorithms. Non monotone algorithms (NMAs) climb on the surrounding hills; i.e., $f(x_{i+1}) > f(x_i)$, and may avoid this undesirable behavior. Moreover, it has been shown that NMAs may jump over local minima [20] and become more fitted to global optimization [4].

A well known non monotone line search strategy was proposed by De Gripo et al [8, Section 3]. An iterate $x_{i+1} = x_i + \lambda_i d_i$ is accepted if

$$f(x_{i+1}) = f(x_i + \lambda_i d_i) \leq \max_{0 \leq j \leq q} f(x_{i-j}) + 0.01\lambda_i \nabla f(x_i)^T d_i. \quad (2)$$

This strategy has been adopted by many researchers in constrained and unconstrained problems with success. Currently many monotone algorithms have a non monotone counterpart [6]. The reader may consult additional material in [8, 9, 12, 16, 17, 19, 21] and references therein.

This paper adapts a sufficient decrease condition that does not require the computation of derivatives [5, 15]. Therefore, it can be used in derivative-free optimization and gradient-related algorithms. Furthermore, line search is not mandatory and can be replaced by trust region or some other technique. Finally, the maximum $f(\cdot)$ value on the previous q iterations is replaced by an upper bound $\varphi_i \geq f(x_i)$, which essentially has to be decreased a certain number of iterations (assumption **A5** below). An iterate x_{i+1} is accepted if

$$f(x_{i+1}) \leq \varphi_i - \phi(||d_i||), \quad (3)$$

where $\phi(\cdot) : I\!R_+ \to I\!R_+$, $\lim_{\tau \downarrow 0} \phi(\tau)/\tau = 0$. Note that a monotone algorithm is recovered when $\varphi_i = f(x_i)$ for all i, and the algorithm will behave as explained before; on the other hand, if the function upper bound φ_i is very loose, the algorithm would trend to stay more often on the hills, which implies extra function evaluations.

Next section describes and proves formally the convergence of our non monotone algorithm. It also includes a new approach that we call MultiLine Search (MLS). Section 3 gathers implementation remarks and report preliminary results to compare the monotone version with its non monotone counterpart.

Our notation is standard with minor peculiarities: all vectors are in the Euclidean space $I\!R^n$, unless otherwise stated. $I\!R^*_+$ are vectors in $I\!R^*$ with non negative components; $x^T y$ is the usual inner product $\sum_{k=1}^n x^k y^k$, and $M = xy^T$ is an $n \times n$ matrix with elements $m^{ij} = x^i y^j$. Lower case Greek letters are real values, capital Latin letters I, J, K are subsets of iteration indices. An infinite sequence is denoted as $\{(\cdot)_i\}_1^\infty$, and a subsequence by $\{(\cdot)_i\}_{i \in J}$. The notation $\{(\cdot)_i\}_{i \in I} \leq \alpha$ means that all elements in the subsequence are real numbers not

Table 1. **Non Monotone Algorithm (NMA)**

Input:	$\mu < 1 \leq \gamma, \epsilon > 0$	constants
	$i = 0$, Choose x_1, τ_1	Initial values
	DO $\quad i = i + 1$	next iteration
	$\quad$ Update φ_i	satisfying A5
	$\quad$ Choose $d_i : 0 < \|\|d_i\|\| \leq \tau_i$	
	$\quad$ IF $f(x_i + d_i) \leq \varphi_i - \phi(\tau_i)$	
	$\qquad x_{i+1} = x_i + d_i$	move
	$\qquad 0 < \tau_{i+1} \leq \gamma \|\|d_i\|\|$	$\|\|d\|\|$ may expand
	$\quad$ ELSE	
	$\qquad x_{i+1} = x_i$	no move
	$\qquad \tau_{i+1} = \mu \tau_i$	line search, trust region
	LOOP UNTIL $\|\|\nabla f(x_i)\|\| < \epsilon$	

bigger than α. Throughout the paper the set $D_i = \{d_{i1}, \ldots, d_{im}\}$ is a set of m unit vectors in $I\!R^n$. The set $\tau D = \{\tau d : d \in D\}$.

2. Non monotone algorithms

Our aim is to propose a non monotone algorithm that generates a converging subsequence $\{x_i\}_{i \in J} \to \bar{x}, \nabla f(\bar{x}) = 0$ under suitable conditions. Table 1 describes the algorithm as close as possible to a gradient related method, including the usual stopping criterium $||\nabla f(x)|| < \epsilon$. This version tries to satisfy (3) for $x_{i+1} = x_i + d_i$. Table 2 describes a more practical version where (3) is *tested* on multiple search directions. Theorem 4 below proves that with a slight modification and some usual extra assumptions on $f(\cdot)$ the non monotone algorithm exhibits a superlinear rate of convergence. We now list the assumptions and prove convergence.

A1: $f(\cdot)$ is bounded below, and $\{x_i\}_1^\infty$ remains in a compact set,

A2: $f(\cdot)$ is Fréchet differentiable, that is, $\nabla f(\cdot) : I\!R^n \to I\!R^n$ is everywhere defined and $f(x + d) = f(x) + \nabla f(x)^T d + o(||d||)$ for all $x, d \in I\!R^n$.

A3: $\exists(\alpha > 0) : \left\{ \frac{\nabla f(x_i)^T d_i}{||\nabla f(x_i)|| \, ||d_i||} \right\}_{i \in I} \leq -\alpha.$

A4: $[\{d_i\}_{i \in I} \to 0] \Rightarrow [\{\nabla f(x_i)\}_{i \in I} \to 0]$

A5: Let J be the index set of *successful* iterations. The sequence of reference values $\{\varphi_i\}_1^\infty$

a) *is an upper bound,* $f(x_i) \leq \varphi_i$, and

b) *decreases sufficiently every "q" successful iterations*, i.e.,

1. $\forall(i \in J)\exists(j \in J, i < j) : \varphi_j \leq \varphi_i - \Phi(||d_i||)$, where $\Phi(\cdot) : I\!R_+ \to I\!R_+$, and for any index subset K $[\{\Phi(||d_i||)\}_{i\in K} \to 0] \Rightarrow [\{||d_i||\}_{i\in K} \to 0]$.

2. Between i and j there are at most "q" successful iterations.

Assumptions **A1-A4** are required by most algorithms that solve smooth problems. **A5** is easy to comply. The sequence $\{\varphi_i\}_1^\infty$ may remain constant except at those iterations where it is forced to decrease. It is easy to show that (2) is a special case. We now prove that the non monotone algorithm is well defined; specifically we have

LEMMA 1 *If* $f(x_j) > \varphi_i - \phi(||d_i||)$ *for all* $i \geq j$, *then* $\nabla f(x_j) = 0$.

Proof: As $||d_{i+1}|| = \mu||d_i||$ we have that $\{||d_i||\}_1^\infty \to 0$. Besides, for all $i \geq j$ we have that $f(x_j + d_i) > \varphi_i - \phi(||d_i||)$; therefore

$$\begin{aligned} \nabla f(x_j)^T d_i = \; & f(x_j + d_i) - f(x_j) - o(||d_i||) \\ > \; & \varphi_i - f(x_j) - \phi(||d_i||) - o(||d_i||) \\ \geq \; & -o(||d_i||) - \phi(||d_i||) \end{aligned}$$

Coupling this inequality with **A3** we assert for $i \in I$ that

$$0 \geq -\alpha||\nabla f(x_j)|| \geq \nabla f(x_j)^T \frac{d_i}{||d_i||} > -\frac{o(||d_i||)}{||d_i||} - \frac{\nu(||d_i||)}{||d_i||}.$$

Since $||d_i|| \to 0$ we deduce that $||\nabla f(x_j)|| = 0$ ∎

LEMMA 2 $\{d_i\}_1^\infty \to 0$.

Proof: If the number of successful iterations is finite then by construction $||d_{i+1}|| = \mu||d_i||$ for all i large enough and the lemma is valid.

If, on the contrary, the number of successful iterations is infinite, let K be the index set where the upper bound actually decreases. For any two consecutive indices $i, j \in K$ we have that $f(x_j) \leq \varphi_j \leq \varphi_i - \Phi(||d_i||)$; hence $\{\Phi(||d_i||)\}_{i\in K} \to 0$, otherwise $\{\varphi_i\}_{i\in K}$ would be unbounded below, which in turn forces $\{f(x_i)\}_{i\in K}$ to be unbounded below contradicting **A1**. By **A4** we deduce that $\{||d_i||\}_{i\in K} \to 0$; but for any $j \notin K, ||d_j|| \leq \gamma^q||d_i||$, for some $i \in K$. Therefore, we conclude that $\{d_i\}_1^\infty \to 0$ ∎

As a direct consequence of the previous lemma we can state

REMARK 3 *If* $\Phi(||d||) = \phi(||d||) = 0.01\, d^T d$, *the convergence of a descent method that satisfies* ***A1-A4*** *is ensured, provided* $||d_{i+1}|| \leq \gamma||d_i||$ *at all successful iterations.*

We now establish the superlinear rate of convergence along the lines given in [14, theorem 4.1.8].

THEOREM 4 (SUPERLINEAR RATE) *Assume that for all i large enough: The Hessian $\nabla^2 f(x_i)$ is uniformly positive definite and $B_i d_i = -\nabla f(x_i)$, where $B_i \in \mathbb{R}^{n\times n}$ is a uniformly positive matrix that satisfies the necessary condition for superlinear rate of convergence $||(\nabla^2 f(x_i) - B_i)d_i|| = o(||d_i||)$.*

The proposed Non monotone algorithm exhibits a superlinear rate of convergence if $\lim_{\tau\downarrow 0} \phi(||\tau||)/\tau^2 = 0$.

Proof: We assume that the proof is asymptotic: it happens for all i large enough. From lemma 2 we obtain that $\{d_i\}_1^\infty \to 0$ and by the assumptions in the theorem we also obtain that $\{\nabla f(x_i)\}_1^\infty \to 0$. The algorithm exhibits a superlinear rate of convergence if and only if $f(x_i + d_i) \leq \varphi_i - \phi(||d_i||)$ [3]. Let $v_i = (\nabla^2 f(x_i) - B_i)d_i$. Note that

$v_i^T d_i = o(d_i^T d_i)$ and $d_i^T \nabla f(x_i) = -d_i^T B_i d_i = v_i^T d_i - d_i^T \nabla^2 f(x_i) d_i$;

therefore

$$\begin{aligned} f(x_i + d_i) &= f(x_i) + d_i^T \nabla f(x_i) + \tfrac{1}{2} d_i^T \nabla^2 f(x_i) d_i + o(d_i^T d_i) \\ &= f(x_i) - \tfrac{1}{2} d_i^T \nabla^2 f(x_i) d_i + v_i^T d_i + o(d_i^T d_i) \\ &= f(x_i) + d_i^T d_i \left(-\frac{1}{2} \frac{d_i^T \nabla^2 f(x_i) d_i}{d_i^T d_i} + \frac{v_i^T d_i + o(d_i^T d_i)}{d_i^T d_i} \right) \end{aligned}$$

Let $\lambda > 0$ be a lower bound of the minimum eigenvalue of $\{\nabla^2 f(x_i)\}$ for all i large enough. When $||d_i||$ is small enough we obtain

$$(|v_i^T d_i| + |o(d_i^T d_i)|)/d_i^T d_i \leq \tfrac{1}{4}\lambda;$$

hence, $f(x_i + d_i) \leq f(x_i) - \frac{\lambda}{4} d_i^T d_i \leq \varphi_i - \phi(||d_i||)$ ∎

2.1 Line Search(LS), MultiLine Search(MLS), Trust Region(TR)

A straightforward implementation of the NMA is by line search (LS); that is: if $f(x_i + d_i) \leq \varphi_i - \phi(||d_i||)$, it generates $x_{i+1} = x_i + d_i$ as its monotone counterpart; otherwise, it simply defines $d_{i+1} = \mu d_i$ and proceeds with the next iteration. The Trust Region (TR) approach is a natural extension of LS. It tries to satisfy (3) on a ball of radios τ_i around x_i. This technique has attracted a lot of interest in the optimization community [1, 2, 13]. Essentially TR replaces the *true* function $f(x_i + d_i)$ by a *model* $\tilde{f}(x_i, d)$ and finds

$$d_i = \arg \min_{||d|| \leq \tau_i} \tilde{f}(x_i, d). \tag{4}$$

$x_{i+1} = x_i + d_i$ is accepted if (3) holds; otherwise, it is rejected. The τ value is adjusted depending upon the proximity of the model value $\tilde{f}(x_i, d_i)$ to the true value $f(x_i + d_i)$. There are various issues that TR must face, mainly

Table 2. **Multiple LineSearch NonMonotone Algorithm (MLSNMA)**

PSEUDOCODE	REMARKS
$\tau = 2; \epsilon = 10^{-6}\sqrt{n}, x \in I\!R^n$	Remark 7
$f_x = f(x), \varphi = \max(f_x/2, 2f_x) + 10$	
success= 0	
DO Generate d	Remark 8
IF $\|\|d\|\| > 500\sqrt{n}\min(0.01, \tau)$	
Contract $d : \|\|d\|\| = 500\sqrt{n}\min(0.01, \tau)$	
ELSEIF $(\|\|d\|\| < \min(10^{-4}, \tau)\,\tau)$	Remark 9
Expand $d : \|\|d\|\| = \min(10^{-4}, \tau)\,\tau$	
$\tau = \|\|d\|\|$	Keep $\|\|d\|\|$
Generate $D = \{d_1, \ldots, d_n\}$	Remark 10
$k = 0$; done= FALSE	
WHILE (NOT DONE) AND $(k \leq n)$	
LINESEARCH (d_k)	Remark 11
$k = k + 1$	Next direction
IF (NOT DONE)	x is blocked
$\tau = 0.2\,\tau$	
UNTIL $(\tau < \epsilon)$	Remark 12

- To define an appropriate model, and
- to solve subproblem 4

Current research offers several options that greatly affect the TR performance. See [1, 11, 18] and references therein to be aware of the difficulties encountered in TR methods. We propose here another technique, which is, computationally, between LS and TR. Instead of solving subproblem (4), we carry out a multiline search (MLS). Specifically, given the iterate $x_i \in I\!R^n, \varphi_i \geq f(x_i)$, a set of m unit directions $D_i = \{d_{i1}, \ldots, d_{im}\}$ and a parameter $\tau_i > 0$, we declare that x_i is *blocked* if

$$\forall (d \in \tau_i D_i) : f(x_i + d) > \varphi_i - \phi(\tau_i), \tag{5}$$

where $\phi(\cdot) : I\!R_+ \to I\!R_+$ and $\lim_{\tau \downarrow 0} \phi(\tau)/\tau = 0$. To try to unblock x_i the algorithm imposes a reduction on the norm of the next search directions; namely $d_{i+1} \in \mu\tau_i D_{i+1}, \mu < 1$. The iteration will be considered successful if $x_{i+1} = x_i + d$ satisfies (3) for some $d \in \tau_i D_i$. It is obvious that under assumption **A6** below the algorithm ensures convergence.

A6: $D_i = \{d_{i1}, \ldots, d_{im}\}$ is a *finite* set of m unit directions and $\exists d \in D_i$ that satisfies **A3**.

This assumption cannot be verified on some practical problems, where no derivative information is at hand; however the following theorem is very useful when D_i positively spans $\mathbb{R}^n$, that is,

$$\forall(x \in \mathbb{R}^n)\exists(\alpha_1 \geq 0, \ldots, \alpha_m \geq 0) : x = \alpha_1 d_{i1} + \cdots + \alpha_m d_{im}.$$

THEOREM 5 *If* $\{D_i\}_1^\infty \to D = \{d_1, \ldots, d_m\}$ *positively span* $\mathbb{R}^n$, *and* $f(\cdot)$ *is strictly differentiable at limit points of* $\{x_i\}_1^\infty$, *then*

1. $m \geq (n+1)$.
2. $\forall(x \in \mathbb{R}^n)\exists(d \in D_i) : x^T d < 0.$
3. $$\left[\lim_{\substack{x_i \to \bar{x}, \tau_i \downarrow 0 \\ d_{ik} \to d_k, k=1,\ldots,m}} \frac{f(x_i + \tau_i d_{ik}) - f(x_i)}{\tau_i} \geq 0\right] \Rightarrow \nabla f(\bar{x}) = 0.$$

Proof: These are known facts. The proof can be found in [4, 5]∎

Based on the previous theorem, we propose the following MLS strategy: Generate d_i, let $\tau_i = ||d_i||$, and generate a set D_i of unit directions that satisfies **A6**, with $(d_i/\tau_i) \in D_i$. If (3) holds for $x_{i+1} = x_i + d$, for some $d \in \tau_i D_i$ the iteration has been successful and we proceed with the next iteration; otherwise, we declare that x_i is blocked and go to the next iteration forcing $||d_{i+1}|| = \mu\tau_i$. We now outline the convergence proof of the algorithm described in table 2 and remark 11, which contains this MLS strategy.

THEOREM 6 *Let* $f(\cdot)$ *be strictly differentiable. Under assumptions* ***A1,A2, A4,A5,A6*** *the algorithm shown in table 2 generates a subsequence* $\{x_i\}_{i\in I}$ *that converges to a point* $\bar{x}$ *satisfying a necessary optimality condition.*

Proof: If the number of blocked points is infinite we use theorem 5, or lemma 1; otherwise, we use lemma 2∎

3. Implementation and numerical results

This section shows up a number of remarks that complement the description of the algorithm given in table 2.

REMARK 7 *The starting point* x, *the stopping value* ϵ, τ, φ, *and the number of iterations* q *where* φ *is constant may be input parameters.*

REMARK 8 *d may be randomly generated when no derivative information is available. Depending upon the amount of information at hand d could even be the Newton direction.*

REMARK 9 *This safeguard prevents a premature stop due for instance to singularities when* $d = -B\nabla f(x)$.

REMARK 10 *The choice of* D *seems to have a tremendous impact on the performance of derivative free optimization algorithms [5]. When derivative information is available we suggest the orthogonal directions* $d_k = -\mathbf{sign}(u^k)(e_k - 2u^k u)$, *where* $u = -\nabla f(x_i)/||\nabla f(x_i)||, e_k$ *is the* $k-th$ *column of the identity matrix, and* $\mathbf{sign}(\alpha) = 1$ *if* $\alpha \geq 0$, $\mathbf{sign}(\alpha) = 0$ *otherwise. Note that* $d_k^T u = |u^k| \geq 0, k = 1, \ldots, n$; *hence we assert that* $\exists d \in D : d^T u \geq 1/\sqrt{n}$, *because otherwise*

$$1 = \sum_{k=1}^{n} |u^k|^2 = \sum_{k=1}^{n} (d_k^T u)^2 < \sum_{k=1}^{n} (1/n) = 1,$$

a contradiction.

REMARK 11 *This procedure assumes* $||d|| = 1$ *and evaluates* $f(x + \tau d)$. *It returns* TRUE *if the iteration is successful. It also updates* x *and* φ. *We use the updating on* φ *suggested above.*

```
LINESEARCH (d)
    z = x + τd; f_z = f(z)
    done= FALSE
    IF (f_z ≤ φ − τ min(10^-4, τ^2))
        x = z; f_x = f_z              Accept z
        success= success+ 1
        IF (success= q)
            φ = f_x; success = 0      Update φ
        done= TRUE
end of linesearch
```

REMARK 12 *We have chosen this termination criterium because it is also valid for derivative free optimization.*

We carried out preliminary numerical tests with functions from the Moré, Garbow and Hillstrom collection. The MatLab code was taken from [10] and run on a Pentium 4 desk computer. We used the quasi Newton direction $d = -\mathbf{sign}(\nabla f(x)^T B \nabla f(x)) B \nabla f(x)$, and B was updated with the symmetric formula $B = B + (s - By)(s - By)^T / (s - By)^T y$, where $s = x_{i+1} - x_i, y = \nabla f(x_{i+1}) - \nabla f(x_i)$. Table 3 shows the number of function evaluations needed for functions SINGX, ROSENX, which have an adjustable number of variables. For $q \in \{1, 5, 10, 20\}$ it was observed that the algorithm's performance generally improves for $q > 1$. It was also observed in tests not reported here that LS was superior to MLS on the steepest descent method. These results are by no way conclusive, and a more complete numerical test must be carried out in future research.

Table 3. **# of Function evaluations**

	singx				rosenx			
	φ constant (q)				φ constant (q)			
variables	**1**	**5**	**10**	**20**	**1**	**5**	**10**	**20**
8	321	283	159	154	533	362	343	343
12	647	583	366	232	988	1006	990	793
24	930	930	844	844	3949	2772	2838	2463

References

[1] B. Addis, S. Leiffer. A trust region algorithm for global optimization. Preprint ANL/MCS-P1190-0804, Argonne National Laboratory, Il, USA, 2004.

[2] A.R. Conn, N.I.M. Gould, P.L. Toint. *Trust region methods.* MPS-SIAM Series on Optimization, Philadelphia, ISBN 0-89871-460-5, 2000.

[3] J.E. Dennis, J.J. Moré. A characterization of superlinear convergence and its application to quasi-Newton methods. *Mathematics of Computation* 28:549-560, 1974.

[4] U.M. García-Palomares, F.J. González-Castaño, J.C. Burguillo-Rial. A combined global & local search (CGLS) approach to global optimization. *Journal of Global Optimization* To appear, 2006.

[5] U.M. García-Palomares, J.F. Rodríguez. New sequential and parallel derivative-free algorithms for unconstrained optimization. *SIAM Journal on Optimization* 13:79-96, 2002.

[6] N.I.M. Gould, D. Orban, Ph.L. Toint. GALAHAD a library of thread-safe Fortran 90 packages for large scale nonlinear optimization. *Transactions of the ACM on Mathematical Software* 29-4:353-372, 2003.

[7] N.I.M. Gould, C. Sainvitu, Ph.L. Toint. A filter-trust-region method for unconstrained minimization. Report 04/03, Rutherford Appleton Laboratory, England, 2004.

[8] L. Grippo, F. Lampariello, S. Lucidi. A nonmonotone line search technique for Newton's method. *SIAM Journal Numerical Analysis* 23-4:707-716, 1986.

[9] L. Grippo, M. Sciandrone. Nonmonotone globalization techniques for the Barzilai-Borwein gradient method. *Computational Optimization and Applications* 23:143-169, 2002.

[10] C. Gurwitz, L. Klein, M. Lamba. A MATLAB library of test functions for unconstrained optimization. Report 11/94, Brooklin College, USA, 1994.

[11] W.W. Hager. Minimizing a quadratic over a sphere. *SIAM Journal on Optimization* 12:188-208, 2001.

[12] J. Han, J. Sun, W. Sun. Global convergence of non-monotone descent methods for unconstrained optimization problems. *Journal of Computational and Applied Mathematics* 146-1:89-98, 2002.

[13] P.D. Hough, J.C. Meza. A class of trust region methods for parallel optimization. *SIAM Journal on Optimization* 13-1:264-282, 2002.

[14] C.T. Kelley. Iterative methods for optimization. *SIAM Frontiers in Applied Mathematics* ISBN 0-89871-433-8, 1999.

[15] S. Lucidi, M. Sciandrone. On the global convergence of derivative free methods for unconstrained optimization. *SIAM Journal on Optimization* 13-1:119-142, 2002.

[16] V.P. Plagianakos, G.D. Magoulas, M.N. Vrahatis. Deterministic nonmonotone strategies for effective training of multilayer perceptrons. *IEEE Transactions on Neural Networks* 13-6:1268-1284, 2002.

[17] M. Raydan. The Barzilai and Borwein gradient method for the large scale unconstrained minimization problem. *SIAM Journal on Optimization* 7:26-33, 1997.

[18] M. Rojas, S. Santos, D. Sorensen. LSTRS: Matlab software for large-scale trust-region subproblems and regularization. Technical Report 2003-4, Department of mathematics, Wake Forest University, NC, USA, 2003.

[19] P.L. Toint. An assesment of non-monotone linesearch techniques for unconstrained optimization. *SIAM Journal on Scientific and Statistical Computing* 8-3:416-435, 1996.

[20] P.L. Toint. Non-monotone trust region algorithms for nonlinear optimization subject to convex constraints. *Mathematical Programming* 77:69-94, 1997.

[21] J.L. Zhang, X.S. Zhang. A modified SQP method with nonmonotone linesearch technique. *Journal of Global Optimization* 21:201-218, 2001.

ON THE EFFICIENCY OF THE ε-SUBGRADIENT METHODS OVER NONLINEARLY CONSTRAINED NETWORKS

E. Mijangos,[1]

[1] *University of the Basque Country, Department of Applied Mathematics, Statistics and Operations Research, Spain, eugenio.mijangos@ehu.es*

Abstract The efficiency of the network flow techniques can be exploited in the solution of nonlinearly constrained network flow problems by means of approximate subgradient methods. In particular, we consider the case where the side constraints (non-network constraints) are convex. We propose to solve the dual problem by using ε-subgradient methods given that the dual function is estimated by minimizing *approximately* a Lagrangian function with only network constraints. Such Lagrangian function includes the side constraints. In order to evaluate the efficiency of these ε-subgradient methods some of them have been implemented and their performance computationally compared with that of other well-known codes. The results are encouraging.

keywords: Nonlinear Programming, Approximate Subgradient Methods, Network Flows.

1. Introduction

Consider the nonlinearly constrained network flow problem **(NCNFP)**

$$\begin{aligned} \underset{x}{\text{minimize}} \quad & f(x) && (1)\\ \text{subject to} \quad & x \in \mathcal{F} && (2)\\ & c(x) \leq 0, && (3) \end{aligned}$$

where:

- The set $\mathcal{F}$ is

$$\mathcal{F} = \{x \in \mathbf{R}^n \mid Ax = b,\ 0 \leq x \leq \overline{x}\},$$

where A is a node-arc incidence $m \times n$-matrix, b is the production/demand m-vector, x are the flows on the arcs of the network represented by A, and $\overline{x}$ are the capacity bounds imposed on the flows of each arc.

Please use the following format when citing this chapter:
Author(s) [insert Last name, First-name initial(s)], 2006, in IFIP International Federation for Information Processing, Volume 199, System Modeling and Optimization, eds. Ceragioli F., Dontchev A., Furuta H., Marti K., Pandolfi L., (Boston: Springer), pp. [insert page numbers].

- The side constraints (3) are defined by $c : \mathbf{R}^n \rightarrow \mathbf{R}^r$, such that $c = [c_1, \cdots, c_r]^t$, where $c_i(x)$ is linear or nonlinear and twice continuously differentiable on the feasible set $\mathcal{F}$ for all $i = 1, \cdots, r$.
- $f : \mathbf{R}^n \rightarrow \mathbf{R}$ is nonlinear and twice continuously differentiable on $\mathcal{F}$.

Many nonlinear network flow problems (in addition to the balance constraints on the nodes and the capacity constraints on the arc flows) have nonlinear side constraints. These are termed nonlinearly constrained network flow problems, **NCNFP**. In recent works [11, 12], **NCNFP** has been solved using partial augmented Lagrangian methods with quadratic penalty function and superlinear-order multiplier estimates (ALM).

In this work we focus on the primal problem **NCNFP** and its dual problem

$$\text{maximize} \qquad q(\mu) = \min_{x \in \mathcal{F}} l(x, \mu) = \min_{x \in \mathcal{F}} \{f(x) + \mu^t c(x)\} \tag{4}$$

$$\text{subject to:} \qquad \mu \in \mathcal{M}, \tag{5}$$

where $\mathcal{M} = \{\mu \mid \mu \geq 0,\ q(\mu) > -\infty\}$. We assume throughout this paper that the constraint set $\mathcal{M}$ is closed and convex, and q is continuous on $\mathcal{M}$, and for every $\mu \in \mathcal{M}$ some vector $x(\mu)$ that minimizes $l(x, \mu)$ over $x \in \mathcal{F}$ can be calculated, yielding a subgradient $c(x(\mu))$ of q at μ. We propose to solve **NCNFP** by using primal-dual methods, see [2].

The minimization of the Lagrangian function $l(x, \mu)$ over $\mathcal{F}$ can be performed by means of efficient techniques specialized for networks, see [21].

Since $q(\mu)$ is approximately computed, we consider *approximate subgradient methods* [13] in the solution of this problem. Author's purpose is to improve the efficiency obtained by using the multiplier methods with *asymptotically exact minimization* [11, 12]. Moreover, these methods allow us to solve problems of the kind of **NCNFP** where the dual function might be nondifferentiable in spite of having a differentiable Lagrangian function, as this could happen if the conditions given by the Proposition 6.1.1 in [2] are not fulfilled. The basic difference between these methods and the classical subgradient methods is that they replace the subgradients with inexact subgradients.

Different ways of computing the stepsize in the approximate subgradient methods have been considered. The diminishing stepsize rule (DSR) suggested by Correa and Lemaréchal in [4]. A dynamically chosen stepsize rule based on an estimation of the optimal value of the dual function by means of an adjustment procedure (DSAP) similar to that suggested by Nedić and Bertsekas in [18] for incremental subgradient methods. A dynamically chosen stepsize whose estimate of the optimal value of the dual function is based on the relaxation level-control algorithm (DSRLC) designed by Brännlund in [3] and analyzed by Goffin and Kiwiel in [9]. The convergence of these methods was studied in the cited papers for the case of exact subgradients. The convergence of the corresponding approximate (inexact) subgradient methods is analyzed in [13], see also [10].

The main aim of this work is to evaluate the efficiency of the approximate subgradient methods when we use DSR, DSAP, and DSRLC over **NCNFP** problems, for which we compare it with that of the augmented Lagrangian method (ALM) when it uses superlinear-order estimates [11, 12], and with that of the well-known codes filterSQP [7] and MINOS [17]. Also, we compare the quality of the computed solution by the ε-subgradient methods.

This paper is organized as follows: Section 2 presents the approximate subgradient methods; Section 3, the solution to the nonlinearly constrained network flow problem; and Section 4 puts forward the numerical tests.

2. Approximate subgradient methods

When, as happens in this work, for a given $\mu \in \mathcal{M}$, the dual function value $q(\mu)$ is calculated by minimizing approximately $l(x, \mu)$ over $x \in \mathcal{F}$ [see (4)], the subgradient obtained, as well as the value of $q(\mu)$, will involve an error.

In order to put forward such methods, it is useful to introduce a notion of approximate subgradient [2, 20]. In particular, given a scalar $\varepsilon \geq 0$ and a vector $\overline{\mu}$ with $q(\overline{\mu}) > -\infty$, we say that c is an *ε-subgradient at $\overline{\mu}$* if

$$q(\mu) \leq q(\overline{\mu}) + \varepsilon + c^t(\mu - \overline{\mu}), \qquad \forall \mu \in \mathbf{R}^r. \tag{6}$$

The set of all ε-subgradients at $\overline{\mu}$ is called the *ε-subdifferential at $\overline{\mu}$* and is denoted by $\partial_\varepsilon q(\overline{\mu})$. Note that every subgradient at a given point is also an ε-subgradient for all $\varepsilon > 0$. Generally, however, an ε-subgradient need not be a subgradient, unless $\varepsilon = 0$.

An approximate subgradient method is defined by

$$\mu^{k+1} = [\mu^k + s_k c^k]^+, \tag{7}$$

where c^k is an ε_k-subgradient at μ^k, $[\cdot]^+$ denotes the projection on the closed convex set $\mathcal{M}$, and s_k is a positive stepsize.

In our context, we minimize approximately $l(x, \mu^k)$ over $x \in \mathcal{F}$, thereby obtaining a vector $x^k \in \mathcal{F}$ with

$$l(x^k, \mu^k) \leq \inf_{x \in \mathcal{F}} l(x, \mu^k) + \varepsilon_k. \tag{8}$$

As is shown in [2, 13], the corresponding constraint vector, $c(x^k)$, is an ε_k-subgradient at μ^k. If we denote $q_{\varepsilon_k}(\mu^k) = l(x^k, \mu^k)$, by definition of $q(\mu^k)$ and using (8) we have

$$q(\mu^k) \leq q_{\varepsilon_k}(\mu^k) \leq q(\mu^k) + \varepsilon_k \qquad \forall k. \tag{9}$$

2.1 Stepsize rules

Throughout this section, we use the notation

$$q^* = \sup_{\mu \in \mathcal{M}} q(\mu), \quad \mathcal{M}^* = \{\mu \in \mathcal{M} \mid q(\mu) = q^*\},$$

and $\|\cdot\|$ denotes the standard Euclidean norm.

In this work, three kinds of stepsize rules have been considered.

2.1.1 Diminishing stepsize rule (DSR). The convergence of the subgradient method using a diminishing stepsize was shown by Correa and Lemaréchal, see [4]. Next, we consider the special case where c^k is an ε_k-subgradient.

In a recent work [13], the following proposition is proved.

PROPOSITION 1 *Let the optimal set* $\mathcal{M}^*$ *be nonempty. Also, assume that the sequences* $\{s_k\}$ *and* $\{\varepsilon_k\}$ *are such that*

$$s_k > 0, \ \sum_{k=0}^{\infty} s_k = \infty, \ \sum_{k=0}^{\infty} s_k^2 < \infty, \ \sum_{k=0}^{\infty} s_k \varepsilon_k < \infty. \tag{10}$$

Then, the sequence $\{\mu^k\}$*, generated by the* ε*-subgradient method, where* $c^k \in \partial_{\varepsilon_k} q(\mu^k)$ *(with* $\{\|c^k\|\}$ *bounded), converges to some optimal solution.*

An example of such a stepsize is

$$s^k = 1/\widehat{k}, \tag{11}$$

for $\widehat{k} = \lfloor k/m \rfloor + 1$. In this work we use by default $m = 5$.

An interesting alternative for the ordinary subgradient method is the *dynamic stepsize rule*

$$s_k = \gamma_k \frac{q^* - q(\mu^k)}{\|c^k\|^2}, \tag{12}$$

with $c^k \in \partial q(\mu^k)$ and $0 < \underline{\gamma} \le \gamma_k \le \overline{\gamma} < 2$, which was introduced by Poljak in [19] (see also Shor [20]).

Unfortunately, in most practical problems q^* and $q(\mu^k)$ are unknown. Then, the latter can be approximated by $q_{\varepsilon_k}(\mu^k) = l(x^k, \mu^k)$ and q^* replaced with an estimate q^k_{lev}. This leads to the stepsize rule

$$s_k = \gamma_k \frac{q^k_{lev} - q_{\varepsilon_k}(\mu^k)}{\|c^k\|^2}, \tag{13}$$

where $c^k \in \partial_{\varepsilon_k} q(\mu^k)$ is bounded for $k = 0, 1, \ldots$.

2.1.2 Dynamic stepsize with adjustment procedure (DSAP). An option to estimate q^* is to use the *adjustment procedure* suggested by Nedić and Bertsekas [18], but fitted for the ε-subgradient method, its convergence is analyzed by Mijangos in [13], see also [10].

In this procedure q_{lev}^k is the best function value achieved up to the kth iteration, in our case $\max_{0\leq j\leq k} q_{\varepsilon_j}(\mu^j)$, plus a positive amount δ_k, which is adjusted according to algorithm's progress.

The adjustment procedure obtains q_{lev}^k as follows:

$$q_{lev}^k = \max_{0\leq j\leq k} q_{\varepsilon_j}(\mu^j) + \delta_k,$$

and δ_k is updated according to

$$\delta_{k+1} = \begin{cases} \rho\delta_k, & \text{if } q_{\varepsilon_{k+1}}(\mu^{k+1}) \geq q_{lev}^k, \\ \max\{\beta\delta_k, \delta\}, & \text{if } q_{\varepsilon_{k+1}}(\mu^{k+1}) < q_{lev}^k, \end{cases}$$

where δ_0, δ, β, and ρ are fixed positive constants with $\beta < 1$ and $\rho \geq 1$.

2.1.3 Dynamic stepsize with relaxation-level control (DSRLC). Another choice to compute an estimate q_{lev}^k for (13) is to use a dynamic stepsize rule with relaxation-level control, which is based on the algorithm given by Brännlund [3], whose convergence was proved by Goffin and Kiwiel in [9] for $\varepsilon_k = 0$ for all k.

Mijangos in [13] has fitted this method to the dual problem of **NCNFP** (4-5) for $\{\varepsilon_k\} \to 0$ and analized its convergence, see also [10].

In this case, in contrast to the *adjustment procedure*, q^* is estimated by q_{lev}^k, which is a target level that is updated only if a sufficient ascent is detected or when the path long done from the last update exceeds a given upper bound B.

ALGORITHM 1

Step 0 (*Initialization*): Select μ^0, $\delta_0 > 0$, and $B > 0$.
Set $\sigma_0 = 0$ and $q_{rec}^{-1} = \infty$.
Set $k = 0$, $l = 0$ and $k(l) = 0$, where $k(l)$ will denote the iteration k when the lth update of q_{lev}^k occurs. Then $k(l) = k$ will be set.

Step 1 (*Function evaluation*): Compute $q_{\varepsilon_k}(\mu^k)$ and $c^k \in \partial_{\varepsilon_k} q(\mu^k)$.
If $q_{\varepsilon_k}(\mu^k) > q_{rec}^{k-1}$, set $q_{rec}^k = q_{\varepsilon_k}(\mu^k)$.
Otherwise set $q_{rec}^k = q_{rec}^{k-1}$.

Step 2 (*Stopping rule*): If $\|c^k\| = 0$, terminate with $\mu^* = \mu^k$.

Step 3 (*Sufficient ascent detection*): If $q_{\varepsilon_k}(\mu^k) \geq q_{rec}^{k(l)} + \frac{1}{2}\delta_k$, set $k(l+1) = k$, $\sigma_k = 0$, $\delta_{l+1} = \delta_l$, $l := l+1$, and go to Step 5.

Step 4 (*Oscillation detection*): If $\sigma_k > B$, set $k(l+1) = k$, $\sigma_k = 0$, $\delta_{l+1} = \frac{1}{2}\delta_l$, $l := l+1$.

Step 5 (*Iterate update*): Set $q_{lev}^k = q_{rec}^{k(l)} + \delta_l$. Choose $\gamma \in [\underline{\gamma}, \overline{\gamma}]$ and compute μ^{k+1} by means of (7) with the stepsize s_k obtained by (13).

Step 6 (*Path long update*): Set $\sigma_{k+1} = \sigma_k + s_k \|c^k\|$, $k := k+1$, and go to Step 1.

Note that q_{rec}^k keeps the record of the highest value attained by the iterates that are generated so far; i.e., $q_{rec}^k = \max_{0 \leq j \leq k} q_{\varepsilon_j}(\mu^j)$. Moreover, the algorithm uses the same target level $q_{lev}^k = q_{rec}^{k(l)} + \delta_l$ for $k = k(l), k(l)+1, k(l+2), \ldots, k(l+1)-1$. In [13] we analize the convergence of the ε-subgradient method with the stepsize (13) for q_{lev} given by this algorithm.

3. Solution to NCNFP

An algorithm is given below for solving **NCNFP**. This algorithm uses the approximate subgradient method described in Section 2.

The value of the dual function $q(\mu^k)$ is estimated by minimizing approximately $l(x, \mu^k)$ over $x \in \mathcal{F}$ (the set defined by the network constraints) so that the optimality tolerance, τ_x^k, becomes more rigorous as k increases; i.e., the minimization will be *asymptotically exact* [1]. In other words, we set $q_{\varepsilon_k}(\mu^k) = l(x^k, \mu^k)$, where x^k minimizes approximately the nonlinear network subproblem $\mathbf{NNS_k}$

$$\underset{x \in \mathcal{F}}{\text{minimize}}\ l(x, \mu^k)$$

in the sense that this minimization stops when we obtain a x^k such that

$$\|Z^t \nabla_x l(x^k, \mu^k)\| \leq \tau_x^k$$

where $\lim_{k \to \infty} \tau_x^k = 0$ and Z represents the reduction matrix whose columns form a base of the null subspace of the subspace generated by the rows of the matrix of active network constraints of this subproblem, see [16]. Let $\overline{x}^k$ be the minimizer of this subproblem approximated by x^k. Then, it can be proved (see [13]) that there exists a positive w, such that $l(x^k, \mu^k) \leq l(\overline{x}^k, \mu^k) + w\tau_x^k$ for $k = 1, 2, \ldots$. If we set $\varepsilon_k = \omega\tau_x^k$, this inequality becomes (8). Moreover, as

$$\tau_x^{k+1} = \alpha\tau_x^k, \qquad \text{for a fixed} \quad \alpha \in (0,1),$$

then $\sum_{k=1}^{\infty} \varepsilon_k < \infty$, and so $\lim_{k \to \infty} \varepsilon_k = 0$. Consequently, we can denote $q_{\varepsilon_k} = l(x^k, \mu^k)$, which holds the inequality (9), and we may use the methods described in Section 2. In this work, $\alpha = 10^{-1}$ by default. Note that in this case, $\varepsilon_k = \tau_x^k \omega = 10^{-k-1}\omega$.

ALGORITHM 2 (APPROXIMATE SUBGRADIENT METHOD FOR **NCNFP**)

Step 0 *Initialize.* Set $k = 1$, N_{max}, τ_x^1, ϵ_μ and τ_μ. Set $\mu^1 = 0$.

Step 1 *Compute* the dual function estimate, $q_{\varepsilon_k}(\mu^k)$, by solving **NNS**$_k$, so that if $\|Z^t \nabla_x l(x^k, \mu^k)\| \leq \tau_x^k$, then $x^k \in \mathcal{F}$ is an approximate solution, $q_{\varepsilon_k}(\mu^k) = l(x^k, \mu^k)$, and $c^k = c(x^k)$ is an ε_k-subgradient of q in μ^k.

Step 2 *Check the stopping rules* for μ^k.

T_1: Stop if $\max_{i=1,\ldots,r} \left\{ \dfrac{(c_i^k)^+}{1+(c_i^k)^+} \right\} < \tau_\mu$, where $(c_i^k)^+ = \max\{0, c_i(x^k)\}$.

T_2: Stop if $\left| \dfrac{q^k - (q^{k-1} + q^{k-2} + q^{k-3})/3}{1 + q^k} \right| < \epsilon_\mu$, where $q^n = q_{\varepsilon_n}(\mu^n)$.

T_3: Stop if $\dfrac{1}{4} \sum_{i=0}^{3} \|\mu^{k-i} - \mu^{k-i-1}\|_\infty < \epsilon_\mu$.

T_4: Stop if k reaches a prefixed value N_{max}.

If μ^k fulfils one of these tests, then it is optimal, and the algorithm stops. Without a duality gap, (x^k, μ^k) is a primal-dual solution.

Step 3 *Update* the estimate μ^k by means of the iteration

$$\mu_i^{k+1} = \begin{cases} \mu_i^k + s_k c_i^k, & \text{if } \mu_i^k + s_k c_i^k > 0 \\ 0, & \text{otherwise} \end{cases}$$

where s_k is computed using some stepsize rule. Go to Step 1.

In Step 0, for the checking of the stopping rules, $\tau_\mu = 10^{-5}$, $\epsilon_\mu = 10^{-5}$ and $N_{max} = 200$ have been taken. In addition, $\tau_x^0 = 10^{-1}$ by default.

Step 1 of this algorithm is carried out by the code PFNL, described in [14] (downloadable from website http://www.ehu.es/~mepmifee/).

In Step 2, alternative heuristic tests have been used for practical purposes. T_1 checks the feasibility of x^k, as if it is feasible the duality gap is zero, and then (x^k, μ^k) is a primal-dual solution for **NCNFP**. T_2 and T_3 mean that μ does not improve for the last N iterations. Note that $N = 4$.

To obtain s_k in Step 3, we have used the iteration (7) with the three rules considered in Section 2 for computing the stepsize:

- ▷ Diminishing stepsize rule DSR given by (11), which holds (10), for the ϵ_k given above.

▷ Dynamic stepsize with adjustment procedure DSAP, using $\rho = 2$, $\beta = 1/\rho$, $\gamma_k = 1$ for all k, $\delta_0 = 0.05\|(c^1)^+\|$, and $\delta = 10^{-5}|l(x_0, \mu_0)|$, where x^0 is the initial feasible point for Step 1 and $k = 0$.

▷ Dynamic stepsize with relaxation level-control DSRLC, replacing B in Algorithm 1 with $B_l = \max\{\overline{B}, B/l\}$ when an oscillation is detected, for $B = 10^{-3}\|x^1 - x^0\|$ and $\overline{B} = 0.01$. As can be seen, $\sum_{l=1}^{\infty} B_l = \infty$. In addition, we set $\gamma_k = 1$ for all k, $\delta = 10^{-5}|l(x_0, \mu_0)|$, $\delta_0 = 0.5\|(c^1)^+\|B$

The values given above have been heuristically chosen. The implementation in Fortran-77 of the previous algorithm, termed PFNRN05, was designed to solve large-scale nonlinear network flow problems with nonlinear side constraints.

4. Numerical tests

In order to obtain an evaluation of PFNRN05, some computational tests are performed, which consist in solving nonlinear network flow problems with nonlinear side constraints using this code and comparing the results with those obtained using PFNRN (with ALM), filterSQP and MINOS (see Section 1). These last two solvers are available on the NEOS server [5] with AMPL input [8]. (See the site `http://www-neos.mcs.anl.gov/`.) PFNRN is executed on a Sun Sparc 10/41 work station under UNIX (which has a similar speed to that of the NEOS machines).

Table 1. Test problems.

problem	# arcs	# nodes	# side const.	# actives	# sb. arcs
D12e2	1524	360	180	5	75
D13e2	1524	360	360	10	89
D14e2	1524	360	36	31	151
D12n1	1524	360	180	22	685
D13n1	1524	360	360	38	681
D14n1	1524	360	36	31	596
D21e2	5420	1200	120	3	30
D22e2	5420	1200	120	18	45
D23e2	5420	1200	120	17	238
D31e1	4008	501	5	1	63
D31e2	4008	501	5	1	60

The problems used in these tests were created by means of the following DIMACS-random-network generators: Rmfgen and Gridgen, see [6]. These generators provide linear flow problems in networks without side constraints. The inequality nonlinear side constraints for the DIMACS networks were generated through the *Dirnl* random generator described in [14]. The last two

Table 2. Comparison of the computed solution.

	DSR		DSAP		DSRLC	
Problem	e/f^*	$\|\|c\|\|_\infty$	e/f^*	$\|\|c\|\|_\infty$	e/f^*	$\|\|c\|\|_\infty$
D12e2	10^{-4}	10^{-3}	0.	10^{-11}	0.	10^{-11}
D13e2	10^{-4}	10^{-3}	10^{-7}	10^{-8}	10^{-7}	10^{-8}
D14e2	10^{-4}	10^{-3}	10^{-6}	10^{-6}	10^{-6}	10^{-6}
D12n1	10^{-6}	10^{-3}	0.	10^{-8}	10^{-7}	10^{-6}
D13n1	10^{-6}	10^{-3}	10^{-7}	10^{-6}	10^{-7}	10^{-7}
D14n1	10^{-4}	10^{-5}	10^{-5}	10^{-6}	0.	10^{-10}
D21e2	10^{-4}	10^{-2}	10^{-7}	10^{-7}	0.	10^{-11}
D22e2	10^{-3}	10^{-2}	10^{-6}	10^{-6}	0.	10^{-7}
D23e2	10^{-5}	10^{-1}	10^{-6}	10^{-7}	10^{-6}	10^{-7}
D31e1	10^{-7}	10^{-7}	0.	10^{-11}	0.	10^{-15}
D31e2	10^{-8}	10^{-6}	0.	10^{-14}	0.	10^{-7}

Table 3. Comparison of the efficiency.

Problem	filterSQP	MINOS	ALM	DSR	DSAP	DSRLC
D12e2	3.3	0.5	1.0	0.3	0.6	0.4
D13e2	5.0	0.7	2.3	0.6	0.5	0.6
D14e2	22.7	–	18.7	4.1	3.3	4.2
D12n1	409.6	31.1	46.2	48.3	58.3	39.6
D13n1	560.8	47.9	60.5	73.3	54.0	46.6
D14n1	–	–	162.3	330.3	178.3	103.3
D21e2	38.7	3.7	2.2	1.4	1.4	1.6
D22e2	63.5	6.8	5.4	2.5	1.9	1.9
D23e2	112.4	239.3	20.0	5.8	6.2	5.9
D31e1	28.7	45.6	2.0	1.6	1.6	1.6
D31e2	22.7	14.1	1.3	0.9	1.0	1.0

letters indicate the type of objective function that we have used: Namur functions, **n***, and EIO1 functions, **e***. The EIO1 family creates problems with a moderate number of superbasic variables (i.e., dimension of the null space) at the solution (# sb. arcs). By contrast, the Namur functions [21] generates a high number of superbasic arcs at the optimizer, see Table 1. More details about these problems can be found in [14, 15].

In Table 2, e/f^* represents the relative error in the computation of the optimum value of the objective function, whereas $\|\bar{c}\|_\infty$ represents the maximum violation of the side constraints in the optimal solution; that is, it offers information about the feasibility of this solution and, hence, about its duality gap. The results point out that the quality of the solution computed by PFNRN05 when it uses DSR is lower than that obtained when using dynamic stepsizes, such as DSAP or DSRLC.

In Table 3, for each method used to compute the stepsize the efficiency is evaluated by means of the run-times in CPU-seconds. The efficiency for the three stepsizes is very similar and, for these tests, the subgradient methods were more efficient than the quadratic multiplier method, ALM (see Section 1), filterSQP, and MINOS. Moreover, with default values, filterSQP was more accurate than MINOS.

These results encourage to carry out further experimentation, which also includes real problems, and to analyze more carefully the influence of some parameters over the performance of this code.

References

[1] D.P. Bertsekas. *Constrained Optimization and Lagrange Multiplier Methods*. Academic Press, New York, 1982.

[2] D.P. Bertsekas, *Nonlinear Programming*. 2nd ed. Athena Scientific, Belmont, Massachusetts, 1999.

[3] U. Brännlund. On relaxation methods for nonsmooth convex optimization. Doctoral Thesis, Royal Institute of Technology, Stockholm, Sweden, 1993.

[4] R. Correa, C. Lemarechal. Convergence of some algorithms for convex minimization. *Mathematical Programming*, 62:261–275, 1993.

[5] J. Czyzyk, M.P. Mesnier, J. Moré. The NEOS server. *IEEE Computational Science and Engineering*, 5(3):68–75, 1998.

[6] DIMACS. The first DIMACS international algorithm implementation challenge : The bench-mark experiments. Technical Report, DIMACS, New Brunswick, NJ, USA, 1991.

[7] R. Fletcher and S. Leyffer. User manual for filterSQP, University of Dundee Numerical Analysis Report NA\181, 1998.

[8] R. Fourer, D.M. Gay, B.W. Kernighan. *AMPL a modelling language for mathematical programming*. Boyd and Fraser Publishing Company, Danvers, MA 01293, USA, 1993.

[9] J.L. Goffin, K. Kiwiel. Convergence of a simple subgradient level method. *Mathematical Programming*, 85:207–211, 1999.

[10] K. Kiwiel. Convergence of approximate and incremental subgradient methods for convex optimization. *SIAM Journal on Optimization*, 14(3):807–840, 2004.

[11] E. Mijangos. An implementation of Newton-like methods on nonlinearly constrained networks. *Computers and Operations Research*, 32(2):181–199, 2004.

[12] E. Mijangos. An efficient method for nonlinearly constrained networks. *European Journal of Operational Research*, 161(3):618–635, 2005.

[13] E. Mijangos, Approximate subgradient methods for nonlinearly constrained network flow problems. To appear in *Journal of Optimization Theory and Applications*, 128(1), 2006.

[14] E. Mijangos, N. Nabona. The application of the multipliers method in nonlinear network flows with side constraints. Technical Report 96/10, Dept. of Statistics and Operations Research, Universitat Politècnica de Catalunya, Barcelona, Spain, 1996.

[15] E. Mijangos, N. Nabona. On the first-order estimation of multipliers from Kuhn-Tucker systems. *Computers and Operations Research*, 28:243–270, 2001.

[16] B.A. Murtagh, M.A. Saunders. Large-scale linearly constrained optimization. *Mathematical Programming*, 14:41–72, 1978.

[17] B.A. Murtagh, M.A. Saunders. MINOS 5.5. User's guide. Report SOL 83-20R, Department of Operations Research, Stanford University, Stanford, CA, USA, 1998.

[18] A. Nedić, D.P. Bertsekas. Incremental subgradient methods for nondifferentiable optimization. *SIAM Journal on Optimization*, 12:109–138, 2001.

[19] B.T. Poljak. Minimization of unsmooth functionals, *Z. Vyschisl. Mat. i Mat. Fiz.*, 9:14–29, 1969.

[20] N.Z. Shor. *Minimization methods for nondifferentiable functions*. Springer-Verlag, Berlin, 1985.

[21] Ph.L. Toint, D. Tuyttens. On large scale nonlinear network optimization. *Mathematical Programming*, 48:125–159, 1990.

PRECONDITIONED CONJUGATE GRADIENT ALGORITHMS FOR NONCONVEX PROBLEMS WITH BOX CONSTRAINTS

R. Pytlak,[1] and T. Tarnawski,[2]

[1]*Faculty of Cybernetics, Military University of Technology, 00-908 Warsaw, Poland, rpytlak@isi.wat.edu.pl* [2]*Faculty of Cybernetics, Military University of Technology, 00-908 Warsaw, Poland, tarni@isi.wat.edu.pl*

Abstract The paper describes a new conjugate gradient algorithm for large scale nonconvex problems with box constraints. In order to speed up the convergence the algorithm employs a scaling matrix which transforms the space of original variables into the space in which Hessian matrices of functionals describing the problems have more clustered eigenvalues. This is done efficiently by applying limited memory BFGS updating matrices. Once the scaling matrix is calculated, the next few iterations of the conjugate gradient algorithms are performed in the transformed space. The box constraints are treated by the projection as previously used in [R. Pytlak, The efficient algorithm for large-scale problems with simple bounds on the variables, SIAM J. on Optimization, Vol. 8, 532-560, 1998]. We believe that the preconditioned conjugate gradient algorithm gives more flexibility in achieving balance between the computing time and the number of function evaluations in comparison to a limited memory BFGS algorithm. The numerical results show that the proposed method is competitive to L-BFGS-B procedure.

keywords: bound constrained nonlinear optimization problems, conjugate gradient algorithms, quasi-Newton methods.

1. Introduction

In this paper we consider algorithms for the problem:

$$\min_{x \in \mathcal{R}^n} f(x) \tag{1}$$

$$\text{s. t. } \; l \leq x \leq u, \tag{2}$$

where $l,\ u \in \mathcal{R}^n$.

In [9] (see also [4]) a new family of conjugate gradient algorithms has been introduced whose direction finding subproblem is given by

$$d_k = -\mathbf{Nr}\{g_k, -\beta_k d_{k-1}\}, \tag{3}$$

Please use the following format when citing this chapter:
Author(s) [insert Last name, First-name initial(s)], 2006, in IFIP International Federation for Information Processing, Volume 199, System Modeling and Optimization, eds. Ceragioli F., Dontchev A., Furuta H., Marti K., Pandolfi L., (Boston: Springer), pp. [insert page numbers].

where $\mathbf{Nr}\{a, b\}$ is defined as the point from a line segment spanned by the vectors a and b which has the smallest norm, i.e.,

$$\| \mathbf{Nr}\{a, b\} \| = \min\{\| \lambda a + (1 - \lambda) b \| : 0 \leq \lambda \leq 1\}, \tag{4}$$

$\| \cdot \|$ is the Euclidean norm and $g_k = \nabla f(x_k)$.

Notice that if $\beta_k = 1$ then we have the Wolfe–Lemaréchal algorithm ([7], [13]). In [9] it was shown that the Wolfe–Lemaréchal algorithm is in fact the Fletcher–Reeves algorithm when directional minimization is exact. Moreover, the sequence $\{\beta_k\}$ was constructed in such way that directions generated by (3) are equivalent to those provided by the Polak–Ribiére formula (under the assumption that directional minimization is exact). This sequence

$$\beta_k = \frac{\|g_k\|^2}{|\langle g_k - g_{k-1}, g_k \rangle|} \tag{5}$$

has striking resemblance to the Polak–Ribiére formula.

2. General preconditioned conjugate gradient algorithm

The idea behind preconditioned conjugate gradient algorithm is to transform the decision vector by linear transformation D such that after the transformation the nonlinear problem is *easier* to solve. If $\hat{x}$ is transformed x:

$$\hat{x} = Dx \tag{6}$$

then our minimization problem will become

$$\min_{\hat{x}} \left[\hat{f}(\hat{x}) = f(D^{-1}\hat{x}) \right] \tag{7}$$

and for this problem the search direction will be defined as follows

$$\hat{d}_k = -\mathbf{Nr}\{\nabla \hat{f}(\hat{x}_k), -\hat{\beta}_k \hat{d}_{k-1}\} \tag{8}$$

Notice that

$$\nabla \hat{f}(\hat{x}) = D^{-T} \nabla f(\hat{x}) \tag{9}$$

therefore we can write

$$\hat{d}_k = -\mathbf{Nr}\{D^{-T} \nabla f(D^{-1}\hat{x}_k), -\hat{\beta}_k \hat{d}_{k-1}\}. \tag{10}$$

If we multiply both sides of (10) by D^{-1} we will get

$$d_k = -\lambda_k D^{-1} D^{-T} \nabla f(x_k) + (1 - \lambda_k)\, \hat{\beta}_k d_{k-1}. \tag{11}$$

where $0 \leq \lambda_k \leq 1$ and either

$$\hat{\beta}_k = 1 \tag{12}$$

for the Fletcher-Reeves version, or

$$\hat{\beta}_k = \frac{\|\hat{g}_k\|^2}{|\langle \hat{g}_k - \hat{g}_{k-1}, \hat{g}_k\rangle|} = \frac{g_k^T D^{-1} D^{-T} g_k}{|(g_k - g_{k-1})^T D^{-1} D^{-T} g_k|}$$

for the Polak-Ribiere version.

The equation (11) can be stated as

$$d_k = -\lambda_k H \nabla f(x_k) + (1 - \lambda_k) \beta_k d_{k-1}. \tag{13}$$

where $H = D^{-1}D^{-T}$. This suggests that D should be chosen in such a way that $D^T D$ is an approximation to $\nabla^2_{xx} f(\bar{x})$ where $\bar{x}$ is a solution of problem (1).

Moreover, if D is an upper triangular matrix then at each iteration of the algorithm we will have to solve the system of linear equations

$$D^T \hat{g}_k = g_k, \qquad D d_k = \hat{d}_k. \tag{14}$$

It is worthwhile to notice that the following holds (see [11]):

$$\langle g_k, d_k\rangle \leq -\|\hat{d}_k\|^2 \text{ and } \langle g_k, d_k\rangle = -\|\hat{d}_k\|^2, \tag{15}$$

if $0 < \lambda_k < 1$.

If box constraints (2) are present in our problem then we can tackle them by using the projection procedure proposed initially in [1] (see also [10]).

In the rest of the paper we consider, for the simplicity of presentation, the problems with simpler constraints $x \geq 0$. We define the set of indices I_k^+

$$I_k^+ := \{i \in \overline{1,n} : (x_k)_i \leq \varepsilon_k \text{ and } \nabla_{x_i} f(x_k) > 0\}, \tag{16}$$

where $\{\varepsilon_k\}$ is such that $\varepsilon_k > 0$ and

$$\lim_{k \in K} \|x_k - P[x_k - \nabla f(x_k)]\| = 0 \quad \Leftrightarrow \quad \lim_{k \in K} \varepsilon_k = 0. \tag{17}$$

for any subsequence $\{x_k\}_{k \in K}$. Here, by $P[\cdot]$ we denote the projection operator on the set $\{x \in \mathcal{R}^n : l \leq z \leq u\}$ ([1]).

The sets I_k^+ are used to modify the direction finding subproblem. Instead of solving problem (3) we find a new direction according to the rule

$$d_k = -\mathbf{Nr}\{\nabla f(x_k), -\beta_k d_{k-1}^+\}. \tag{18}$$

Here d_{k-1}^+ is defined by

$$(d_{k-1}^+)_i := \begin{cases} (d_{k-1})_i & \text{if } i \notin I_k^+ \\ -\nabla_{x_i} f(x_k)/\beta_k & \text{if } i \in I_k^+ \end{cases}. \tag{19}$$

To complete the description of the main components of our algorithm we have to show how to use scaling matrices in its preconditioned version. Having the set of indices I_k^+ we do not scale variables corresponding to them and we apply general scaling to the others. Therefore, we use the scaling matrix of the form

$$D_k = \begin{bmatrix} D & 0 \\ 0 & I_{n_k} \end{bmatrix}.$$

where $n_k = |I_k^+|$.

In order to describe the line search procedure notice that the function $f(P[x_k + \alpha d_k])$ can be interpreted as a composition of two functions: the first one is Lipschitzian and the second one continuously differentiable. If we define

$$x_k(\alpha) = x_k + d_k(\alpha), \quad \text{where } (d_k(\alpha))_i := \begin{cases} \alpha(d_k)_i & \text{if } \alpha \leq \alpha_k^i \\ \alpha_k^i(d_k)_i & \text{if } \alpha > \alpha_k^i \end{cases} \tag{20}$$

and the breakpoints $\{\alpha_k^i\}_1^n$ are calculated as follows

$$\alpha_k^i := -\frac{(x_k)_i}{(d_k)_i}, \quad i = 1, \ldots, n \tag{21}$$

(assuming that if $(d_k)_i \geq 0$ then $\alpha_k^i = \infty$), then our directional minimization rule can be stated as follows.

R1 find the largest positive number α_k from the set $\{\theta^k : k = 0, 1, \ldots, \theta \in (0,1)\}$ such that for $\mu \in (0,1)$ we have

$$f(x_k(\alpha_k)) - f(x_k) \leq -\mu \left[\alpha_k \sum_{i \notin I_k^+} (d_k)_i^2 + \sum_{i \in I_k^+} \frac{\partial f(x_k)}{\partial x^i} \left[x_k^i - x_k^i(\alpha_k) \right] \right],$$

Notice that in the rule *R1* we employ d_k instead of $\hat{d}_k$ as (15) would imply. Since we assume that $D_k^T D_k$ are uniformly bounded from below and above (in the sense of condition (26)) there exist constants $0 < c_1 < c_2 < +\infty$ such that $c_1 \|d_k\| \leq \|\hat{d}_k\| \leq c_2 \|d_k\|$. Thus the use of d_k on the left on inequality in the rule *R1* is justified (it corresponds to appropriately choosing the coefficient μ). It is worthwhile to observe that our descent direction rule allows for such inaccurate directional minimization search.

Our general algorithm is as follows:

Algorithm Parameters: $\mu \in (0,1)$, $\epsilon > 0$, $\{\hat{\beta}_k\}_1^\infty$, $\{D_k\}_1^\infty$, $D_k \in R^{n \times n}$ nonsingular matrix, $T \in R^{n \times n}$ nonsingular diagonal matrix.
Data: x_0

1) Set $k = 0$, compute $d_k = -g_k$, go to Step 3).

2)Compute: $w_k = x_k - P[x_k - T\nabla f(x_k)]$, $\varepsilon_k = \min(\varepsilon, \|w_k\|)$,

$$D_k^T \hat{g}_k = g_k \quad (22)$$

$$\hat{d}_k = -\mathbf{Nr}\{\hat{g}_k, -\hat{\beta}_k \hat{d}_{k-1}^+\} \quad (23)$$

$$D_k d_k = \hat{d}_k \quad (24)$$

If $w_k = 0$ then STOP.

3) Find a positive number α_k according to the rule *R1*.

4) Substitute $P[x_k + \alpha_k d_k]$ for x_{k+1}, increase k by one, go to Step 2.

We can prove the lemma

LEMMA 1 *Assume that x_k is a noncritical point, $D_k^T D_k$ is positive definite and $d_k \neq 0$ is calculated in* Step 2 *of* Algorithm. *Then there exists a positive α_k such that the condition stated in the rule* R1 *is*

$$\lim_{\alpha \to \infty} f(P[x_k + \alpha d_k]) = -\infty. \quad (25)$$

To investigate the convergence of *Algorithm* we begin by providing a crucial lemma which requires the following assumptions.

ASSUMPTION 1 *There exists $L < \infty$ such that*

$$\|\nabla f(y) - \nabla f(x)\| \leq L\|y - x\|$$

for all x, y from a bounded set.

ASSUMPTION 2 *There exist d_l, d_u such that $0 < d_l < d_u < +\infty$ and*

$$d_l\|u\|^2 \leq u^T D_k^T D_k u \leq d_u\|u\|^2 \quad (26)$$

for all $u \in \mathcal{R}^n$ and k.

LEMMA 2 *Suppose that* Assumptions 1–2 *hold, the direction d_k is determined by* (22)–(24) *and the step–size coefficient α_k is calculated according to the rule* R1. *Then, for any bounded subsequence $\{x_k\}_{k\in K}$ either*

$$\lim_{k\in K} \|x_k - P[x_k + d_k]\| = 0, \quad (27)$$

or

$$\lim_{k\in K} \|x_k - P[x_k - \nabla f(x_k)]\| = 0. \quad (28)$$

For the convenience of future notations we assume that variables $(x)_i$ have been reordered in such a way that d_k can be partitioned into two vectors (d_k^1, d_k^2) where the first vector d_k^1 is represented by

$$d_k^1 := \{(d_k)_i\}_{i \notin I_k^+}.$$

The same convention applies to other vectors.

THEOREM 3 *Suppose that* Assumptions 1–2 *are satisfied. Moreover, assume that for any convergent subsequence* $\{x_k\}_{k\in K}$ *whose limit is not a critical point*

i) $\{\hat{\beta}_k\}$ *is such that*

$$\liminf_{k\to\infty}\left(\hat{\beta}_k\|d^1_{k-1}\|\right) \geq \nu_1 \liminf_{k\to\infty}\|\nabla^1 f(x_k)\| \tag{29}$$

where ν_1 *is some positive constant,*

ii) there exists a number ν_2 *such that* $\nu_2\|D_k^{-T}\|_2\|D_{k-1}\|_2 \in (0,1)$ *and*

$$\langle\nabla^1 f(x_k), d^1_{k-1}\rangle \leq \nu_2\|\nabla^1 f(x_k)\|\|d^1_{k-1}\|,\ whenever\ \lambda_k \in (0,1). \tag{30}$$

Then $\lim_{k\to\infty} f(x_k) = -\infty$*, or every accumulation point of the sequence* $\{x_k\}_0^\infty$ *generated by* Algorithm *is a critical point.*

Our global convergence result is as follows.

THEOREM 4 *Suppose that* Assumptions 1–2 *are satisfied. Then* Algorithm *generates* $\{x_k\}$ *such that every accumulation point of* $\{x_k\}$ *satisfies necessary optimality conditions for problem (1)–(2) provided that:*

i) $\hat{\beta}_k$ *is given by*

$$\hat{\beta}_k = \frac{\|\nabla^1\hat{f}((\hat{x}^1_k,\hat{x}^2_{k-1}))\|^2}{|\langle\nabla^1\hat{f}((\hat{x}^1_k,\hat{x}^2_{k-1})) - \nabla^1\hat{f}(\hat{x}_{k-1}), \nabla^1\hat{f}((\hat{x}^1_k,\hat{x}^2_{k-1}))\rangle|}, \tag{31}$$

ii) there exists $M < \infty$ *such that* $\alpha_k \leq M$, $\forall k$.

3. Scaling matrices based on the compact representation of BFGS matrices

In the previous section we showed that for a given nonsingular matrix $H^{-1} = D^T D$ the preconditioned conjugate gradient algorithm is globally convergent. The use of constant scaling matrix is likely to be inefficient since the function f we minimize is nonlinear. Therefore, we are looking at the sequence of matrices $\{H_k\}$ such that each H_k^{-1} is as close as possible to the Hessian $\nabla^2_{xx} f(x_k)$ and can be easily factorized as $D_k^{-1}D_k^{-T}$ where D_k is a nonsingular matrix. We assume, for the simplicity of presentation, that $n_k \equiv 0$.

In the paper we present the preconditioned conjugate gradient algorithm based on the BFGS updating formula. To this end we recall compact representations of quasi–Newton matrices described in [8]. Suppose that the k vector pairs $\{s_i, y_i\}_{i=0}^{k-1}$ satisfy $s_i^T y_i > 0$ for $i = k - m - 1, \ldots, k - 1$. If we assume that $B_0 = \gamma_k I$ and introduce matrices $M_k = [\gamma_k S_k \; Y_k]$,

$$S_k = [s_{k-m-1}, \ldots, s_{k-1}], \; Y_k = [y_{k-m-1}, \ldots, y_{k-1}] \tag{32}$$

where $s_i = x_{i+1} - x_i$ and $y_i = g_{i+1} - g_i$, then LBFGS approximation to the Hessian matrix is

$$B_k = \gamma_k I - M_k W_k M_k^T \tag{33}$$

and $W_k \in \mathcal{R}^{m \times m}$ is nonsingular ([3]).

In order to transform the matrix B_k to the form $D_k^T D_k$ we do the QR factorization of the matrix M_k^T:

$$M_k^T = Q_k R_k \tag{34}$$

where Q_k is $n \times n$ orthogonal matrix and R_k the $n \times m$ matrix which has zero elements except the elements constituting the upper $m \times m$ submatrix. Taking into account that $Q_k^T Q_k = I$ we can write (33) as

$$B_k = Q_k^T \left[\gamma_k I - R_k^T W_k R_k\right] Q_k. \tag{35}$$

Notice that the matrix $R_k^T W_k R_k$ has zero elements except those lying in the upper left $m \times m$ submatrix. We denote this submatrix by T_k and we can easily show that it is a positive definite matrix. If we compute the Cholesky decomposition of the matrix $\gamma_k I_k - T_k$, $\gamma_k I_k - T_k = C_k^T C_k$ then eventually we come to the relation

$$B_k = Q_k^T F_k^T F_k Q_k \tag{36}$$

with

$$F_k = \begin{bmatrix} C_k & 0 \\ 0 & \sqrt{\gamma_k} I_{n-k} \end{bmatrix}. \tag{37}$$

The desired decomposition of the matrix B_k is thus given by

$$B_k = D_k^T D_k, \; D_k = F_k Q_k \tag{38}$$

where the matrix D_k is nonsingular provided that $s_i^T y_i > 0$ for $i = k - m - 1$, $\ldots, k - 1$. Notice that the matrix Q_k does not have to be stored since it can be easily evaluated from the Householder vectors which have been used in the QR factorization.

Recall the relation (14) which now can be written as

$$Q_k^T F_k^T \hat{g}_k = g_k, \qquad F_k Q_k d_k = \hat{d}_k. \tag{39}$$

4. Scaling matrices - the reduced Hessian approach

The approach is based on the limited memory reduced Hessian method proposed by Gill and Leonard ([5],[6], see also [12])

Suppose that $\mathcal{G}_k = \text{span}\{g_0, g_1, \ldots, g_k\}$ and let $\mathcal{G}_k^{\perp}$ denote the orthogonal complement of $\mathcal{G}_k$ in $\mathcal{R}^n$. If $B_k \in \mathcal{R}^{n \times r_k}$ have columns that define the basis of $\mathcal{G}_k$ and

$$H_k = Q_k T_k$$

is the QR decomposition of B_k then

$$Q_k^T B_k Q_k = \begin{pmatrix} Z_k^T B_k Z_k & 0 \\ 0 & \sigma I_{n-r_k} \end{pmatrix} \tag{40}$$

where $Q_k = (Z_k\ W_k)$ and $\text{range}(B_k) = \text{range}(Z_k)$. (40) follows from the theorem which was stated, among others, in [6]:

THEOREM 5 *Suppose that the BFGS method is applied to a general nonlinear function. If $B_0 = \sigma I_n$ and*

$$B_k d_k = -g_k,$$

then $d_k \in \mathcal{G}_k$ for all k. Furthermore, if $z \in \mathcal{G}_k$ and $w_k \in \mathcal{G}_k^{\perp}$, then $B_k z \in \mathcal{G}_k$ and $B_k w = \sigma w$.

From (40) we have

$$B_k = D_k^T D_k = Q_k \begin{pmatrix} Z_k^T B_k Z_k & 0 \\ 0 & \sigma I_{n-r_k} \end{pmatrix} Q_k^T$$

Therefore, it follows that we can take as D_k:

$$D_k = \begin{pmatrix} R_k & 0 \\ 0 & \sqrt{\sigma} I_{n-r_k} \end{pmatrix} Q_k^T = G_k Q_k^T$$

At every iteration we have to solve equations

$$Q_k G_k^T \hat{g}_k\,, \qquad = g_k G_k Q_k^T d_k = \hat{d}_k.$$

Solving these equations requires multiplication of vectors in $\mathcal{R}^n$ by the orthogonal matrix Q_k (or Q_k^T), and this can be achieved by the sequence of m multiplications of the Householder matrices H_k^i, $i = 1, \ldots, m$ such that $Q_k = H_k^1 H_k^2 \cdots H_k^m$. The cost of these multiplications is proportional to n. Furthermore, we have to solve the set on n linear equations with the upper triangular matrix G_k, or its transpose.

5. Numerical experiments

In order to verify the effectiveness of our algorithm we have tested it on problems from the CUTE collection ([2]). We tried it on problems with various dimension although its application is recommended for solving large scale problems.

Algorithm has been implemented in C on Intel PC under Linux operating system. We compared our algorithm with the L-BFGS-B code which is the benchmark procedure for problems with box constraints for which evaluating the Hessian matrix is too expensive. L-BFGS-B code was used with the parameter $m = 5$ and we applied $m = 5$ and we recalculated matrices D_k every five iterations in *Algorithm*. The stopping criterion was $\|\nabla f(x)\|/\max(1, \|x\|) \leq 10^{-7}$. We used the scaling matrices as described in Section 3.

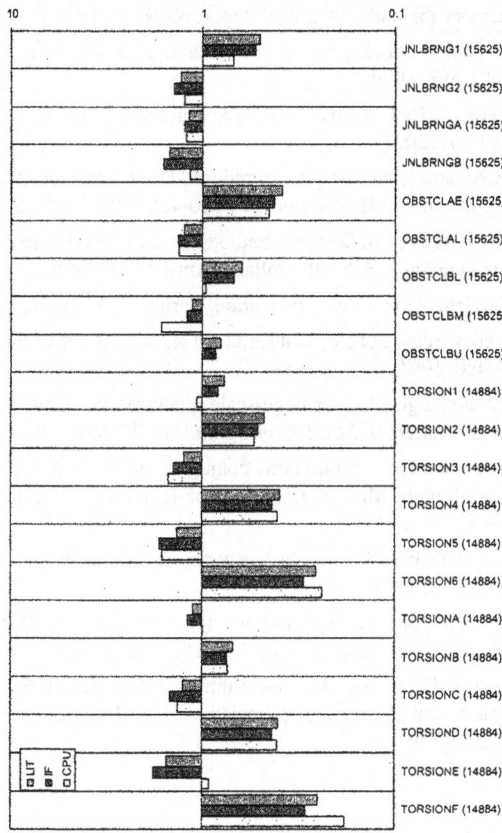

Figure 1. Performance comparison of *Algorithm* against the L-BFGS-B code.

The performance comparison of *Algorithm* is given in Figure 1. where we compare it with the code L-BFGS-B presented in [14]. For each problem the bars represent the ratio of the number of iterations (LIT), number of function evaluations (IF) and computing time (CPU) needed by the *Algorithm* divided by those from the executions of the L-BFGS-B code. Therefore values above one testify in favor of the L-BFGS-B and below one – in favor of our algorithm.

References

[1] D.P. Bertsekas, Projected Newton methods for optimization problems with simple constraints. *SIAM J. Control and Optimiz.* 20:221-245. 1982.

[2] I. Bongartz, A.R. Conn, N.I.M. Gould, Ph.L. Toint, CUTE: Constrained and Unconstrained Testing Environment. Research Report RC 18860, IBM T.J. Watson Research Center, Yorktown Heights, NY, USA, 1994.

[3] R. Byrd, J. Nocedal, R. B. Schnabel, Representations of quasi–Newton matrices and their use in limited memory methods. Technical Report NAM-03, 1996.

[4] Y. Dai, Y. Yuan, Global convergence of the method of shortest residuals. *Numerische Mathematik* 83:581-598, 1999.

[5] P.E. Gill, M.W. Leonard, Reduced–Hessian methods for unconstrained optimization. *SIAM J. Optimiz.* 12:209-237, 2001.

[6] P.E. Gill, M.W. Leonard, Limited–memory reduced hessian methods for large–scale unconstrained optimization. *SIAM J. Optimiz.* 14:380-401, 2003.

[7] C. Lemaréchal, An Extension of Davidon methods to nondifferentiable Problem. In *Mathematical Programming Study 3.* North-Holland, Amsterdam, 1975.

[8] J. Nocedal, S.J. Wright, *Numerical optimization.* Springer–Verlag, New York, 1999.

[9] R. Pytlak, On the convergence of conjugate gradient algorithms. *IMA Journal of Numerical Analysis.* 14:443-460, 1994.

[10] R. Pytlak, An efficient algorithm for large-scale nonlinear programming problems with simple bounds on the variables. *SIAM J. on Optimiz.* 8:532-560, 1998.

[11] R. Pytlak, T. Tarnawski, The preconditione conjugate gradient algorithm for nonconvex problems. Research Report, Military University of Technology, Faculty of Cybernetics, N.-1, 2005.

[12] D. Siegel, Modifying the BFGS update by a new column scaling technique. *Mathematical Programming.* 66:45-78, 1994.

[13] P. Wolfe, A Method of Conjugate Subgradients for Minimizing Nondifferentiable Functions. In *Mathematical Programming Study 3.* North-Holland, Amsterdam, 1975.

[14] C. Zhu, R.H. Byrd, P. Lu, J. Nocedal, Algorithm 778: L-BFGS-B, FORTRAN subroutines for large scale bound constrained optimization. *ACM Transactions on Mathematical Software.* 23:550-560, 1997.

MULTIOBJECTIVE OPTIMIZATION FOR RISK-BASED MAINTENANCE AND LIFE-CYCLE COST OF CIVIL INFRASTRUCTURE SYSTEMS

D. M. Frangopol,[1] and M. Liu [2]

[1] *Department of Civil, Environmental, and Architectural Engineering, University of Colorado, Boulder, CO 80309-0428, U.S.A. dan.frangopol@colorado.edu,* [2] *Formerly, Department of Department of Civil, Environmental, and Architectural Engineering, University of Colorado, Boulder, CO 80309-0428, U.S.A. minliu@illinoisalumni.org**

Abstract Reliability and durability of civil infrastructure systems such as highway bridges play a very important role in sustainable economic growth and social development of any country. The bridge infrastructure has been undergoing severe safety and condition deterioration due to gradual aging, aggressive environmental stressors, and increasing traffic loads. Maintenance needs for deteriorating highway bridges, however, have far outpaced available scarce funds highway agencies can provide. Bridge management systems (BMSs) are thus critical to cost-effectively allocate limited maintenance resources to bridges for achieving satisfactory lifetime safety and performance. In existing BMSs, however, visual inspections are the most widely adopted practice to quantify and assess bridge conditions, which are unable to faithfully reflect structural capacity deterioration. Failure to detect structural deficiency due to, for example, corrosion and fatigue, and inability to accurately assess real bridge health states may lead to unreliable bridge management decisions and even enormous safety and economic consequences. In this paper, recent advances in risk-based life-cycle maintenance management of deteriorating civil infrastructure systems with emphasis on highway bridges are reviewed. Methods of predicting lifetime safety and performance of highway bridges with and without maintenance are discussed. Treatment of various uncertainties associated with the complex deterioration processes due to time-dependent loading, environmental stressors, structural resistances, and maintenance actions are emphasized. The bridge maintenance management is formulated as a nonlinear, discrete, combinatorial optimization problem with simultaneous consideration of multiple and conflicting objectives, which address bridge safety and performance as well as long-term economic consequences. The effectiveness of genetic algorithms as a numerical multiobjective optimizer for

*The authors gratefully acknowledge the partial financial support of the U.K. Highways Agency, the U.S. National Science Foundation through grants CMS-9912525 and CMS-0217290, and the Colorado Department of Transportation. The opinions and conclusions presented in this paper are those of the writers and do not necessarily reflect the views of the sponsoring organizations.

Please use the following format when citing this chapter:
Author(s) [insert Last name, First-name initial(s)], 2006, in IFIP International Federation for Information Processing, Volume 199, System Modeling and Optimization, eds. Ceragioli F., Dontchev A., Furuta H., Marti K., Pandolfi L., (Boston: Springer), pp. [insert page numbers].

producing Pareto-optimal tradeoff solutions is demonstrated. The proposed probabilistic multiobjective optimization BMS is applied at project-level for similar bridges and at network-level for a group of different bridges that form a highway network.

Keywords: System reliability, optimization, civil infrastructure, bridges, genetic algorithms.

1. Introduction

Future sustained economic growth and social development of any country is intimately linked to the reliability and durability of its civil infrastructure systems such as highway bridges, which are the most critical but vulnerable elements in highway transportation systems. Highway bridges have been and are constantly subject to aggressive environments and ever-increasing traffic volumes and heavier truckloads, which degrade at an alarming rate the long-term bridge performance. In the United States, nearly 30% of the 600,000 existing bridges nationwide are structurally deficient or functionally obsolete; the associated costs of maintenance, repair, and replacement are enormous [14].

Deteriorating civil infrastructure leads to increased direct and indirect costs for business and users. Catastrophic failure of civil infrastructures due to natural hazards (e.g. earthquakes, hurricanes, and floods) and manmade disasters (e.g. vehicular collision and explosive blasts due to terrorists' attacks) [15] can cause widespread social and economic consequences. Therefore, timely and adequate maintenance interventions become indispensable to enhance resilience of civil infrastructure to adverse circumstances. This can substantially increase a country's economic competitiveness. In addition to development of advanced inspection and maintenance technologies, methodologies for cost-effective allocation of limited budgets to maintenance management of aging and deteriorating civil infrastructure over the life-cycle are urgently needed in order to optimally balance the lifetime performance and life-cycle cost while ensuring structure safety above acceptable levels.

A variety of practical bridge management systems (BMSs) have been developed and implemented in the United States for achieving desirable management solutions to maintain satisfactory bridge infrastructure performance, including BRIDGIT [20] and Pontis [30]. Most existing BMSs, however, utilize the least long-term economic cost criterion [28]. Recently, practicing bridge managers showed that this approach may not necessarily result in satisfactory long-term bridge performance [29]. Additionally, visual inspection is the most widely used practice to determine the condition and performance deterioration of bridges [1]. This highly subjective evaluation technique leads to significant variability in condition assessment [27]. More importantly, the actual level of structure safety against sudden failure and progressive degradation risks cannot be faithfully or accurately described by visual inspection-based bridge condition assessment [13]. Accordingly, maintenance decisions made solely on visual inspection results are not necessarily cost-effective and may cause tremendous safety and economic consequences if inadequate or unnecessary maintenance interventions are performed.

In order to resolve the above problems, all necessary long-term performance and expense considerations need to be incorporated into the maintenance management decision-making process. These multiple aspects include bridge performance such as visual inspection-based condition states, computation-based safety and reliability indices, and life-cycle costs such as agency cost and user cost. Unlike the traditional cost minimization approach, the above multiple criteria should be treated simultaneously so that a multiobjective optimization formulation is generated. As a result, the proposed risk/reliability-based maintenance management methodology leads to a group of optimized management solution options, exhibiting tradeoff between reducing life-cycle cost and improving structure performance. This significantly enables bridge managers to actively and preferably compromise structure safety/reliability and other conflicting objectives under budget and/or performance constraints.

In order to make rational decisions in preservation of deteriorating civil infrastructure, it is imperative that sources of uncertainty associated with the deterioration process with and without maintenance be addressed appropriately. These include imperfect description of mechanical loadings and environmental stressors as well as inexact prediction of deteriorating structure performance. There are two general types of uncertainty: aleatory and epistemic. The aleatory uncertainty is caused by inherent variation of structure deterioration due to combined effects of complex traffic loadings and environmental stressors as well as physical aging. The epistemic uncertainty stems from the randomness caused by subjective assumption in evaluating demand and load-carrying capacity of bridges or insufficient knowledge in understanding, for example, deterioration mechanisms. This type of uncertainty may be reduced provided more information is available [12]. Probable maintenance actions over the life cycle add further uncertainty to accurate prediction of time-varying structure performance.

In this paper, recent advances in application of multiobjective optimization techniques to risk-based maintenance management of civil infrastructure, in particular, highway bridges are reviewed. The multiple and competing objective functions of interest include condition, safety and life-cycle cost. Uncertainties associated with the deterioration process with and without maintenance interventions are treated by Monte Carlo simulation and/or structural reliability theory. The basic theory and effectiveness of evolutionary computation techniques such as genetic algorithms (GAs) in solving multiobjective optimization problems are discussed. Two application examples of GA-based bridge maintenance management are provided. The first example deals with project-level maintenance management of preserving a large population of similar deteriorating highway reinforced concrete crossheads. The second example is concerned with network-level bridge maintenance management for a number of different bridges that form a highway transportation network.

2. Multiobjective Optimization Algorithms

Because bridge management involves scheduling of different maintenance strategies to different bridges at discrete years, it can be readily formulated as a combinatorial optimization problem for which multiple and usually conflicting objectives need to be considered. In this section, the basic concept of multiobjective optimization is presented, the techniques of genetic algorithms (GAs) are discussed, and the application of GAs to the civil infrastructure management problems is emphasized.

2.1 General Formulation

A generic multiobjective optimization problem can be stated as

$$\begin{array}{c} \text{Optimize } \mathbf{f}\,(\mathbf{x}) \equiv [f_1(\mathbf{x}), f_2(\mathbf{x}), \cdots, f_m(\mathbf{x})] \\ \text{Subject to } \mathbf{C}\,(\mathbf{x}) \equiv [C_1(\mathbf{x}), C_2(\mathbf{x}), \cdots, C_n(\mathbf{x})] < 0 \end{array} \tag{1}$$

where $\mathbf{f}$ is a set of objective functions that are usually conflicting in nature; $\mathbf{C}$ is a set of constraints that define the valid solution space; $\mathbf{x}$ = a vector of design variables. Unlike optimization problems with single objectives, there are no unique solutions that can optimize all objectives simultaneously for a multiobjective optimization problem. Instead, a group of *Pareto-optimal* or *nondominated* solutions are present, which exhibit the optimized tradeoff in compromising these objectives. A solution $\mathbf{x}^*$ is Pareto-optimal if and only if there does not exist another solution that is no worse in all objectives and is strictly better in at least one objective. If all objectives are to be minimized, this can be stated mathematically as

$$\begin{array}{c} f_i(\mathbf{x}) \leq f_i(\mathbf{x}^*), \text{ for i = 1,2, ..., m; and} \\ f_k(\mathbf{x}) \leq f_k(\mathbf{x}^*), \text{ for at least } k\text{th objective.} \end{array} \tag{2}$$

2.2 Genetic Algorithms

Most traditional optimization algorithms are problem-dependent and single-objective oriented. Gradients are usually utilized to guide the search process and continuous design variables are often assumed. These pose significant difficulties to practical maintenance management problems. In contrast, heuristic algorithms based on evolutionary strategies such as GAs [19], simulated annealing [25], and tabu search [18] are very suitable for practical maintenance scheduling problems. In particular, GAs are stochastic search and optimization engines that follow the survival-of-the-fitness theory from the biological sciences. Since their inception in the 1960's, GAs have been successfully used in a wide array of applications due to their ease of implementation and robust performance for difficult engineering and science problems of vastly different natures. GAs are general-purpose numerical tools and gradients are no longer needed and discrete-valued design variables can be handled without difficulty. More importantly, GAs can handle multiple objectives simultaneously.

GAs usually operate on solutions that are encoded as genotypic representations (i.e. chromosomes) from their original phenotypic representations (i.e. actual data values). GAs start with a set of initial solutions (population) that is randomly generated in the search space. For each solution in the current population, objective functions defining the optimization problem are evaluated and a fitness value is assigned to reflect its (relative) merit standing in the population. Based on the fitness values, GAs perform a selection operation that reproduces a set of solutions with higher fitness values from the previous generation to fill a mating pool. A crossover operation is then pursued with which two parent solutions in the mating pool are randomly selected and interchange, with a prescribed probability, their respective string components at randomly selected bit locations referred to as cross sites. The resulting new solutions are called children or offspring. This step is meant to hopefully combine better attributes from the parent solutions so that child solutions with improved merits could be created. The next operation in GA is mutation that changes the genotype value at one or more randomly selected bit locations in a child solution with another prescribed probability. This operation serves to possibly recover useful information that could by no means be accessible through selection or crossover operation and therefore encourages search into a completely new solution space. After these three basic operations, a new generation is created. The search process continues until prescribed stopping criteria are met.

A successful multiobjective GA must have the ability to obtain a nondominated set of solutions close to the global Pareto-optimal front, and to have this solution set as diverse as possible, that is, to prevent solution clustering from occurring. Note that the selection operation is based on the relative fitness measures of solutions. Unlike single-objective problems where the objective function itself may be used as the fitness measure, after scaling and constraint-handling treatment, a multiobjective GA needs a single fitness measure that reflects the overall merit of multiple objectives. Multiobjective GAs have been fruitfully studied and developed in the last decade [7], many of which adopt Goldberg's nondominated sorting technique [19] to rank all solutions in a population, as discussed in the following.

For a given population of solutions, a nondominated subset is first identified according to the definition of Pareto optimality as defined previously. All solutions in this nondominated subset are assigned a rank of one and are then

Figure 1. Population ranking based on nondominated sorting

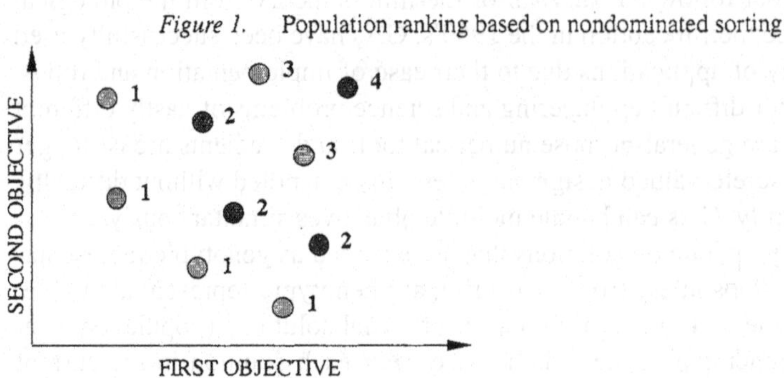

temporarily deleted from the population. The nondominated subset of the remaining solutions is identified and assigned a rank of two. This procedure continues until all solutions in the population have been assigned appropriate ranks. A solution in the front of a lower-numbered rank is assigned a higher fitness than that of a solution in the front of a higher-numbered rank. As a result, solutions closer to the global Pareto-optimal front have higher fitness values. As an illustration, consider a generic problem that has two objectives to be minimized. Fig. 1 indicates, in the solution space, ten solutions that are classified into four fronts with varied ranks. To assign fitness values to solutions with the same rank, niching strategies are used to determine relative fitness values by, for example, a crowding distance measure [10]. This measure is taken as an average distance of the two solution points on either side of the current solution along each of the objectives and thus serves as an estimate of the density of solutions surrounding a particular solution in the population.

Constraints in the GA-based optimization must be handled appropriately in GAs [6]. One possible approach is to assign to constraint-violating solutions dummy fitness values, which are defined in terms of degrees of constraint violation and are always less than those of valid solutions in the population. Thus the original constrained optimization problems are equivalently converted into unconstrained problems. Alternatively, constraint-violation may be considered by modifying genetic operators instead of assigning fictitious fitness values to invalid solutions. Deb [10] proposed a constrained binary tournament selection scheme that determines from two randomly picked solutions in the population the better solution based on three rules: (i) if both solutions are feasible, the one with higher fitness wins; (ii) if one solution is feasible and the other is infeasible, the feasible one always wins; (iii) if two solutions are both infeasible, the one with less degree of constraint-violation wins.

In multiobjective GAs, elitists usually refer to the generation-wise nondominated solutions. It is beneficial to retain elitist solutions in the subsequent generations for evolution operations due to their excellent genetic properties.

The stochastic nature of GAs, however, may disturb this ideal situation especially at early generations when the number of elitists is much smaller than the population size. To solve this problem, the elitist strategy may be adopted by forcibly inserting nondominated solutions (elitists) from the last generation back to its offspring population after basic genetic operations (i.e. selection, cross, and mutation) are performed. An updated set of elitists is then identified based on the population augmented by the elitists from the last generation. Previous studies have shown that the elitist scheme plays a crucial role in improving the optimization results [10].

2.3 GA-Based Maintenance Management

Research on use of multiobjective optimization techniques in maintenance management of civil infrastructure has appeared recently in the literature. Multiple and conflicting performance indicators such as condition, safety, durability along with life-cycle cost are simultaneously considered as separate criteria [21,24,26,17]. Interestingly almost all these research activities are conducted using GAs as numerical optimizers. This is because the practical maintenance management problems can be best posed as combinatorial optimization [18]. Due to their inherent features as previously discussed, GAs are very effective for solving these kinds of problems.

Many GAs work with a fixed population size. As generations evolve, the nondominated solutions fill most solution slots in a population, which may make it very difficult for dominated solutions to enter the population for genetic operations. As a result, the diversity of nondominated solutions in the subsequent generations may not be fully explored due to lack of information from valid yet dominated solutions. In this study, the initial GA population consists of 1,000 randomly generated trial solutions and each of the subsequent generations contains 200 offspring solutions plus the nondominated solutions from the previous (i.e. parent) generation. In addition, the fitness value is determined according to Goldberg's nondominated sorting plus Deb's crowding distance measure; Deb's constrained binary tournament selection scheme is adopted; a uniform crossover is applied with a probability of 50%; a uniform mutation is performed with a probability of 5%. Although this is a relatively high rate of mutation, by using elitism to preserve the nondominated solutions at each generation, mutation tends not to be very disruptive; sometimes a high level of mutation is used to avoid premature convergence.

3. Bridge Maintenance Management at Project-Level

Much effort has been devoted by researchers and practitioners to develop methodologies for long-term maintenance management of deteriorating bridges [8, 12, 13, 16, 17] Most previous research can be categorized as project-level types because only individual bridges or a group of similar bridges are considered. In this section, the time-dependent bridge performance deterioration with and without maintenance interventions is predicted by a continuous computational model [16]. This model describes the performance profiles without maintenance by a curve characterized by an initial performance level, time to damage initiation, and a deterioration curve governed by appropriate functions and, in the simplest form, a linear function with a constant deterioration rate. Effects of a generic maintenance action include prompt performance improvement, deterioration suppression for a prescribed period of time, deterioration severity reduction, and prescribed duration of maintenance effect. Epistemic uncertainties associated with the deterioration process are considered in terms of respective probabilistic distributions of the controlling parameters of this computational model. Monte Carlo simulation is used to account for these uncertainties by obtaining statistical performance profiles of deteriorating structures.

3.1 Problem Statement

The GA-based management procedure is used to prioritize maintenance needs for deteriorating reinforced concrete highway crossheads through simultaneous optimization of both structure performance and life-cycle maintenance cost. The maintenance management problem is thus posed as a combinatorial multiobjective optimization problem in that, for any year over the specified time horizon, at most one maintenance strategy may be carried out. Time-dependent performances of these structures are described using appropriate indicators in terms of condition and safety states.

For reinforced concrete elements under corrosion attack in the United Kingdom, Denton [11] categorized visual inspection-based condition states into four discrete levels, denoted as 0, 1, 2, and 3, that represent no chloride contamination, onset of corrosion, onset of cracking, and loose concrete/significant delamination, respectively. A value larger than 3 indicates an unacceptable condition state. As a subjective measure, however, the condition index may not faithfully reflect the true load-carrying capacity of structural members. According to bridge specifications in the United Kingdom, the safety index is defined as the ratio of available to required live load capacity [9]. It is considered unacceptable structure performance if the value of safety index drops below 0.91.

The goal is to obtain a set of sequences of maintenance actions applied over the specified time horizon that, in an optimized tradeoff manner, (i) decrease the largest (i.e. worst) lifetime condition index value, (ii) increase the smallest (i.e. worst) lifetime safety index value, and (iii) decrease the present value of life-cycle maintenance cost. The constraints are enforced such that the condition index value must be always less than 3.0 and the safety index value must be always greater than 0.91.

Five maintenance strategies are considered: replacement of expansion joints, silane, cathodic protection, minor concrete repair, and rebuilding [11,22]. Replacement of expansion joints is statistically the least costly. It does not improve performance or delay deterioration but alleviates deterioration severity of both condition and safety performance. The silane treatment reduces chloride penetration but does not correct existing defects or replace deteriorated structural components. Statistically speaking, silane reduces deterioration of condition more efficiently than replacement of expansion joints while having the same effects on safety deterioration. Cathodic protection replaces anodes and thus suppresses corrosion of reinforcing bars almost completely. It postpones deterioration of both condition and safety for 12.5 years upon application. The minor concrete repair strategy is applied to replace all cover concrete with visual defects but not corroded reinforcing bars. The rebuilding strategy improves both condition and safety levels to those values typical of a new structural component.

3.2 Numerical Results

In the numerical implementation, Monte Carlo simulation with a sample size of 1,000 is used to consider effects of uncertainty on prediction of both structure performance and life-cycle maintenance cost. All three objective functions are evaluated in terms of sample mean values. The service life is considered 50 years and the monetary discount rate is 6%. A number of different optimized maintenance planning solutions are generated. These solutions represent the optimized tradeoff among the condition, safety, and life-cycle maintenance cost objectives. Three representative maintenance solutions with different levels of performance enhancement and maintenance needs are shown in Fig. 2. Detailed information can be found in [22].

4. Bridge Maintenance Management at Network-Level

Compared to the above project-level maintenance management, a transportation network-oriented methodology provides more rational solutions because the ultimate objective of maintenance management is to improve performance of the entire transportation network instead of merely that of individual structures in the network. In this section, performance evaluation of deteriorating bridge networks is discussed and network-level maintenance management is presented and illustrated with numerical examples.

5. Problem Statement

The network reliability measures the level of satisfactory network performance. Most studies on assessment of reliability for transportation highway

Figure 2. Tradeoff of three project-level maintenance-scheduling solutions

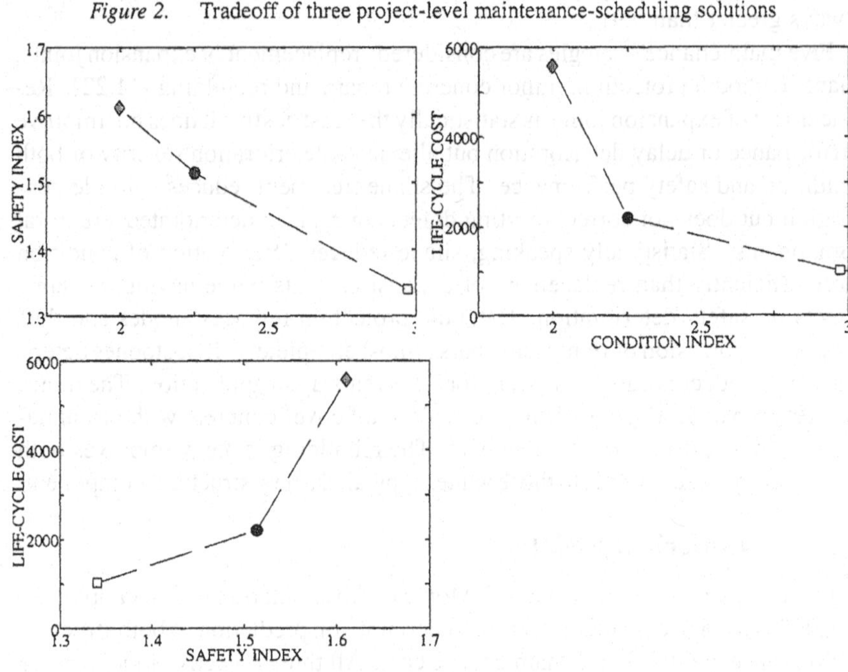

infrastructures have focused on maintenance management of deteriorating road networks for which a travel path consists of multiple links (i.e. roadways between any two nodes) with binary states (either operational or failed). There are three network reliability measures with ascending levels of sophistication: connectivity reliability, travel time reliability, and capacity reliability [4,5]. The connectivity reliability is associated with the probability that nodes in a highway network are connected; in particular, the terminal connectivity refers to the existence of at least one operational path that connects the origin and destination (OD) nodes of interest. The travel time reliability indicates the probability that a successful trip between a specified OD pair can be made within given time interval and level-of-service. Based on this reliability measure, the appropriate level of service that should be maintained in the presence of network deterioration can be determined. The third measure is the capacity reliability, which reflects the possibility of the network to accommodate given traffic demands at a specified service level. In this formulation, link capacities may be treated as random variables to consider the time-dependent probabilistic capacity deterioration. Inherent in the last two reliability measures are the determination of risk-taking route choice models for simulating travelers' behavior in the presence of both perception error and network uncertainty [5].

For maintenance management of deteriorating highway networks, it is also very important to use economic terms as a measure of the overall network performance. There are two basic types of costs: agent cost and user cost. The agent cost is composed of direct material and labor expenses needed to perform routine and preventive maintenance, rehabilitation, and replacement of existing transportation facilities. The indirect user costs are caused by loss of adequate service due to, for example, congestion and detour. In some situations the user cost may be a dominating factor in evaluating the overall life-cycle costs for a transportation network. The uncertainty associated with capacity degradation and demand variation should be integrated in the analysis in order to obtain a reliable cost measure.

In this study, the goal of network-level maintenance management is to prioritize maintenance needs to bridges that are of most importance to the network performance and over the specified time horizon. The overall goal is to satisfy the following two requirements in a simultaneous and balanced manner: (i) the overall bridge network performance, which is measured by the lowest level of the lifetime reliability of connectivity between the origin and destination locations, is improved, and (ii) the present value of total life-cycle maintenance cost is reduced.

Four different maintenance strategies are considered herein for enhancing bridge network performance in terms of reliability levels of deteriorating reinforced concrete bridge deck slabs: resin injection, slab thickness increasing, steel plate attaching, and complete replacement [17]. Resin injection is the least costly maintenance type among the four options. It injects epoxy resin into voids and seals cracks in concrete, which repairs the aging deck slabs by reducing the corrosion of reinforcement due to exposure to the open air. The reduction rate in reliability deterioration is assumed 0.03/year for 15 years. The other three maintenance strategies instantly improve the bridge reliability level by various amounts upon application. Increasing slab thickness and attaching steel plate increase the system reliability indices by a maximum of 0.7 and 2.0, respectively, with unit costs being US\$300/m^2 and US\$600/m^2, respectively. The complete replacement option restores the structural system to the initial reliability level with a unit cost of US\$900/m^2.

5.1 Numerical Results

The network-level maintenance management is illustrated using a real bridge network in Colorado [3]. This network consists of thirteen bridges of different types. The network performance is evaluated in terms of the terminal reliability for connectivity between two designated locations. Flexure failure of bridge slabs is considered as the only failure mode [2]. Deterioration of reinforcement is caused by deicing chemicals related corrosion. The life cycle is consid-

Figure 3. Tradeoff of three network-level maintenance-scheduling solutions

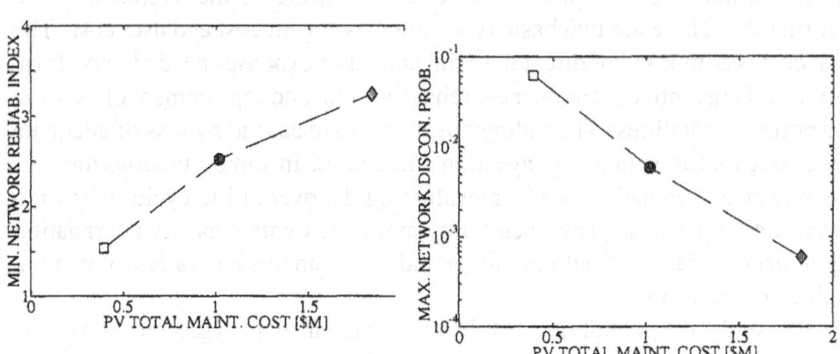

ered 30 years and the discount rate is 6%. The optimized solutions by GA represent a wide spread tradeoff between the conflicting network connectivity (equivalently network disconnectivity probability) and the total maintenance cost objectives. Tradeoff of three representative solutions is plotted in Fig. 3. Detailed information can be found in [23].

6. Monitoring-Integrated Maintenance Management

It is interesting and challenging to integrate the recent developments of structural health monitoring (SHM) technologies into intelligent maintenance management of civil infrastructure systems. Utilizing advanced sensing/information technology and structural modeling/identification schemes, SHM detects, locates, and quantifies structural damages caused by catastrophic natural or man-made events as well as by long-term deterioration. These data assist bridge managers in assessing the health of existing bridges and thus in determining immediate or future maintenance needs for safety consideration and lifespan extension. Most existing research and practice in BMS and SHM, however, are carried out in a disconnected manner. Therefore, a unified framework is necessary to bridge this gap between these two research areas.

Research in these areas represents a crucial step toward improving the traditional approach to BMSs by providing bridge managers with an efficient tool to make timely and intelligent decisions on monitoring, evaluation, and maintenance of deteriorating highway bridges. This can be achieved by exploring the interaction between SHM and BMS strategies in terms of whole-life costing and structural safety/health/reliability. Prediction of time-dependent bridge performance with monitoring is essential in this endeavor. With monitored data, the time-dependent performance will be more reliably estimated and the mainte-

Figure 4. Bridge performance profiles with and without monitoring-integrated maintenance interventions

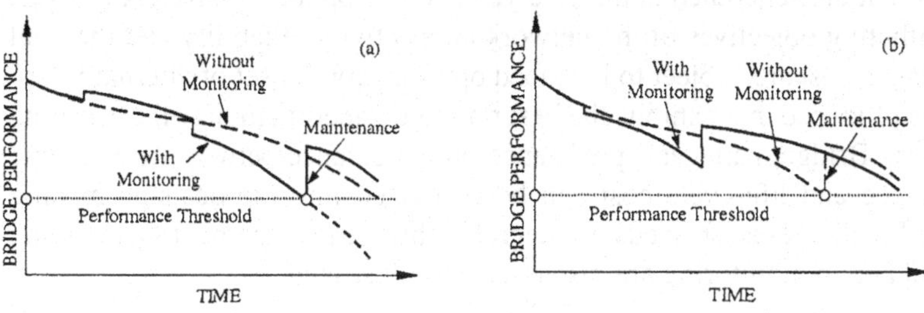

nance interventions will be more accurately applied than in the case without monitoring. Fig. 4 schematically illustrates the influence of monitoring actions on the prediction of bridge performance and on the ensuing maintenance interventions. In Fig. 4(a), with sensed data, earlier reaching the prescribed performance threshold is predicted, which incurs a timely maintenance intervention. Otherwise, if based on the non-monitoring performance prediction, the maintenance would not have been applied, which would cause tremendous risk concerns and consequences due to failure occurrence. Fig. 4(b) indicates another situation where the monitoring-enriched performance prediction makes unnecessary the maintenance actions predicted by the non-monitoring profile; in this case savings of maintenance costs can be enormous. Therefore, interactions among maintenance, monitoring, and management must be accurately analyzed in order to maintain bridges in timely and economical manners.

7. Conclusions

This paper reviews recent developments of risk-based maintenance management of civil infrastructure systems especially of highway bridges, emphasizing simultaneous consideration of multiple criteria related to long-term structure performance and life-cycle cost. Sources of uncertainty associated with the deterioration process are considered in probabilistic performance prediction of structures with and without maintenance interventions. The usefulness of genetic algorithms in solving the posed combinatorial multiobjective optimization problems is discussed. Two illustrative numerical examples are provided. The first example deals with project-level maintenance scheduling for a group of deteriorating reinforced concrete crossheads over a specified time horizon. Structure performance measures, in terms of visually inspected condition and computed load-carrying safety indices, and the present value of long-term maintenance cost are treated as competing objectives. The second example is associated with network-level bridge maintenance management, in which a group of spatially distributed bridges that form a highway network is studied. The overall network performance is assessed in terms of the terminal connectivity reliability. A maintenance solution contains a sequence of maintenance interventions that are scheduled at discrete years to be applied to different bridges. The conflicting objectives of the network connectivity reliability and the total maintenance cost are subject to balanced optimization. A set of alternative solutions is produced that exhibits the best possible tradeoff among all competing objectives. Bridge managers' preference on the balance between the lifetime performance and life-cycle cost can be integrated into the decision-making process. Finally, research needs of integrating bridge management systems and structural health monitoring are discussed and illustrated.

References

[1] AASHTO. *Manual for Condition Evaluation of Bridges.* 2nd edn. American Association of State Highway and Transportation Officials, Washington, D.C., 1994.

[2] AASHTO. *Standard Specifications for Highway Bridges.* 16th edn. American Association Of State Highway And Transportation Officials, Washington, D.C., 1996.

[3] F. Akgül, D. M. Frangopol. Rating and Reliability of Existing Bridges in a Network. *Journal of Bridge Engineering, ASCE* 8(6), 383-393, 2003.

[4] M. G. H. Bell, Y. Iida. *Transportation Network Analysis.* Wiley, Chichester, 1997.

[5] A. Chen, W. W. Recker. Considering Risk Taking Behavior in Travel Time Reliability, Report No. UCI-ITS-WP-00-24, Institute of Transportation Studies, University of California, Irvine, California, 2000.

[6] C. A. Coello Coello Theoretical and Numerical Constraint-Handling Techniques Used with Evolutionary Algorithms: a Survey of the State of the Art." *Computer Methods in Applied Mechanics and Engineering* 191, 1245-1287, 2002.

[7] C:A. Coello Coello, D.A. van Veldhuizen, G.B. Lamont. *Evolutionary Algorithms for Solving Multi-Objective Problems.* Kluwer Academic Publishers, New York, 2002.

[8] P.C. Das. Prioritization of Bridge Maintenance Needs. In: Frangopol, D.M. (ed.): *Case Studies in Optimal Design and Maintenance Planning Of Civil Infrastructure Systems. ASCE* Reston, 26-44, 1999.

[9] DB12/01. The Assessment of Highway Bridge Structures. Highways Agency Standard for Bridge Assessment, London, 2001.

[10] K. Deb. *Multi-Objective Optimization Using Evolutionary Algorithms.* Wiley, Chichester, 2001.

[11] S. Denton. Data Estimates for Different Maintenance Options for Reinforced Concrete Cross Heads (Personal communication). Parsons Brinckerhoff Ltd, Bristol, 2002.

[12] Enright, M. P., Frangopol, D. M. Condition Prediction of Deteriorating Concrete Bridges Using Bayesian Updating. *Journal of Structural Engineering, ASCE* 125(10) (1999) 1118-1125.

[13] A.C. Estes, D.M. Frangopol. *Updating Bridge Reliability Based on Bridge Management Systems Visual Inspection Results. Journal of Bridge Engineering, ASCE* 8(6), 374-382, 2004.

[14] FHWA. The Status of The Nation's Highways, Bridges, and Transit: Conditions and Performance. U.S. Federal Highway Administration, Washington, D.C., 2002.

[15] FHWA. Recommendations for Bridge and Tunnel Security. U.S. Federal Highway Administration, Washington, D.C., 2003.

[16] D. M. Frangopol, J S. Kong, E. S.Gharaibeh. Reliability-Based Life-Cycle Management of Highway Bridges. *Journal of Computing In Civil Engineering, ASCE* 15(1), 27-34, 2001.

[17] H. Furuta, T. Kameda, Y. Fukuda, D.M. Frangopol. Life-Cycle Cost Analysis for Infrastructure Systems: Life Cycle Cost vs. Safety Level vs. Service Life. In: Frangopol, D.M., Brühwiler, E., Faber, M.H., Adey B. (eds.): *Life-Cycle Performance of Deteriorating Structures: Assessment, Design and Management. ASCE* Reston, Virginia, 19-25, 2004.

[18] F. Glover, M. Laguna. *Tabu Search.* Kluwer Academic Publishers, New York, 1997.

[19] D. E. Goldberg. *Genetic Algorithms in Search, Optimization and Machine Learning.* Addison-Wesley, Reading, 1989.

[20] H. Hawk, E. P. Small. The BRIDGIT Bridge Management System. *Structural Engineering International, IABSE* 8(4), 309-314, 1998.

[21] C. Liu, A. Hammad, Y. Itoh,: Multiobjective Optimization of Bridge Deck Rehabilitation Using a Genetic Algorithm. *Computer-Aided Civil and Infrastructure Engineering*, 12 (1 431-443, 1997.

[22] M. Liu, D. M. Frangopol. Bridge Annual Maintenance Prioritization under Uncertainty by Multiobjective Combinatorial Optimization. *Computer-Aided Civil and Infrastructure Engineering*, 20(5), 343-353, 2005.

[23] M. Liu, D. M. Frangopol. Balancing the Connectivity Reliability of Deteriorating Bridge Networks and Long-Term Maintenance Cost Using Optimization. *Journal of Bridge Engineering, ASCE* 10(4), 468-481, 2005.

[24] Z. Lounis, D. J. Vanier. A Multiobjective and Stochastic System for Building Maintenance Management. *Computer-Aided Civil and Infrastructure Engineering.* 15 (2000) 320-329.

[25] R. Marett, M. Wright. A Comparison of Neighborhood Search Techniques for Multiobjective Combinatorial Problems. *Computers & Operations Research.* 23(5), 465-483, 1996.

[26] A. Miyamoto, K. Kawamura, H. Nakamura. Bridge Management System and Maintenance Optimization for Existing Bridges. *Computer-Aided Civil and Infrastructure Engineering.* 15, 45-55, 2000.

[27] M. E. Moore, B. M. Phares, B. A. Graybeal, D. D. Rolander, G. A. Washer. Reliability of Visual Inspection for Highway Bridges (Report Nos. FHWA-RD-01-020 and FHWA-RD-01-21). U.S. Federal Highway Administration, Washington, D.C., 2001.

[28] NCHRP. Bridge Life-Cycle Cost Analysis (Report 483). National Cooperative Highway Research Program, Transportation Research Board, Washington, D.C., 2003.

[29] R. W. Shepard, M. B. Johnson, W. E. Robert, A .R. Marshall. Modeling Bridge Network Performance - Enhancing Minimal Cost Policies. In: Watanabe, E., Frangopol, D.M., Utsunomiya, T. (eds.): *Bridge Maintenance, Safety, Management and Cost.* A. A. Balkema Publishers, Leiden ,Book and CD-ROM, 2004.

[30] P.D. Thompson, E.P. Small, M. Johnson, A.R. Marshall. The Pontis Bridge Management System. *Structural Engineering International, IABSE.* 8(4),303-308, 1998.

APPLICATION OF MULTI-OBJECTIVE GENETIC ALGORITHM TO BRIDGE MAINTENANCE

H. Furuta,[1] and T. Kameda[2]

[1]*Department of Informatics, Kansai University, 2-1-1 Ryozenji-cho, Takatsuki, Osaka 569-1095, Japan, furuta@res.kutc.kansai-u.ac.jp* [2]*Graduate School of Informatics, Kansai University, 2-1-1 Ryozenji-cho, Takatsuki, Osaka 569-1095, Japan*

Abstract In order to establish a rational bridge management program, it is necessary to develop a cost-effective decision-support system for the maintenance of bridges. In this paper, an attempt is made to develop a new multi-objective genetic algorithm for the bridge management system that can provide practical maintenance plans. Several numerical examples are presented to demonstrate the applicability and efficiency of the proposed method.

keywords: Bridge Maintenance, Genetic Algorithm, Life-Cycle Cost, Multi-Objective Optimization, Repair

1. INTRODUCTION

In this paper, an attempt is made to apply a Multi-Objective Genetic Algorithm (MOGA) for the establishment of optimal planning of existing bridge structures. In order to establish a rational maintenance program for the bridge structures, it is necessary to evaluate the structural performance of existing bridges in a quantitative manner. So far, several structural performance indices have been developed, some of which are reliability, durability, and damage indices. However, it is often necessary to discuss the structural performance from the economic and/or social points of view.

Life-Cycle Cost (LCC) is one of useful measures for evaluating the structural performance from another standpoint. Using these measures, the overall cost can be reduced and an appropriate allocation of resources can be achieved. In general, LCC optimization is to minimize the expected total cost which includes the initial cost involving design and construction, routine or preventive maintenance cost, inspection, repair and failure costs. Then, the optimal strategy obtained by LCC optimization can be different according to the prescribed level of structural performance and required service life.

Please use the following format when citing this chapter:
Author(s) [insert Last name, First-name initial(s)], 2006, in IFIP International Federation for Information Processing, Volume 199, System Modeling and Optimization, eds. Ceragioli F., Dontchev A., Furuta H., Marti K., Pandolfi L., (Boston: Springer), pp. [insert page numbers].

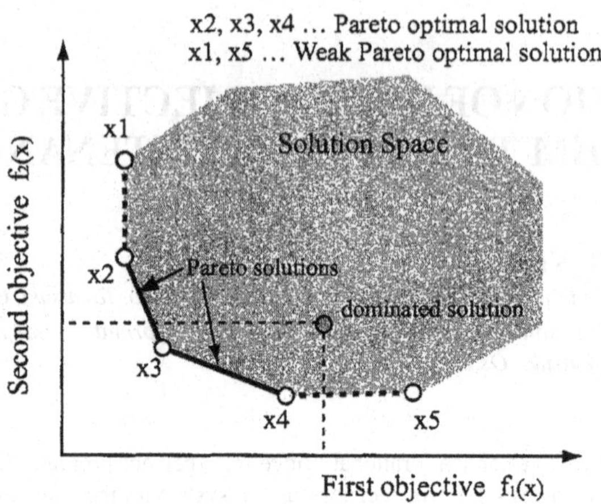

Figure 1. Pareto solutions

In this paper, an attempt is made to discuss the relationships among several performance measures and provide rational balances of these measures by using MOGA. Furthermore, another attempt is made to develop a new multi-objective genetic algorithm for the bridge management problem that have a lot of constraints. Several numerical examples are presented to demonstrate the applicability and efficiency of the proposed method.

2. MULTI-OBJECTIVE GENETIC ALGORITHM

2.1 Multi-Objective Problem

A multiple-objective optimization problem has two or more objective functions that cannot be integrated into a single objective function [1]. In general, the objective functions cannot be simultaneously minimized (or maximized). It is the essence of the problem that trade-off relations exist among the objective functions. The concept of "Pareto optimum" becomes important in order to balance the trade-off relations. The Pareto optimum solution is a solution that cannot improve an objective function without sacrificing other functions (Figures 1 and 2). A dominated, also called non-dominant, solution is indicated in Figure 1.

2.2 Multi-Objective Genetic Algorithm (MOGA)

Genetic Algorithm (GA) is an evolutionary computing technique, in which candidates of solutions are mapped into GA space by encoding. The following steps are employed to obtain the optimal solutions [2]: a) initialization, b)

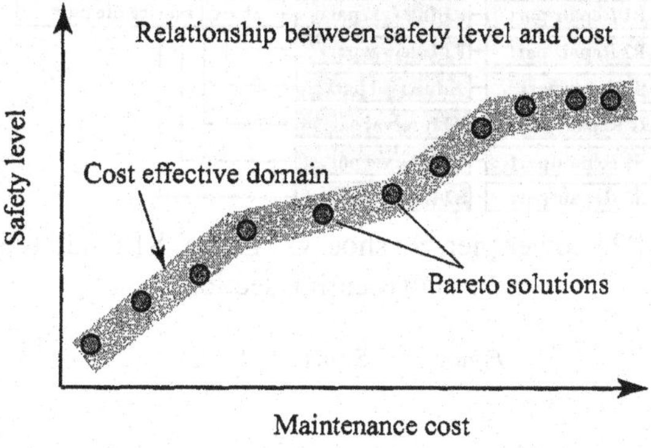

Figure 2. Cost-effective domain including Pareto solutions

crossover, c) mutation, d) natural selection, and e) reproduction. Individuals, which are solution candidates, are initially generated at random. Then, steps b, c, d, and e are repeatedly implemented until the termination condition is fulfilled. Each individual has a fitness value to the environment. The environment corresponds to the problem space and the fitness value corresponds to the evaluation value of objective function. Each individual has two aspects: Gene Type (GTYPE) expressing the chromosome or DNA and Phenomenon Type (PTYPE) expressing the solution. GA operations are applied to GTYPE and generate new children from parents (individuals) by effective searches in the problem space, and extend the search space by mutation to enhance the possibility of individuals other than the neighbor of the solution. GA operations that generate useful children from their parents are performed by crossover operation of chromosome or genes (GTYPE) without using special knowledge and intelligence. This characteristic is considered as one of the reasons of the successful applications of GA [3].

3. FORMULATION OF BRIDGE MAINTENANCE PLANNING

It is desirable to determine an appropriate life-cycle maintenance plan by comparing several solutions for various conditions [4][5]. A new decision support system is developed here from the viewpoint of multi-objective optimization, in order to provide various solutions needed for the decision-making.

In this study, LCC, safety level and service life are used as objective functions. LCC is minimized, safety level is maximized, and service life is maximized. There are trade-off relations among the three objective functions. For example, LCC increases when service life is extended, and safety level and service life

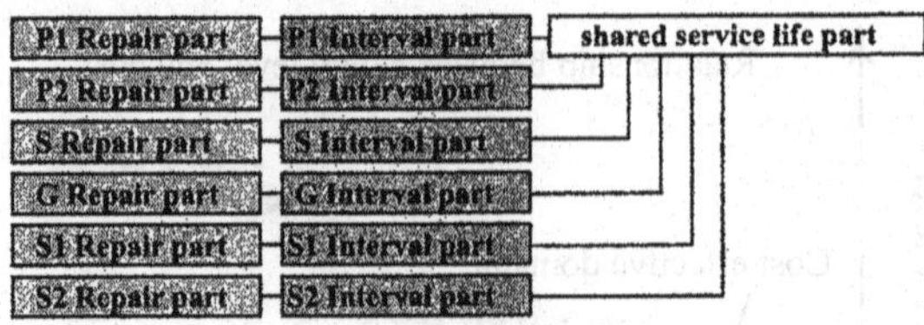

P1 : upper pier, P2 : lower pier, S : shoe, G : girder, S1 : slab (central section), S2 : slab (bearing section)

Figure 3. Structure of DNA

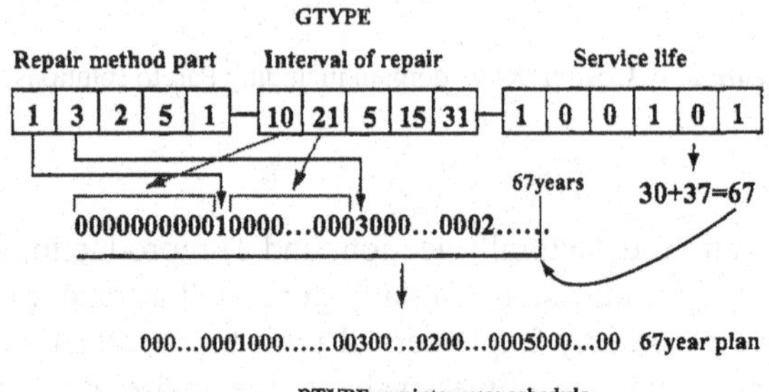

Figure 4. Coding rule

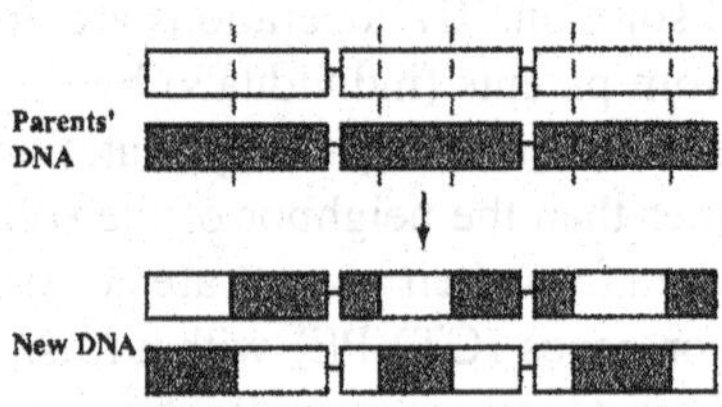

Figure 5. Crossover

decrease due to the reduction of LCC. Then, multi-objective optimization can provide a set of Pareto solutions that can not improve an objective function without making other objective functions worse [6].

In the proposed system, DNA structure is constituted as shown in Figure 3, in which DNA of each individual consists of three parts such as repair method, interval of repair, and shared service life (Figure 4). In this figure, service life is calculated as the sum of repairing years and their interval years. In Figure 4, service life is obtained as 67 years which is expressed as the sum of 30 years and 37 years. The repair part and the interval part have the same length. Gene

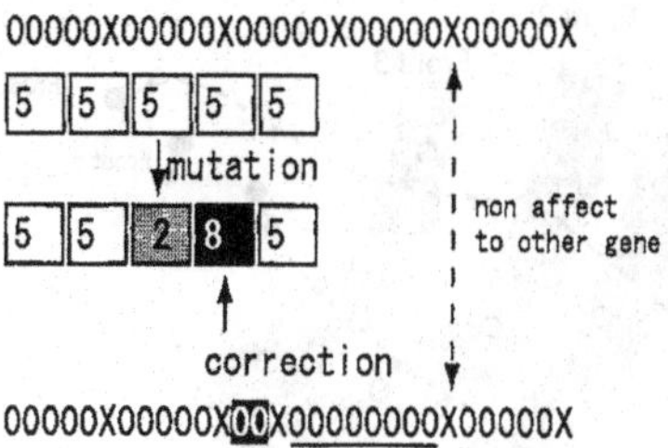

Figure 6. Mutation

of repair part has ID number of repair method. The interval part has enough length to consider service life. In this system, ID 1 means surface painting, ID 2 surface coating, ID 3 section restoring, ID 4 desalting (re-alkalization) or cathodic protection, and ID 5 section restoring with surface covering. DNA of service life part has a binary expression with six bits and its value is changed to decimal number.

In crossover, the system generates new candidates by using the procedure shown in Figure 5. For mutation, the system shown in Figure 6 is used.

Then, objective functions are defined as follows:

$$Objective function1 : C_{total} = \sum LCC_i \rightarrow min \tag{1}$$

where LCC_i = LCC for $bridge_i$

$$Objective function2 : Y_{total} = \sum Y_i \rightarrow max \tag{2}$$

Constraints : $Y_i > Y_{required}$
where Y_i = Service life of $bridge_i$, $Y_{required}$ = Required service life

$$Objective function3FP_{total} = \sum P_i \rightarrow max \tag{3}$$

Constraints : $P_i > P_{target}$
where P_{target} = Target safety level

The above objective functions have trade-off relations to each other. Namely, the maximization of safety level or maximization of service life cannot be realized without increasing LCC. On the other hand, the minimization of LCC can be possible only if the service life and/or the safety level decreases.

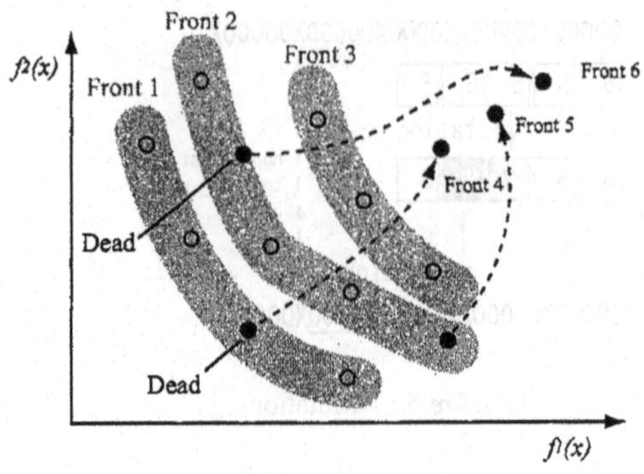

Figure 7. Concept of Front

4. NEW MULTI-OBJECTIVE GENETIC ALGORITHM

As mentioned before, the formulation of bridge maintenance planning has several constraint conditions. In usual, it is not easy to solve multi-objective optimization problems with constraints by applying the multi-objective genetic algorithm.

In this study, an improvement is made on the selection process by introducing the sorting technique. The selection is performed using so-called sorting rules which arrange the list of individuals in the order of higher evaluation values. Then, the fitness values are assigned to them by using the linear normalization technique. In usual, if the fitness values are calculated directly according to the evaluation values, the differences among every individuals decrease so that the effective selection can not be done. On the other hand, the linear normalization technique enables to keep the selection pressure constant so that it may continue the selection well. In this study, the selection procedure is improved coupling the linear normalization technique and the sorting technique. Using the evaluation values, the individuals are reordered and given the new fitness values. Figure 7 presents the process of the selection proposed here. The individuals of satisfying the constraints are arranged first according to the evaluation values and followingly the individuals of unsatisfying the constraints are arranged according to the degree of violating the constraints. Accordingly, all the individuals are given the fitness values using the linear normalization technique.

In order to apply the sorting rules to the multi-objective optimization problems, the non- dominated sorting method is used. In the non-dominated sorting method, the Pareto solutions are defined as Front1. Then, Front2 is determined by eliminating the Front1 from the set of solution candidates. Repeating the process, the new Front is pursued until the solution candidates diminish. Further, the Fronts are stored in the pool of the next generation. If the pool is full, the individuals in the Front are divided into the solutions to survive or die based on the degree of congestion.

In this study, the individuals are divided into the group of satisfying the constraints and the group without satisfying the constraints. The former is called as live individual, and the latter dead individual. While the live individuals are given the fitness values according to the evaluation values after the non-dominated sorting, the dead individuals are given the same fitness value. When implementing the non-dominated sorting, the Pareto front may not exist at the initial generation, because there remain a lot of dead individuals after the non-dominated sorting. Then, the dead individuals are arranged in the order of degree of violating the constraints and some of them are selected for the next generation. Thus, the multi-objective optimization problem with constraints are transformed into the minimization problem of violation of constraints. The elite preserve strategy is employed for the selection of survival individuals.

When the generation is progressed, live individuals appear and then both the live individuals forming the Pareto front and the dead individuals arranged in the order of violation degree exist together. In this case, appropriate numbers of live and dead individuals are selected for the next generation. If the number of individuals involved in the Pareto front increases, only the individuals are selected for the next generation.

5. NUMERICAL EXAMPLES

Figures 5 through 9 present the results calculated for B01 model (Bridge 1) with the constraint that safety level should be greater than 0.6 under the environment that neutralization is dominant. From Figure 5, it is seen that the number of live individuals immediately increases after 20 generations and reaches to 1000 at 40 generations. Similarly, the number of Pareto solutions increases with some fluctuation and converges after 125 generations. As the generation proceeds from 100 to 3000, the Pareto solutions can be improved smoothly. Figure 5 shows that the proposed method can provide solutions satisfying the safety requirement very well. Figure 10 shows the repair methods given by the proposed method. Figure 11 presents the results of B04, which gives the safety for every structural element that is greater than the prescribed target value 0.6, though the safety levels are different. Figure 12 shows the optimal combination of repair methods.

6. CONCLUSIONS

In this paper, an attempt was made to formulate the optimal maintenance planning as a multi-objective optimization problem. By considering LCC, safety level, and service life as objective functions, it is possible to obtain

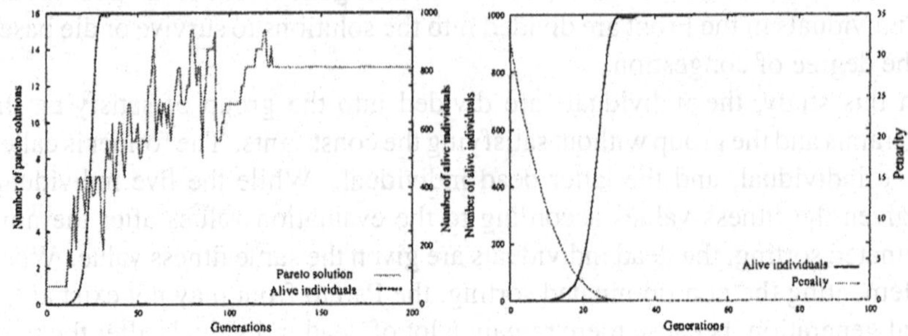

Figure 8. Calculated result for bridge 1 under the neutralization environment -relation between number of Pareto solutions and generation (left) and live solutions and generation (right)

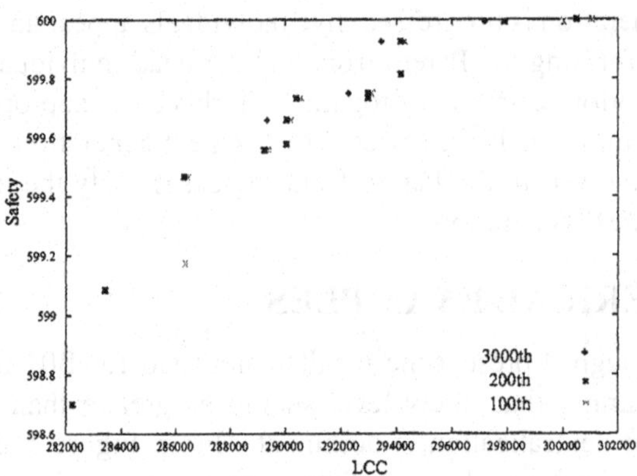

Figure 9. Calculated result for bridge 1 under the neutralization environment -relation between safety level and LCC-

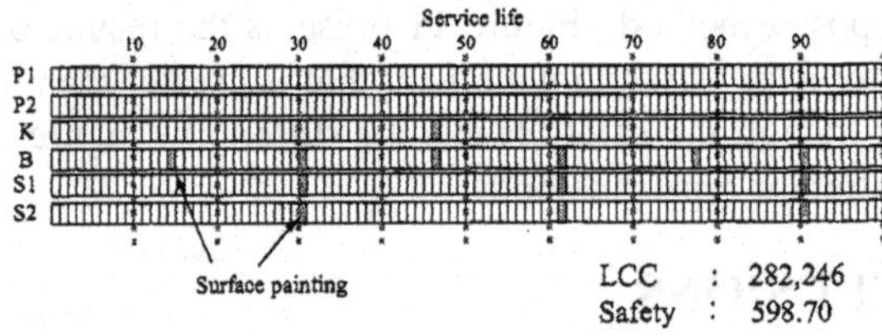

Figure 10. Repair methods for bridge 1

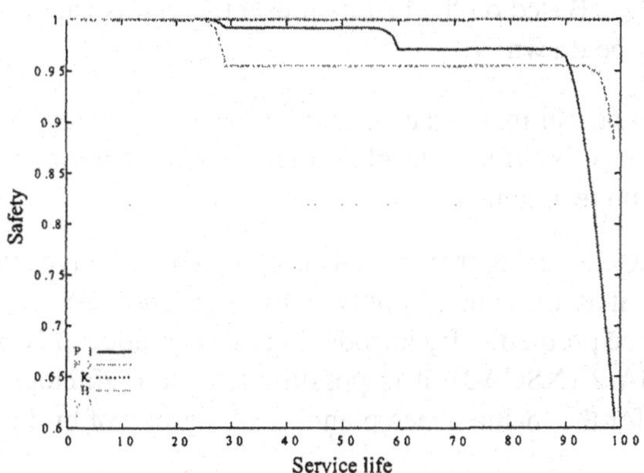

Figure 11. Change of safety level of bridge 4

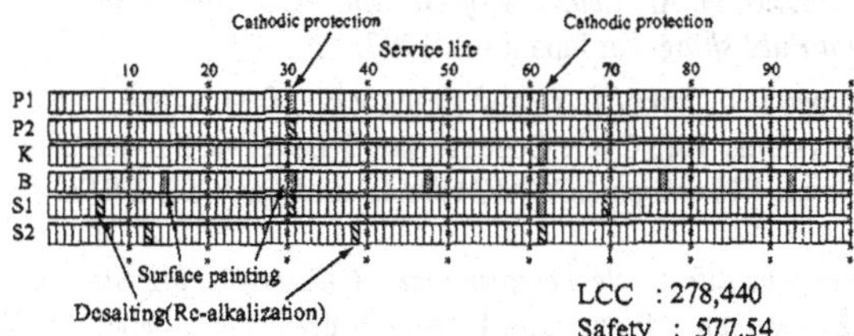

Figure 12. Repair methods for bridge 4

the relationships among these three performance indicators and provide bridge maintenance engineers with various maintenance plans with appropriate allocations of resources. Furthermore, another attempt was made to develop a new multi-objective genetic algorithm for the bridge management problem that has a lot of constraints. Based on the results presented in this paper, the following conclusions may be drawn:

- Since the optimal maintenance problem is a very complex combinatorial problem, it is difficult to obtain reasonable solutions by the current optimization techniques.
- Although Genetic Algorithm (GA) is applicable to solve multi-objective problems, it is difficult to apply it to large and very complex bridge maintenance problems. By introducing the technique of Non-Dominated Sorting GA-2 (NSGA2), it is possible to obtain efficient near-optimal solutions for the maintenance planning of a group of bridge structures.
- Furthermore, the new GA method proposed here can much more improve the convergence and efficiency in the optimization procedure, by introducing the sorting based selection.
- The proposed method can provide many near-optimal maintenance plans with various reasonable LCC values, safety levels and service lives.

References

[1] Iri, M. Konno, H. (eds.). *Multi-Objective Optimization, Tokyo, Sangyo-tosho (in Japanese).* 1982

[2] Goldberg, D.E. *Genetic Algorithms in Search, Optimization and Machine Learning,* Addison Wesley Publishing, 1989.

[3] Furuta, H. Sugimoto, H. *Applications of Genetic Algorithm to Structural Engineering, Tokyo, Morikita Publishing* (in Japanese). 1997.

[4] Furuta, H., Kameda, T., Fukuda, Y., and Frangopol, D. M. Life-Cycle Cost Analysis for Infrastructure Systems: Life Cycle Cost vs. Safety Level vs. Service Life, *Proceedings of Joint International Workshops LCC03/IABMAS and fip/JCSS, EPFL, Lausanne, March 24-26,2003* ,
(keynote lecture); in Life-Cycle Performance of Deteriorating Structures: Assessment, Design and Management , Frangopol, D. M., Bruhwiler, E., Faber, M. H. and Adey, B. (eds.), ASCE, Reston, Virginia, 19-25.

[5] Furuta, H., Kameda, T. and Frangopol, D. M. Balance of Structural Performance Measures, *Proc. of Structures Congress, Nashville, Tennessee, ASCE, May, CD-ROM. 2004*

[6] Liu, M. and Frangopol, D.M. (2004b). Optimal Bridge Maintenance Planning Based on Probabilistic Performance Prediction, *Engineering Structures, Elsevier* , 26(7), 991-1002.

A METHOD FOR THE MIXED DISCRETE NON-LINEAR PROBLEMS BY PARTICLE SWARM OPTIMIZATION

S. Kitayama,[1] M. Arakawa,[2] and K. Yamazaki[1]

[1]*Kanazawa University, Kakuma-machi, Kanazawa, 920-1192, JAPAN, kitagon@t.kanazawa-u.ac.jp* [2]*Kagawa University, 2217-20 Hayashi-cho, Takamatsu, Kagawa, 761-0396, JAPAN*

Abstract An approach for the Mixed Discrete Non-Linear Problems (MDNLP) by Particle Swarm Optimization is proposed. The penalty function to handle the discrete design variables is employed, in which the discrete design variables are treated as the continuous design variables by penalizing at the intervals. By using the penalty function, it is possible to handle all design variables as the continuous design variables. Through typical benchmark problem, the validity of proposed approach for MDNLP is examined.

keywords: Global Optimization, Particle Swarm Optimization, Mixed Discrete Non-Linear Problems

1. Introduction

Particle Swarm Optimization (PSO), which mimics the social behavior, is an optimization technique developed by Kennedy et. al. [1]. It has been reported that PSO is suitable for the minimization of the non-convex function of the continuous design variables through many numerical examples. Few researches of PSO have been reported about the discrete optimizaton [2]. These researches handle the discrete design variables as the continuous ones, directly. That is, firstly all design variables may be handled as the continuous ones, and optimized. Finally, the round-off or cut-off are applied to get the discrete optimum. These approaches may be valid in a sense, but some problems are included as shown in Fig.1(a), (b).

Fig.1(a) shows a case. x_L represents the optimum of the continuous design variables. Point A and B represent the discrete design variables close to x_L. In this case, Point B is chosen as the neighborhood of x_L by the round-off. However, the objective function at Point B makes a change of the function value worse, when compared with the objective function at Point A. [3] Another

Please use the following format when citing this chapter:
Author(s) [insert Last name, First-name initial(s)], 2006, in IFIP International Federation for Information Processing, Volume 199, System Modeling and Optimization, eds. Ceragioli F., Dontchev A., Furuta H., Marti K., Pandolfi L., (Boston: Springer), pp. [insert page numbers].

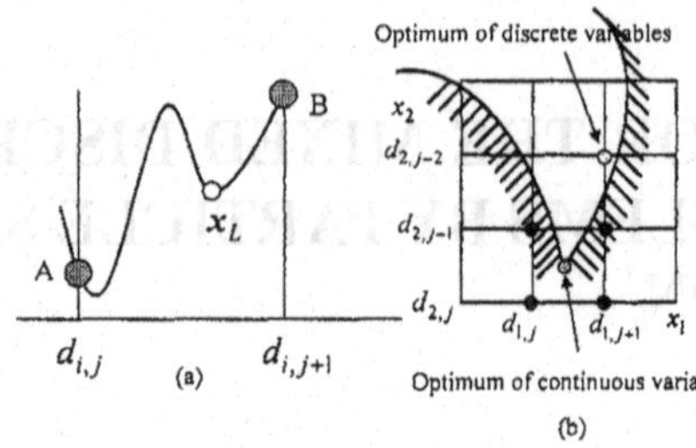

Figure 1. The nature of discrete optimization

case shown in Fig.1(b) is well known. That is, the optimum obtained by the round-off or the cut-off does not satisfy all feasibilities [4].

PSO is suitable for the global optimization of the non-convex function of the continuous design variables. Therefore, all design variables should be handled as the continuous ones whenever PSO is applied to the mixed or discrete design variables problems.

In this paper, the penalty function approach to handle the discrete design variables is proposed. By using the penalty function for the discrete design variables, it is possible to handle the discrete design variables as the continuous ones. Through typical MDNLP, the validity of proposed approach is examined.

2. Particle Swarm Optimization

Particle Swarm Optimization (PSO) is one of the global optimization methods for the continuous design variables [6]. PSO does not utilize the gradient information of function like Genetic Algorithm. In PSO, each particle has the position and velocity, and they are updated by a simple addition and subtraction of vectors during search process.

The position and velocity of particle d are represented by $\boldsymbol{x}_d^k$ and $\boldsymbol{v}_d^k$, respectively. k represents iteration. The position and velocity of particle d at k+1 th iteration are calculated by following equations.

$$\boldsymbol{x}_d^{k+1} = \boldsymbol{x}_d^k + \boldsymbol{v}_d^{k+1} \tag{1}$$

$$\boldsymbol{v}_d^{k+1} = w \times \boldsymbol{v}_d^k + c_1 r_1 \times (\boldsymbol{p}_d^k - \boldsymbol{x}_d^k) + c_2 r_2 \times (\boldsymbol{p}_g^k - \boldsymbol{x}_d^k) \tag{2}$$

in which the coefficient w is called as inertia term, and r_1 and r_2 denote random number between [0,1). The weighting coefficients c_1 and c_2 are parameters. In general, $c_1 = c_2 = 2$ is often used. $\boldsymbol{p}_d^k$ called as *pbest*, represents the best position of particle d till k th iteration, and $\boldsymbol{p}_g^k$ called as *gbest*, represents the best position in the swarm till k th iteration. That is, $\boldsymbol{p}_g^k$ is chosen among $\boldsymbol{p}_d^k$.

The inertia term in Eq.(2) gradually decreases during the search iteration.

$$w = w_{max} - \frac{w_{max} - w_{min}}{k_{max}} \times k \tag{3}$$

in which, w_{max} and w_{min} represent the maximum and minimum value of inertia. k_{max} represents the maximum number of search iteration. In general, w_{max} = 0.9 and w_{min}=0.4 are recommended [7].

2.1 PSO as an Optimization Technique

From Eq.(1) and Eq.(2), the following equation can be obtained.

$$x_d^{k+1} = x_d^k + w \times v_d^k + \alpha(q - x_d^k) \tag{4}$$

in which α and q are represented as follows, respectively.

$$\alpha = c_1 r_1 + c_2 r_2 \tag{5}$$

$$q = \frac{c_1 r_1 p_d^k + c_2 r_2 p_g^k}{c_1 r_1 + c_2 r_2} \tag{6}$$

From Eq.(4), it is possible to interpret that $q - x_d^k$ represents the search direction when we imagine the similarity to the gradient methods. α in Eq.(4) also may be regarded as stochastic step-size, in which its lower and upper bounds are 0 and c_1+c_2, and the mean value is $(c_1 + c_2)/2$. From these relationships, it is possible to consider that PSO has a search direction vector and stochastic step-size even though PSO does not utilize the gradient information of function.

3. Penalty Function Approach for MDNLP by PSO

3.1 Problem Definition

In general, the Mixed Discrete Non-Linear Problem (MDNLP) is described as follows [5]:

$$f(x) \rightarrow min \tag{7}$$

$$x_i^L \leq x_i \leq x_i^U \quad i = 1, 2, \cdots, m \tag{8}$$

$$x_{m+i} \in D_i \quad D_i = \{d_{i,1}, d_{i,2}, \cdots, d_{i,q}\} \quad i = 1, 2, \cdots, n \tag{9}$$

$$g_k(x) \leq 0 \quad k = 1, 2, \cdots, ncon \tag{10}$$

where x represents the design variables, which consist of the continuous and discrete design variables. $f(x)$ is the objective function, and $g_k(x)$ is the behavior constraints. $ncon$ represents the number of behavior constraints. x_i denotes the continuous design variables, and m is the total number of continuous design variables. x_i^L and x_i^U denote the lower and upper bounds of continuous design variables, respectively. On the other hand, n is the total number of discrete design variables. D_i is the set of discrete values for the i-th discrete design variable. $d_{i,j}$ is the j-th discrete value for the i-th discrete design variables. q represents the number of discrete values.

3.2 Penalty Function

In this paper, the following penalty function suggested by [8] is adopted.

$$\phi(\boldsymbol{x}) = \sum_{i=1}^{n} \frac{1}{2} sin[\frac{2\pi \times \{x_{m+i}^c - 0.25(d_{i,j+1} + 3d_{i,j})\}}{d_{i,j+1} - d_{i,j}} + 1] \qquad (11)$$

where x_{m+i}^c is the continuous design variables between $d_{i,j}$ and $d_{i,j+1}$. Then the augmented objective function $F(\boldsymbol{x})$ is constructed by using above penalty function as follows:

$$F(\boldsymbol{x}) = f(\boldsymbol{x}) + s \times \phi(\boldsymbol{x}) + r \sum_{k=1}^{ncon} max[0, g_k(\boldsymbol{x})] \qquad (12)$$

in which s and r denote the penalty parameters for Eq.(11) and Eq.(10), respectively. Finally, MDNLP is transformed into the following continuous design variables problem.

$$F(\boldsymbol{x}) \rightarrow min \qquad (13)$$

$$x_i^L \leq x_i \leq x_i^U \quad i = 1, 2, \cdots, m \qquad (14)$$

$$d_{i,1} \leq x_{m+i}^c \leq d_{i,q} \quad i = 1, 2, \cdots, n \qquad (15)$$

For the simplicity, the design variables are supposed as the discrete design variables in the following discussion. In the case of the mixed design variables, we discuss at section 3.7 separately.

3.3 Characteristics of Penalty Function

The value of Eq.(11) becomes small around the neighborhood of discrete value. On the other hand, the value of Eq.(11) becomes large, turning from discrete value. When $\boldsymbol{p}_g^k$ satisfies the following equation, the discrete value resides around the neighborhood of $\boldsymbol{p}_g^k$.

$$\phi(\boldsymbol{p}_g^k) \leq \varepsilon \qquad (16)$$

ε in Eq.(16) represents small positive value. As a result, the penalty parameter s in Eq.(12) must be updated so that Eq.(16) is satisfied. In order to examine the effect of the penalty parameter s, let us consider a following simple problem.

$$f(x) = x^4 - \frac{8}{3} - 2x^2 + 8x \rightarrow min \tag{17}$$

$$x \in \{-1, 0, 1, 2\} \tag{18}$$

In this simple example, the objective $f(x)$ and the augmented objective function $F(x)$ are shown in Fig.2.

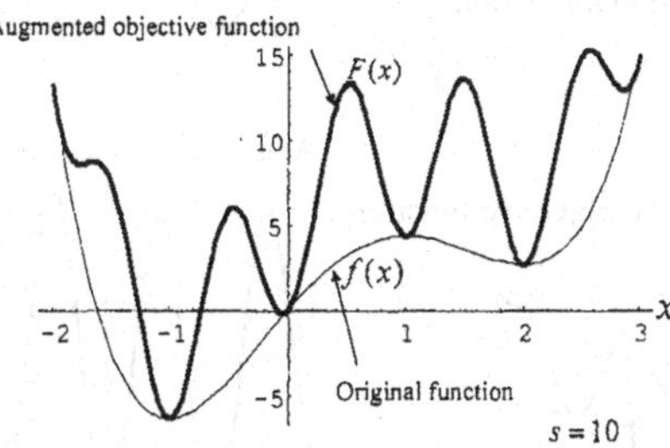

Figure 2. Behavior of the augmented objective function

From Fig.2, it is apparent that $F(x)$ becomes non-convex and continuous. Additionally many local minima are generated around the neighborhood of the discrete values. As a result, the problem to find the discrete optimum is transformed into finding global minimum of $F(x)$. Additionally, the discrete values are given at the point, where the relative error between $f(x)$ and $F(x)$ becomes small. The following equation is utilized as terminal criteria.

$$\frac{|F(\boldsymbol{p}_g^k) - f(\boldsymbol{p}_g^k)|}{|F(\boldsymbol{p}_g^k)|} \leq \varepsilon \tag{19}$$

PSO does not use the gradient information of function, so that it is difficult to satisfy Eq.(16) strictly. Then, Eq.(19) is used instead of that.

Behaviors of $F(x)$ for various penalty parameter s are shown in Fig.3. From Fig.3, it is found that to determine a penalty parameter s in advance is very difficult.

3.4 Initial Penalty Parameter s

An initial search point $\boldsymbol{x}_d$ of particle d is determined randomly. Then the value of penalty function in Eq.(11) is calculated for each particle. The penalty parameter s is determined as follows.

$$s_d = 1 + \phi(\boldsymbol{x}_d) \quad d = 1, 2, \cdots, agent \tag{20}$$

where s_d represents the penalty parameter of particle d. $agent$ in Eq.(20) is the total number of particles. And initial penalty parameter $s_{initial}$ is determined

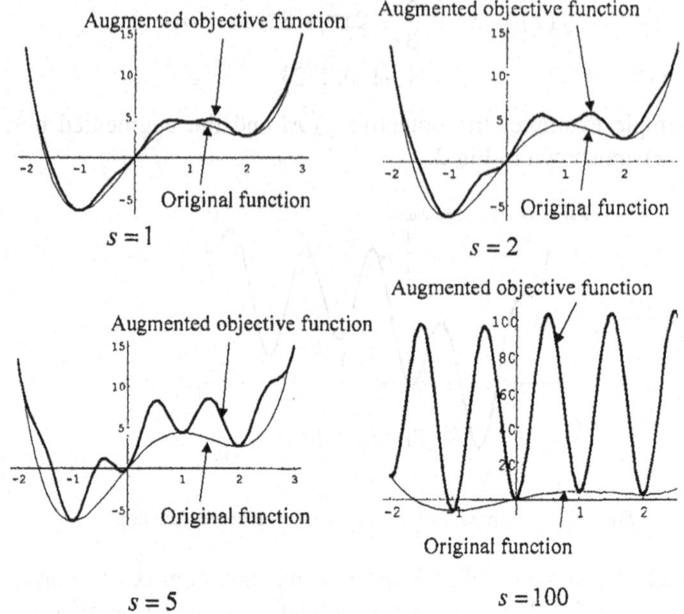

Figure 3. Behavior of the augmented objective function for some parameter

as follows.

$$s_{initial} = min\{s_1, s_2, \cdots, s_{agnet}\} \tag{21}$$

At the initial stage to search the optimum, $F(x)$ is actively transformed into non-convex and continuous function, and many local minima are enerated around the neighborhood of the discrete value.

3.5 Update Scheme of Penalty Parameter s

The following equation is used to update the penalty parameter s.

$$s = s \times exp(1 + \phi(p_g^k)) \tag{22}$$

The behavior of $F(x)$ by updating the penalty parameter s is shown schematically in Fig.4.

In Fig.4, solid line shows $F(x)$ at k th iteration, and dotted line shows $F(x)$ at k+1 th iteration. As shown in Fig.4, $F(x)$ at k+1 th iteration becomes highly non-convex function in comparison with $F(x)$ at k th iteration. For example, point A in Fig.4 corresponds to the point p_g^k at k th iteration. By updating

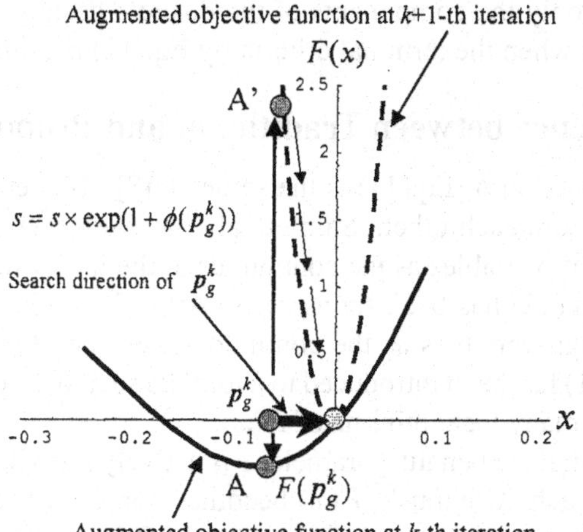

Figure 4. Update of penalty parameter s

penalty parameter s, p_g^k corresponds to the point A' on the dotted line. As discussed in section 2.1, PSO has similar structure to the gradient methods, so that it is expected that p_g^k moves to the direction in Fig.4. Finally, it is also expected that p_g^k will satisfy Eq.(19).

3.6 Initilization of Penalty Parameter s

When Eq.(19) is satisfied, the discrete value around the neighborhood of p_g^k resides. Then an initial penalty parameter by Eq.(21) is utilized in order to find another discrete value, because $F(x)$ becomes highly non-convex function by updating the penalty parameter s. It is assumed that p_g^k fails to escape from local minimum. In such occasions, $F(x)$ is relaxed by using an initial penalty parameter when Eq.(19) is satisfied. As a result, it is expected that p_g^k can escape from local minimum.

3.7 In the Case of Mixed Design Variables

The component of p_g^k can be expressed as follows:

$$p_g^k = (x^{cont}, x^{discrt})^T \tag{23}$$

where x^{cont} and x^{discrt} represent the components of the continuous and discrete design variables, respectively. Then, the components of the continuous design variables x^{cont} in p_g^k are neglected when the terminal criteria by Eq.(19) is applied. That is, only the components of the discrete design variables x^{discrt} in p_g^k are checked when the terminal criteria by Eq.(19) is utilized.

3.8 Difference between Traditional and Proposed Method

The penalty function of Eq.(11) is the same as [8]. However, its approach is very different from each other. Shin et. al. have searched an optimum by regarding all design variables as the continuous at the initial stage, the penalty parameter s in Eq.(12) has been set as zero. After the optimum obtained by regarding all design variables as the continuous has been found, the penalty function of Eq.(11) has been introduced to avoid the search procedure of global minimum among many local minima of $F(x)$.

On the other hand, the penalty parameter s is actively introduced at the initial stage in our approach. Obviously $F(x)$ becomes non-convex and continuous. However, this is not serious problem because PSO is applied to $F(x)$. The new update scheme of penalty parameter s by Eq.(22) is proposed. In the past reports [4, 8], the constant positive number is used to update the penalty parameter. However, the constant positive number depends on the problems. On the other hand, the penalty parameter s may always changes in our approach because the value of $\phi(p_g^k)$ is utilized. It may be expected that flexible applications may be possible. Finally, the initialization of the penalty parameter s is also introcuded in order to relax $F(x)$. As a result, it is expected that p_g^k can escape from local minimum.

Binary PSO is also easy to handle the discrete design variables [9, 10]. However, the search process of binary PSO is stochastic. Additionally, no proof that the objective or design domain is continuous is given. On the other hand, our approach adopted here utilizes the characteristics of PSO and the penalty function of Eq.(11), in order to find optimum. That is, our approach may be deterministic, when compared with binary PSO.

3.9 Algorithm

The proposed algorithm for MDNLP by PSO is shown in Fig.5.

4. Numerical Example

To examine the validity of proposed approach, let us consider the optimum design of pressure vessel as shown in Fig.6.

This problem is one of the most famous benchmark for MDNLP [9, 11–14]. Several results are shown in table 1. From table 1, it is very difficult to find optimum solutions even though this problem consists of only 4 design variables. The design variables are 1) Radius R (continuous design variables: x_1), 2) Length L (continuous design variables: x_2), 3) Thickness T_s (discrete

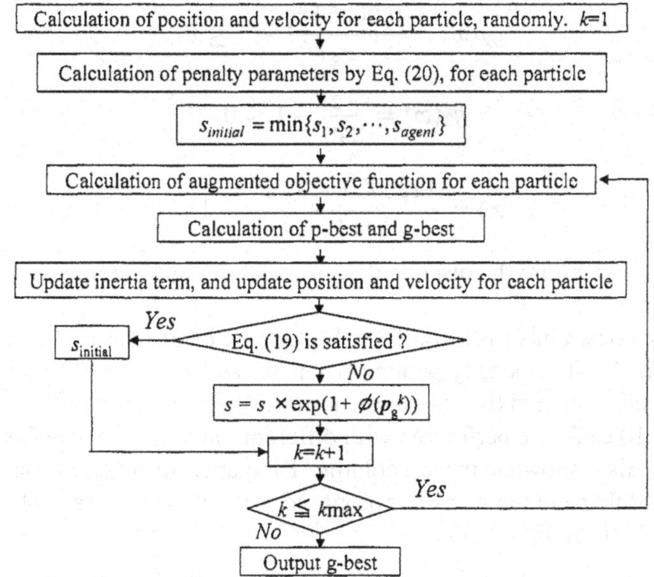

Figure 5. The algorithm for MDNLP by PSO

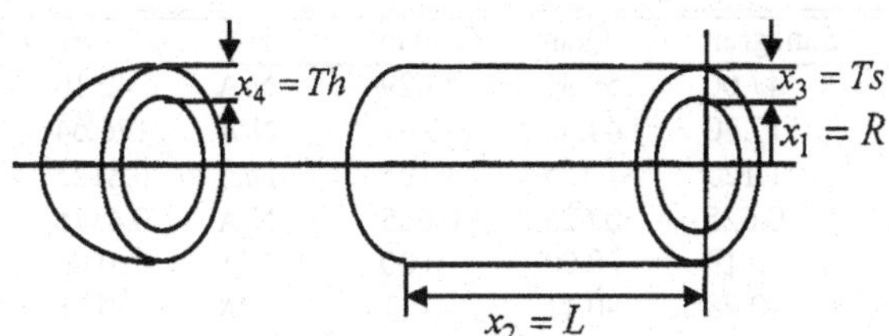

Figure 6. Optimum Design of Pressure Vessel

design variables: x_3), and 4) Thickness T_h (discrete design variables: x_4). An optimization problem is defined as follows.

$$f(x) = 0.6224x_1x_2x_3 + 1.7781x_1^2x_4 + 3.1661x_2x_3^2 + 19.84x_1x_3^2 \rightarrow min \tag{24}$$

$$10 \leq x_1, x_2 \leq 200 \tag{25}$$

$$0.0625 \leq x_3, x_4 \leq 6.1875 \tag{26}$$

$$g_1(x) = \frac{0.0193x_1}{x_3} - 1 \leq 0 \tag{27}$$

$$g_2(\boldsymbol{x}) = \frac{0.00954x_1}{x_4} - 1 \leq 0 \tag{28}$$

$$g_3(\boldsymbol{x}) = \frac{x_2}{240} - 1 \leq 0 \tag{29}$$

$$g_4(\boldsymbol{x}) = \frac{1296000 - \frac{4}{3}\pi x_1^3}{\pi x_1^2 x_2} - 1 \leq 0 \tag{30}$$

in which x_3 and x_4 are the discrete design variables, and are integer multiples of 0.0625 inch.

Behavior constraints from Eq.(27) to Eq.(30) are handled as penalty function by using Eq.(12) The penalty parameter r is set as 1.0×10^8. The number of particle is set as 50, and the maximum number of search iteration k_{max} is also set as 500. 10 trials are performed with different random seed. The best result during 10 trials is shown in the last column "Kitayama" in table 1. From table 1, best result could be obtained by our proposed method. The average of function calls through 10 trials is 22500.

Table 1. Comparison of results

	Sandgren	Qian	Kannan	Hsu	He	Kitayama
R[inch]: x_1	47.00	58.31	58.29	N/A	42.10	42.37
L[inch]: x_2	117.70	44.52	43.69	N/A	176.64	173.42
T_s[inch]: x_3	1.125	1.125	1.125	N/A	0.8125	0.8125
T_h[inch]: x_4	0.625	0.625	0.625	N/A	0.4375	0.4375
$g_1(\mathbf{x})$	-0.19	0.00	0.00	N/A	0.00	0.00
$g_2(\mathbf{x})$	-0.28	-0.11	-0.11	N/A	-0.08	-0.08
$g_3(\mathbf{x})$	-0.51	-0.81	-0.82	N/A	-0.26	-0.28
$g_4(\mathbf{x})$	0.05	-0.02	-1.11	N/A	0.00	0.00
Objective	8129.80	7238.83	7198.20	7021.67	6059.71	6029.87

5. Conclusions

In this paper, PSO has been applied to MDNLP. The penalty function for the discrete design variables is introduced, in order to handle as the continuous design variables. The augmented objective function becomes non-convex function of continuous design variables, by introducing penalty function. As considered that PSO is naturally suitable for the global search of non-convex function of the continuous design variables, our proposed approach may be valid. A method to determine the penalty parameter s and new update scheme of penalty parameter s have been also proposed. Through typical benchmark problem, the validity has been examined.

References

[1] J.Kennedy, R.C.Eberhart. *Swarm Intelligence.* Morgan Kaufmann Publishers, 2001.

[2] G.Venter, J.S.Sobieski. Particle Swarm Optimization. *9-th AIAA/USAF/NASA/ISSMO Symposium on Multidisciplinary Analysis and Optimization Conference,* AIAA2002-1235, 2002.

[3] S.S.Rao. *Engineering Optimization: Theory and Practice.* Wiley InterScience, 1996.

[4] J.F.Fu, R.G.Fenton, W.L.Cleghorn. A Mixed Integer-Discrete Continuous Programming and its Application to Engineering Design Optimization. *Eng. Opt.* 17:263-280, 1996.

[5] J.S.Arora, M.W.Huang. Methods for Optimization of Nonlinear Problems with Discrete Variables: A Review. *Struc. Opt.* 8:69-85, 1994.

[6] K.E.Parsopoulos, M.N.Vrahatis. Recent Approaches to Global Optimization Problems through Particle Swarm Optimization. *Neural Computing.* 1:235-306, 2002.

[7] Y.Fukuyama. A Particle Swarm Optimization for Reactive Power and Voltage Control Considering Voltage Security Assessment. *IEEE Trans. Power Syst.* 15-4:1232-1239. 2000.

[8] D.K.Shin, Z.Gurdal, O.H.Griffin. A Penalty Approach for Nonlinear Optimization with Discrete Design Variables. *Eng. Opt.* 16:29-42, 1990.

[9] S.He, E.Prempain, Q.H.Wu. An Improved Particle Swarm Optimizer for Mechanical Design Optimization Problems. *Eng. Opt.* 35-5:585-605, 2004.

[10] J.Kennedy, R.C.Eberhart. A Discrete Binary Version o f the Particle Swarm Optimization Algorithm. *Proc. of the Conference on Systems. Man, and Cybernetics.* 4104-4109, 1997.

[11] E.Sandgren. Nonlinear and Discrete Programming in Mechanical Design Optimization. *Trans. of ASME, J. of Mech. Des.* 112:pp.223-229, 1990.

[12] B.K.Kannan, S.N.Kramer. An Augmented Lagrange Multiplier Based Method for Mixed Integer Discrete Continuous Optimization and Its Applications to Mechanical Design. *Trans. of ASME, J. of Mech. Des.* 116:405-411, 1994.

[13] Z.Qian, J.Yu, J.Zhou. A Genetic Algorithm for Solving Mixed Discrete Optimization Problems. *DE-Vol.65-1, Advances in Design Automation.* 1:499-503, 1993.

[14] Y.H.Hsu, L.H.Leu. A Two Stage Sequential Approximation Method for Non-linear Discrete Variable Optimization. *ASME/DETC/DAC MA* 197-202, 1995.

OPTIMIZATION OF COOLING PIPE SYSTEM OF PLASTIC MOLDING

T. Matsumori,[1] K. Yamazaki,[2] and Y. Matsui[2]

[1]*Graduated School of Natural Science & Technology, Kanazawa University, Kakuma-machi, Kanazawa, 920-1192, JAPAN, tadayosi@stu.kanazawa-u.ac.jp* [2]*Kanazawa University, Kakuma-machi, Kanazawa, 920-1192, JAPAN*

Abstract In a plastic injection molding, design of cooling pipe system is one of the important problems to reduce internal residual stresses of molded products. If plastic materials in the injection molding die are cooled down uniformly and slowly, generation of the residual stresses can be reduced.

In this paper, a new method to design a cooling pipe system in the plastic injection molding die taking account of coolant flow in the pipe are presented. To consider the effect of the coolant flow, two kinds of models assumed plastic injection die are prepared. And shape optimization techniques are applied to design the cooling pipe system in the models. To evaluate the optimality, two kinds of evaluation functions, one is to obtain uniform temperature distribution and the other is to control cooling rate, are defined.

keywords: Fluid Dynamics, Heat Transfer, Coupled Problem, Finite Element Method, Design Optimization

1. Introductions

Plastic injection molding is used widely to mold complex shapes of industrial products for mass production. In the plastic injection molding, one of the serious problems is the generation of residual strain and stress caused by non-uniform solidification of plastic materials. Cooling velocity of the injected plastic is slower than that of metal because of low heat conductivity of plastics. Therefore, differences of the cooling velocity part by part due to partial temperature gradient is appeared. These differences produce different molding shrinkage and internal residual stresses, which cause warps and cracks in the plastic molded products in some years after the molding. If the plastic materials in the solidification process are controlled well and cooled down uniformly, it is expected that the generation of the residual stress can be reduced much.

Many researchers have tried to optimize the cooling pipe system to reduce the generation of residual stresses by numerical simulation. The automatic

Please use the following format when citing this chapter:
Author(s) [insert Last name, First-name initial(s)], 2006, in IFIP International Federation for Information Processing, Volume 199, System Modeling and Optimization, eds. Ceragioli F., Dontchev A., Furuta H., Marti K., Pandolfi L., (Boston: Springer), pp. [insert page numbers].

design system of cooling pipes used straight line cooling pipe has been given [1, 2]. Koresawa et al. suggested the automatic layout design system adopting complex cooling channels [3]. Recently, the plastic injection die with complex cooling channels can be produced by stereolithography system of metal powder. Therefore the complex layout will be adapted to plastic injection die.

However, in these researches, some factors such as a coolant flow in the cooling pipe have not been taken into consideration. The coolant flow has not been considered when a cooling pipe system is designed and optimized.

In this paper, we discuss about an optimization method of the cooling pipe system by taking coolant flow into account. For this purpose, two kinds of numerical models have been prepared; one is to confirm the validity of the evaluation function of objectives, the other is to consider the effect of the coolant flow in the process of optimization. These models are analyzed by the finite element code, which implements the heat transfer analysis and the fluid analysis. And the temperature distribution and its variation as well as the coolant velocity during the cooling process calculated by the analyses are adopted to optimize the cooling pipe system. Then, to decide the position of the cooling pipe, the basis vector method [4], which is one of the ways to treat a shape change and to control mesh adaptation, is adopted. In the basis vector method, the design variables are set by using the orthogonal array of the design of experiment (DOE). From the numerical analysis results the response surfaces are constructed. Then the optimum shape of the cooling pipe system is obtained by the mathematical programming method.

2. How to optimize a cooling pipe system

2.1 Optimization procedure

In this research, the shape and the position of the cooling pipe system are taken as the design variables. The sampling points are assigned by the orthogonal array in DOE, and the numerical simulations have been implemented. Then the response surface for the evaluation fanctions are constructed and optimized. Figure 1 shows flow chart of the design processes.

2.2 Numerical model

The objective of this paper is to establish an optimization methodology for the cooling pipe system considering the coolant flow effect. To confirm the effect of suggested methodology, a numerical model shown in Figure 2, which is two dimensional model, with two kinds of different conditions, which are called "Case 1" and "Case 2", is prepared. The Case 1 does not consider the coolant flow effects. Therefore only the heat transfer analysis will be done. On the other hand, the Case 2 takes account of the coolant flow, which requires analyzing both the fluid flow and the heat transfer. And we make the numerical model to compare the optimization results between Case 1 and Case 2. If the optimum shape of Case 2 is different from the shape of Case 1, we can confirm the effect of coolant flow.

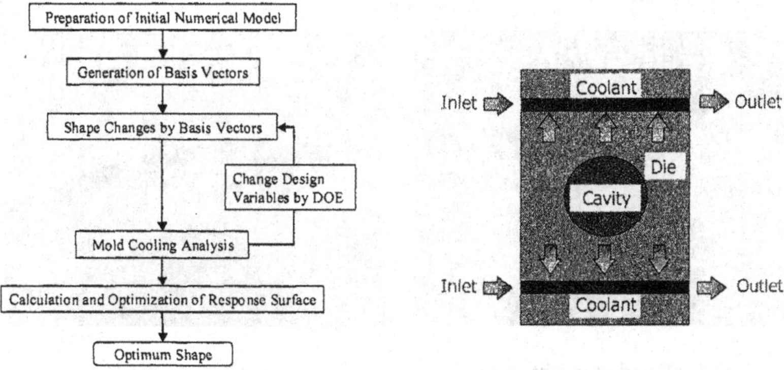

Figure 2. Numerical model

Figure 1. Flow chart to optimize cooling pipe system

In the numerical analysis, we consider only half of the numerical model because the conditions of the model are given symmetric with respect to the x-axis.

2.3 Basis vector method

When an optimization method is applied to the shape design problem, the basis vector method is adopted to decide the shape and position of cooling pipe, and to reduce the number of design variables. The basis vector method can prevent the mesh warping due to the boundary shape changes during optimization. In the basis vector method, an original shape vector and basis vectors need to be prepared. In general, a new coordinate **G** is generated from the following equation,

$$\mathbf{G} = \mathbf{G}_0 + \sum_{i=1}^{n} \alpha_i (\mathbf{V}_i - \mathbf{G}_0) \tag{1}$$

where $\mathbf{G}_0$ is an original shape vector, $\mathbf{V}_i$ is a basis vector, n is the number of basis vectors and α_i is a design variable, respectively.

In this study, the new shape vector is calculated from the following equation,

$$\mathbf{G} = \mathbf{G}_0 + \sum_{i=1}^{n} \alpha_i \sum_{j=1}^{non} \mathbf{V}_j \tag{2}$$

where non is the number of node in a numerical model and $\mathbf{V}_i$ corresponds to $(\mathbf{V}_i - \mathbf{G}_0)$.

Four kinds of basis vectors that include four design variables are prepared. Figure 3 illustrates the shape changes of the numerical model by using the basis vectors.

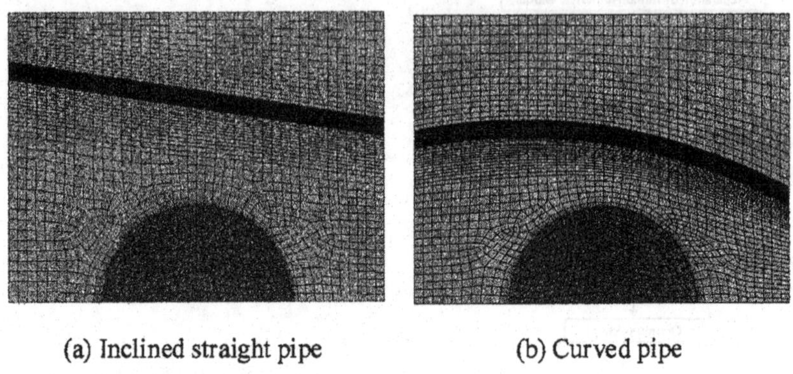

(a) Inclined straight pipe (b) Curved pipe

Figure 3. Examples of shape change using basis vectors

2.4 Response surface approach [5, 6]

In the optimization problem using numerical analyses, when much calculation time is required for one analysis, the design space may be approximated to optimize effectively. Some approximate optimization techniques have been developed, such as the response surface method (RSM), Kriging model, the radial basis function (RBF) and so on. In this study, we have applied the RSM to predict the numerical results without a lot of analyses in optimization process.

When the shape of numerical model is changed by using basis vector method, a few numbers of design variables have to be decided. In this study, the combination of these design variables is determined by an orthogonal array of the design of experiment (DOE) to be calculated efficiently under required reliability. And an orthogonal polynomial calculated by a regression model based on the array is regarded as the response surface.

2.5 Evaluation function

In a plastic injection molding, when the temperature distribution of melted plastic in a cavity is uniform and the cooling rate is slow, the generation of residual stresses is suppressed. Therefore, we introduce two kinds of quantities to evaluate these conditions. One is an average heat quantity Q_{ave} defined by Eq.(3) to obtain uniform temperature distribution along the cavity surface. The other is a deviation of heat quantity Q_{dev} defined by Eq.(4) to measure the cooling rate during t_{end} in the molding die.

$$Q_{\text{ave}} = \int_0^{t_{\text{end}}} q_{\text{ave}} dt = -kA \int_0^{t_{\text{end}}} \frac{\sum_{i=1}^{n} T_i}{n} dt \tag{3}$$

$$Q_{\text{dev}} = \int_0^{t_{\text{end}}} q_{\text{dev}} dt = -kA \int_0^{t_{\text{end}}} \sqrt{\frac{\sum_{i=1}^{n} (T_i - T_{\text{ave}})^2}{n-1}} dt \tag{4}$$

where q_{ave} and q_{dev} are heat flows of the cavity surface nodes at unit time, t_{end} is calculation time, k is thermal conductivity, A is an area of the cavity surface, T_i is a temperature of cavity surface node, T_{ave} is an average temperature of the surface and n is number of the nodes, respectively.

After normalizing these two functions, an objective function summed up these functions with weighted coefficients is introduced to transform this multi-objective optimization problem into a single one.

$$f = w_1 Q_{\text{ave}} + w_2 Q_{\text{dev}} \quad (w_1 \geq 0, w_2 \geq 0, w_1 + w_2 = 1) \tag{5}$$

where w_1 and w_2 are weighted coefficients.

3. Shape optimization problem of cooling pipe system in plastic injection molding

3.1 Optimum shape of cooling pipe system

A simple design model of cooling pipe system (Figure 2) is considered to confirm the coolant flow effects between the different conditions, the Case 1 and the Case 2, as mentioned above. The same initial conditions are assumed for the both cases. As the boundary conditions, a constant cool temperature is given for all area of cooling pipe in the Case 1. On the other hand, in the Case 2 a constant velocity and temperature at inlet, a constant pressure at outlet and no slip condition at cooling pipe wall are assumed to consider the coolant flow in cooling pipe.

The analytical models are made from the combination of the basis vectors based on the orthogonal array $L_{27}(3^4)$. Using 27 kinds of numerical results analyzed these models by the FEM, the response surface is constructed.

The response surfaces in the design space are illustrated in Figs.4. These graphs show the relationship among the average heat quantity, the deviation of heat quantity and the transformed single objective function. By changing the coefficient values of w_1 and w_2 in Eq.(5), a Parato front drawn by red symbol is obtained.

The optimum shapes of cooling pipe systems shown in Figs.5 and 6 are obtained by using the response surfaces (Eq.(5)) and the mathematical programming method under the condition of $w_1 = 0.5$, $w_2 = 0.5$. The optimum

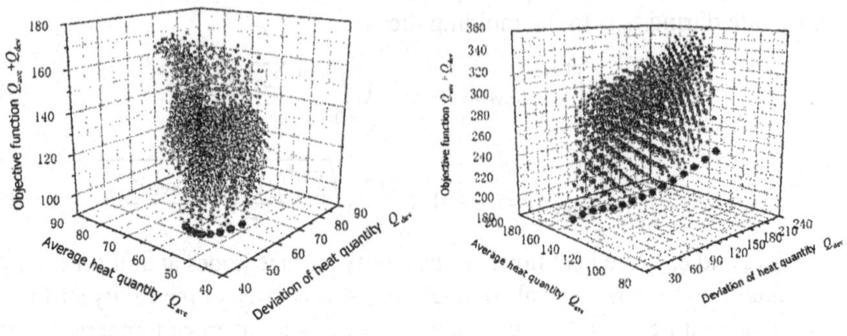

Figure 4. Response surface of Case 1 (left), Case 2 (right)

shape of cooling pipe in the Case 1 shows round and symmetric as shown in Figure 5 because of the symmetry of simulation model. On the other hand, the shape is asymmetry and that the cooling pipe approaches the cavity toward outlet and goes away from the cavity toward inlet in the Case 2 (Figure 6).

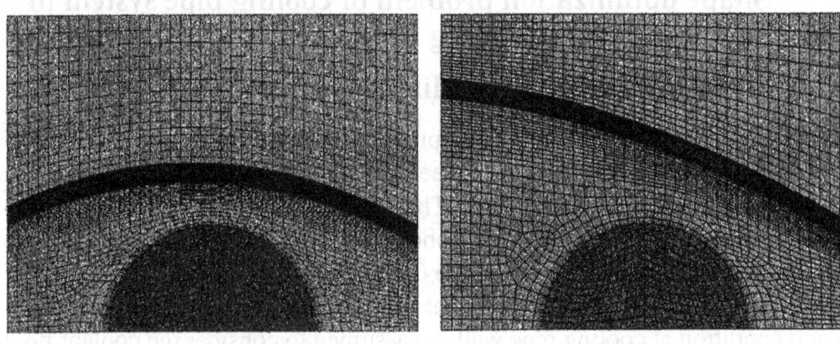

Figure 5. Optimum shape of Case1

Figure 6. Optimum shape of Case2

3.2 Discussions

The optimum shape in Case 1 can be expected to reduce the differences of the temperatures distribution and to cool down fast on the cavity surface. Compared with the optimized model in Case 1, we can find clearly the difference of the inlet and outlet coordinate of the cooling pipe. When the coolant flow is considered, the coolant temperature of inlet side is higher than that of outlet side because of the absorption of heat. This will cause the nonuniform temperature distribution in the die. From this reason, the asymmetry optimum shape in Case 2, has been obtained in order to make a temperature distribution uniformly.

In Figs.5 and 6, the Parato fronts are obtained to optimize the objective function with the different values of the weights. The pipe shapes of these solutions in Figure 6 are illustrates in Figure 7. The trade off relationship between the average heat quantity Q_{ave} and the deviation of heat quantity Q_{dev} is observed. From the shapes of Parato solution, we can recognize that it is necessary to make the position of inlet go away from the cavity for uniform temperature distribution.

When the optimum results shown in Figure 6 are focused, the result of deviation of heat quantity is the minimum at the optimized results as shown Figure 7. In Figure 7, the result of average heat quantity is not the minimum of objective function, because the function value depends on the normalizing standard. However, changing the weighted coefficient, Parato front is obtained. Therefore, we consider the objective function and evaluation functions show a natural phenomenon in plastic injection molding.

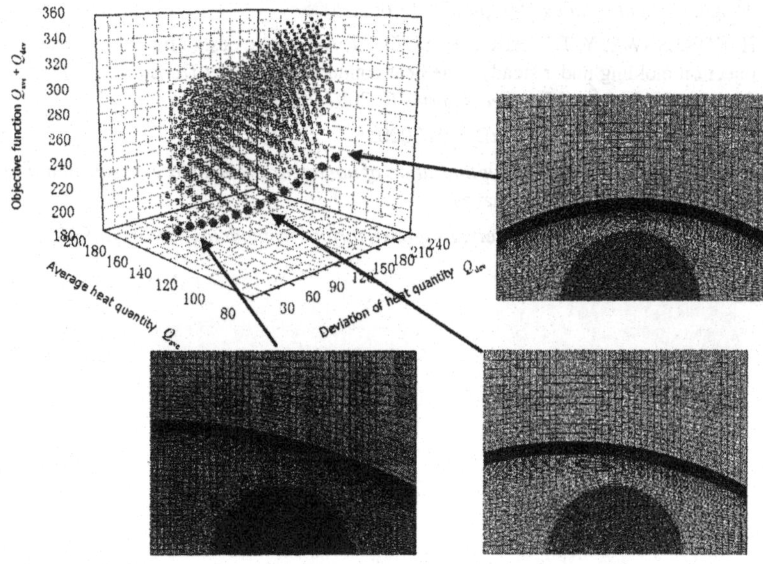

Figure 7. Shapes of parato solutions in Case2

4. Conclusions

A new approach to optimize the cooling pipe system in the plastic injection molding taking account of the coolant flow has been developed. To confirm the coolant flow effect, we prepared a simple cooling pipe system model, and optimized the model with two different conditions. In the optimization processes, we defined the two kinds of quantities to evaluate the cooling process in the cavity of plastic injection die, and adapted the basis vector method to control the shape of cooling pipe system and the response surface method to optimize effectively. As a result, we obtained the different optimum shapes that appear the coolant flow effect. And the response surfaces using the optimizations express the trade off relation ship between the cooling rate and the uniformity of the temperature distribution in the die. However, the optimum shape is lean to keep the uniformity compared with the cooling rate when two evaluation functions are treated at the same rate. However, we can obtain the parato front to change the weight of two functions.

References

[1] C.L.Li, Automatic layout design of plastic injection mould cooling system, *Computer Aided Design* , 33:1073-1090, 2001.

[2] C.L. Li, C.G. Li, A.C.K. Mok, A feature-based approach to injection mould cooling system design, *Computer Aided Design* 37:645-662, 2005.

[3] H. KORESAWA, Y. TOCHIKA, H. SUZUKI, Automatic design for cooling channels in injection molding under steady-state heat conduction analysis,*Transactions of the Japan Society of Mechanical Engineers Series C*, 2087-2093, 1999.

[4] *Genesis user's manual* , VR&D Inc., 2000.

[5] T. KASHIWAMURA et al., *Optimization of nonlinear problem by experimental design* , Asakura shoten, 1998, (in Japanese).

[6] R. H. Myers, D. C. Montgomery, *Response surface methodology* , Wiley, 1995.

OPTIMUM DESIGN OF COOLING PIPE SYSTEMS BY BRANCHING TREE MODEL IN NATURE

K. Yamazaki,[1] and X. Ding[2]

[1] *Dept. of Human & Mechanical Systems Engineering, Kanazawa University, Kakuma-machi, Kanazawa, 920-1192, JAPAN, yamazaki@t.kanazawa-u.ac.jp* [2] *Dept. of Mechanical Engineering University of Shanghai for Science and Technology, 516 Jungong d.,Shanghai 00093, CHINA*

Abstract This paper suggests an innovative design methodology of heat transfer system based on a so-called adaptive growth law, which is an essential optimum growth rule of branch systems in nature. The branch systems in nature can grow by adapting themselves automatically to the growth environments in order to achieve better global functional performances, such as the maximal absorption of nutrition or sunlight in plants and the intelligent blood delivery of a vascular system in animal body. Thus, it can be expected that an optimum layout of heat transfer system would be obtained by the generation method based on the growth mechanism of branch systems in nature. First, the emergent process of branch systems in nature is reproduced in computer model by studying their common growth mechanisms. The branch systems are grown by the control of a so-called nutrient density so as to make it possible that the distribution of branches is dependent on the nutrient distribution. Then, the generation method is applied to the layout design problem for heat transfer systems. Both the conductive heat transfer system and the convective heat transfer system are designed by utilizing the generation method based on the growth mechanisms of branch systems in nature. The effectiveness of the suggested design method is validated by the FEM analysis and by the comparison with other conventional optimum design methods.

keywords: Layout Optimization, Cooling Channel, Branch System, Bionic Design

1. Introductions

Geometric forms (shapes and topologies) of branch systems in nature, such as lungs, vascular tissues, botanical tree (canopies, roots, leaves), etc., always show approximating a globally optimal performance that can minimize the costs of the construction and maintenance of the fluid transportation system under restraints of growth environment. So the branch systems are interested in modeling and visualizing not only by biologists but also by engineers. By studying the growth mechanisms of branch systems in nature, branch systems

Please use the following format when citing this chapter:
Author(s) [insert Last name, First-name initial(s)], 2006, in IFIP International Federation for Information Processing, Volume 199, System Modeling and Optimization, eds. Ceragioli F., Dontchev A., Furuta H., Marti K., Pandolfi L., (Boston: Springer), pp. [insert page numbers].

of plants and animals have been simulated by some kinds of approaches [1]- [4]. On the other hand, development of more intelligent and optimum engineering systems are expected by utilizing the optimality of branch systems in nature, and some effort have been done on this issue. For example, a topology design optimization method to generate stiffener layout pattern for plate and shell structures has been suggested [5], in which a growing and branching tree model is applied. The effectiveness of the method is proved because discrete stiffener layout pattern rather than a vague material distribution can be obtained.

In this paper, an innovative layout design methodology of heat transfer system by utilizing the optimality of branch systems in nature is suggested. The method bases on such essential characteristics of branch systems in nature that the branches can grow by adapting themselves automatically to the growth environments and achieve better global functional performances, such as maximal absorption of nutrition or sunlight in plants and intelligent blood delivery of a vascular system in animal bodies. Thus, it can be expected that an optimum layout of heat transfer system would be obtained by utilizing the generation method based on the growth mechanism of branch systems in nature. First, optimality and growth mechanisms of branch systems in nature are studied, and a reproduction approach of emergent process of branch systems is proposed. Branches are grown by the control of a so-called nutrient density so as to make it possible that the distribution of branches is dependent on the nutrient distribution. The growth of branches also satisfies the hydrodynamic conditions and minimum energy loss principle. If the so-called nutrient density in the generation process of branch system is referred to as the temperature in a heat transfer system, the distribution of branches is responsible to the distribution of cooling channels. Because branch system can grow adaptively corresponding to the nutrient distribution in order to absorb the nutrition to the maximal extent, the cooling channel can be constructed adaptively by the control of the temperature so as to make it possible to achieve comparative uniform temperature distribution of the whole heat transfer system. Having the similar optimality of branch systems in nature, the constructed cooling channel can be designed flexibly under any complex thermal boundary conditions within any shapes of perfusion volumes to be cooled and will achieve good cooling performance. The design problems of both the conductive heat transfer system and the convective heat transfer system are studied, and the cooling performances of the designed heat transfer systems are confirmed being improved by carrying out the FEM analysis and by comparing with the results designed by other conventional design methods.

2. Reproduction of Emergent Process of Branch Systems in Nature

It is necessary to reproduce the emergent process of branch systems in nature in order to apply the optimality of branch systems in nature to engineering design, thus a generation approach of the emergent process for a hierarchical dichotomous branch system being considered as a material or energy transportation system is studied. A certain distributed nutrition density is assigned in advance in the perfusion space to control the growth of branches to make it possible that the distribution of branches is dependent on the distribution of the so-called nutrition density. During the generation process of branch system, the nutrition density in the whole perfusion space decreases by growing branches and its distribution tends to be uniform. The prerequisites of constructing such branch system are briefly described as follows:

1 Branching law stands for the relationship of radii between the parent branch and the daughter branches, which is adopted at every bifurcation point. For a dichotomous branch system, it is formulated as the following Eq.(1).

$$r_0^\lambda = r_1^\lambda + r_2^\lambda \tag{1}$$

where r_0, r_1 and r_2 are radii of the parent branch and the daughter branches, as shown in Figure 1. The bifurcation exponent, λ, is physiologically reasonable when it is in the range of $2 \leq \lambda \leq 3$. Murray's law shows the energy loss for transporting material throughout the whole network can be made minimum when $\lambda = 3$.

2 Growing law relates to the growing direction and the growing velocity for a new branch, which is assumed to be dependent on the local nutrient distribution. A new terminal site is always positioned at the point with the highest nutrient density in the local growth space (vicinity) around the grown branches. If there is more than one such point in the local growth space, a point is selected by a pseudo random number sequence.

3 The hydrodynamic conditions are assumed as that each terminal branch has the same flow and pressure so as to bathe the whole perfusion space evenly. The branches are assumed to be cylinders. Flows in the branches are assumed as fully developed laminar flow, so they obey Poiseuille's law formulated by the following Eq.(2).

$$Q = \frac{\pi r^4}{8\nu} \frac{\Delta P}{l} \tag{2}$$

where Q is the volumetric flow rate, ΔP is the pressure drop, r and l are radius and length of the vessel, and ν is the dynamic viscosity of the fluid.

4 The volume of whole branch system is selected as the cost function, so the branch system is designed in such a way that the volume of it is minimized,

$$V = \sum_{i=1}^{n} r_i^2 l_i \rightarrow \min \tag{3}$$

where r_i, l_i are radius and length of branch i, and n is the total number of branches.

According to the above prerequisites, the growth process can be implemented as follows. A certain nutrition density distribution is assigned in a specified perfusion area in advance. The initial branches are grown, which satisfy the hydrodynamic conditions and a certain branching law. The local nutrition densities nearby the existent branches are updated. Then a new terminal site is selected at a point with the highest nutrition density in the growth area. If there is more than one point having the highest nutrition density, a candidate point is selected at random by a pseudo random number sequence. Therefore the density of branch distribution is dependent on the initial nutrition density of the perfusion area in order to absorb nutrition as much as possible. Next, the new terminal site is attached to the existent branches near it. For each attachable candidate branch, the bifurcation point is selected optimally with the objective of minimum volume of the whole network under the restraint of hydrodynamic conditions, and the new terminal site connects with the bifurcation point tentatively. The radius of the new branch is decided according to Poiseuille's law, and radii of all parents of the new segment are updated according to the branching law. Then the connection is dissolved but the volume of whole network for the connection is recorded. By comparing volumes of all possible tentative connected topologies, a connected topology with the minimum volume is finally adopted permanently. The growth process will stop when the average nutrition density in the perfusion area cannot be decreased anymore.

Figure 2 shows the simulation result of the growth process for a 2-dimensional branch system filled up a circular space. The initial nutrition density is distributed uniformly, as shown in Figure 2(a). Murray's law is adopted at each bifurcation. Figure 2(b)-(d) show the emergent process of the branch system, in which the total branch number of the finally generated branch system is 6971. It is found that the branch system can fill up the whole specified space and the distribution of the branches is dependent on the initial nutrition density in the original space, which is almost uniform. During the generation process, a trunk grows at first, and some boughs grow spreading the whole perfusion area at very initial period. However, a number of twigs grow finally. The simulated branch system has both thicker and thinner size levels, which is similar to the branch systems in nature, in which the thicker ones convey long-distance material transportation, and the thinner ones exchange material with environment.

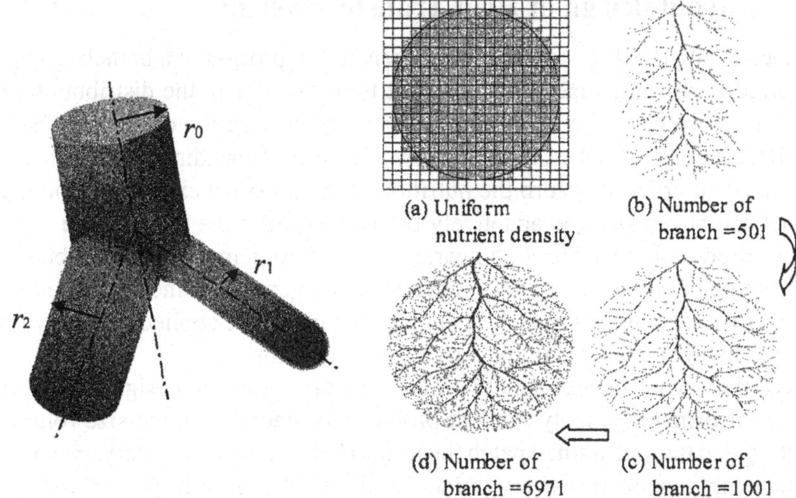

Figure 1. Geometry of a dichotomous branch system

Figure 2. Growth process of branch system under uniformly distributed initial nutrition density (λ=3.0, Volume rate=1.60%)

If the circular perfusion area is considered as a plate with the thickness of 1/14 of its radius, the volume rate of the branch system, i.e., the ratio of the branch volume to the whole plate volume, is 1.60%.

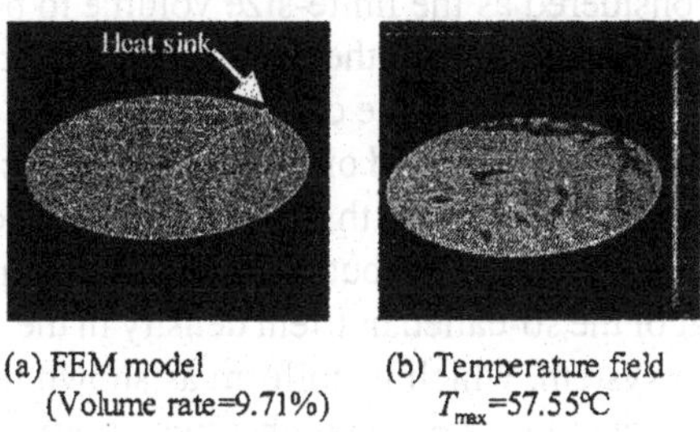

(a) FEM model (Volume rate=9.71%)

(b) Temperature field T_{max}=57.55°C

Figure 3. FEM model and temperature field for a circular plate with a natural branch-like conductive cooling channel under uniform heat-generating rate

In conclusion, it is said that the suggested generation method can grow branch systems to fill up any specified perfusion areas with arbitrarily distributed initial nutrition densities, and the generated branch systems are qualitatively similar to the branch systems in nature.

3. Layout design of Heat Transfer System

If the so-called nutrient density in the generation process of a branch system is referred to as the temperature in a heat transfer system, the distribution of branches can be considered corresponding to the distribution of cooling channels. Because the branch system can grow adaptively depending on the nutrient distribution in order to absorb the nutrition to the maximal extent, the cooling channel can be constructed adaptively by the control of the temperature so as to make it possible to achieve comparatively uniform temperature distribution of the whole heat transfer system. Therefore, the generation method based on the growth mechanism of branch systems in nature can be applied to the layout design of cooling channels in the heat transfer systems.

First, the layout of a conductive heat transfer system is designed and its cooling performance is analyzed. The problem is stated as: "a finite-size volume is to be cooled through a small patch (heat sink) located on its boundary, in which heat is being generated at every point. "A finite amount of high conductivity (k_p) material is available. Determine the optimum distribution of material k_p through the given volume such that the temperature distribution of the whole volume is as uniform as possible." The natural branch-like conductive cooling channel is constructed based on the corresponding original branch system, in which its cross-section is assumed to be a rectangle with the same thickness as the plate. The width of the channel, however, is assumed to be the same diameter as the corresponding branch in the branch system. The ratio of thermal conductivity of the high conductivity material (k_p) to the low conductivity (k_0) is assumed as $\tilde{k} = k_p/k_0 \gg 1$. A circular plate with the ratio of thickness to diameter of 0.01 is considered as the finite-size volume to be cooled, in which the heat generates at every point with the uniform volumetric heat-generating rate $q''' = 10^5 \mathrm{W/m}^3$. The layout of the conductive cooling channel made of a high conductivity material (k_p) is based on the original branch system shown in Figure 2(d), in which some branches with smaller cross-sections are omitted for the simplicity. It is noted that the distribution of the volumetric heat-generating rate is identical to that of the so-called nutrient density in the generation process of the original branch system, which is uniform as shown in Figure 2(a). The ratio of the thermal conductivities of the high conductivity material (k_p) to the low conductivity (k_0) is assumed to be 10^4. The temperature at the heat sink located on the boundary is set as $T_{\min} = 10\,°\mathrm{C}$. The whole structure is insulated from the environment. Figure 3(a) shows the FEM model generated by some imaged section slide by Voxelcon. The volume rate of the cooling channel is 9.71%. Figure 3(b) illustrates the temperature field of the whole plate, in which the red parts stand for the hot spots, while the blue parts stands for the comparative cool spots. The maximal temperature is $T_{\max} = 57.55$ °C. It is found that the hot spots are distributed over the whole volume.

Next, the layout of a convective heat transfer system is designed by the generation method based on the growth mechanisms of branch systems in nature. The difference between the convective and the conductive heat transfer systems is that the coolant flow is available to remove the heat in the convective heat transfer system. The problem can be stated as: "construct an optimum convective cooling channel in a specified volume applied a certain distributed heat flux, in which coolant flows through the channel to remove the heat." A branch-like convective cooling channel is constructed inside a flat plate, which is based on the corresponding original branch system. The diameter of each segment in the cooling channel is assumed to be identical to that of the corresponding branch in the original branch system. A certain distributed heat flux is applied on the top surface of the plate, the bottom surface is insulated from the environment, and the circumference is on the forced air convection. Coolant with a certain volumetric flow rate is flown through the channel to remove the heat. The volume to be cooled is assumed to be made of beryllium-copper alloy, thermal conductivity of which is set as $k = 260$ W/mK. The coolant flowed through the cooling channels is assumed as water, and its temperature at inlet is assumed as 20 °C.

Actually, the problem is a transient heat conduction problem between the solid and coolant in the channel. However, because our goal here is only to confirm the heat conduction efficiency of the heat transfer system with a convective cooling channel, the problem is simplified as a steady-state heat conductive problem. However, it is necessary to consider the energy balance in the system responsible for the fluid convection. Because the temperature of coolant becomes higher and higher by passing through the cooling channel from inlet to outlet, the energy balance due to the fluid convection can be considered approximately as the distribution change of the temperatures at the channel wall. The temperature difference at the channel wall between the inlet and outlet of segment i can be evaluated approximately by the following equation.

$$\Delta T_{\mathrm{wi}} = \frac{q l_{\mathrm{i}}}{\rho_{\mathrm{c}} C_{\mathrm{c}} r_{\mathrm{i}} u_{\mathrm{ci}}} \tag{4}$$

where q is the heat flux applied on the channel, ρ_{c} and C_{c} are density, specific heat of coolant, respectively. And u_{ci} is the coolant average velocity of branch i.

Figure 4 shows the FEM models and temperature fields of three circular plates with branch-like convective cooling channels resulted from different branching laws. The branch-like convective cooling channels in the heat transfer systems are constructed based on the corresponding branch systems shown in Figure 2(d), in which the bifurcation exponents λ are 3.0, 2.5 and 3.5, respectively. The thickness of each plate is assumed to be 1/14 of its radius. The volumetric flow rate of coolant at the inlet is assumed $Q = 8 \times 10^{-5}$ m^3/s , and the heat flux applied at the top surface of the plate is set as $q = 5.0 \times 10^5$ W/m^2. The left figures in Figure 4 are the middle planes of the FEM models, in which the volume rates of cooling channels are 1.77% for $\lambda = 2.5$, 1.28% for $\lambda = 3.0$ and 1.32% for $\lambda = 3.5$, respectively. The right figures in Figure 4 show the temperature fields that are scaled by the maximal temperature $T_{\max} = 44.67\,°\mathrm{C}$ for $\lambda = 2.5$. And the maximal temperatures are $T_{\max} = 33.94\,°\mathrm{C}$ for $\lambda = 3.0$ and $T_{\max} = 34.20\,°\mathrm{C}$ for $\lambda = 3.5$, respectively. It is easy to find that the case of $\lambda = 3.0$ results the most uniform temperature distribution, and the most non-uniform one is the case of $\lambda = 2.5$. It is noted that the volumes of the convective cooling channels from small to large is put in order as λ =3.0, 3.5 and 2.5, which is the same sequence with the uniformities of temperature distributions. As a result, Murray's law ($\lambda = 3.0$) is the most effective branching law for constructing the branch-like convective cooling channel system. The reason is that Murray's law is derived from such principle that the energy loss due to the viscous friction throughout the whole network is minimal, that provides the easiest way for the coolant to pass.

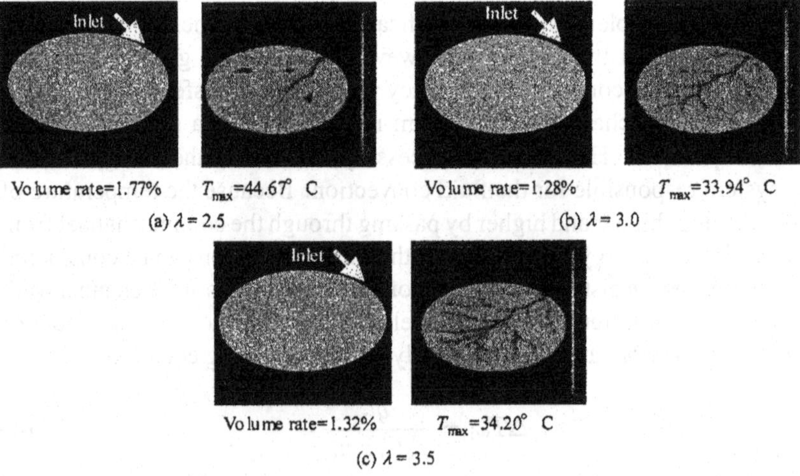

Figure 4. FEM models and temperature fields for heat transfer systems with branch-like convective cooling channels resulted from different branching laws

In order to validate the effectiveness of the suggested generation method, the cooling performances between the natural branch-like cooling channel designed by the suggested generation method and the horizontal-vertical tree-like cooling channels constructed by the constructal theory proposed by Bejan [6] are compared.

The unique principle of the constructal theory for designing the conductive cooling channel is: "every portion of the given volume can have its shape optimized such that its resistance to heat flow is minimal". By utilizing this principle, the cooling channel is determined in a sequence of steps consisting of shape optimization and subsequent construction. It starts from the smallest building block (elemental system) and proceeds toward larger building block (assemblies).

Figure 5(a) shows the FEM model and the corresponding temperature field of a horizontal- vertical tree-like conductive heat transfer system till second construction. The ratio of the thermal conductivities of the high-conductivity material to the low-conductivity material is assumed to be $\tilde{k} = k_p/k_0 = 3333.33$. The volumetric heat-generating rate is set as $q''' = 10^5 \mathrm{W/m^3}$, and is distributed uniformly over the whole volume to be cooled. The volume is a square plate with the ratio of the thickness to the edge length 0.01, which is derived from the sequence of shape optimization and construction by the constructal theory. The temperature at the heat sink end is set as 10 °C, and the whole structure is insulated from the environment. As shown in Figure 5(a), the volume rate of the cooling channel, i.e., the k_p material in which allocated in the whole volume is 9.47%, and the maximum temperature is $T_{\max} = 58.25$ °C.

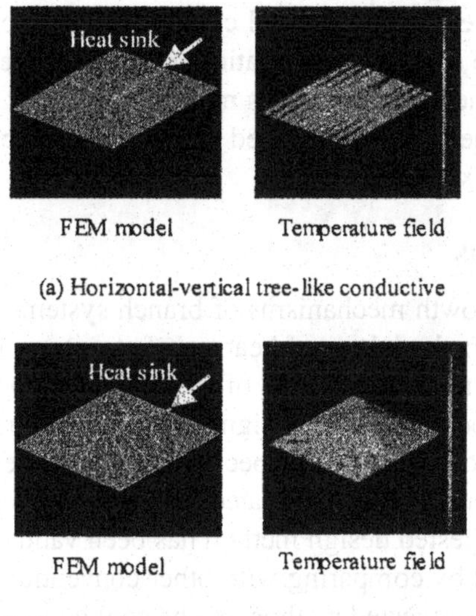

Figure 5. Comparison of horizontal-vertical tree-like and natural branch-like conductive cooling channels

According to the geometries of the heat transfer system with the horizontal-vertical tree-like cooling channel, a corresponding branch system is generated on a square perfusion area applied a uniformly distributed nutrient density. Figure 5(b) shows its FEM model and the corresponding temperature field. The volume rate of the cooling channel is 9.62%, which is very close to that of the horizontal-vertical conductive cooling channel. And the maximum temperature is $T_{\max} = 53.32$ °C, which is a little lower than that of the horizontal-vertical tree-like cooling channel. It is found that hot spots are distributed over the whole volume to be cooled in both cases, so it is said that both conductive cooling channels can achieve good cooling performances. However, it should be noted it is just because the thermal boundary conditions are very simple (uniformly distributed heat-generating rates are applied), the simple and regular distributed conductive cooling channel, i.e., the horizontal-vertical tree-like conductive cooling channel, is available and effective. If the heat-generating rate is applied non-uniformly, it is difficult for the constructal theory to design effective conductive cooling channel, while the flexible natural branch-like conductive cooling channel can be designed adapting to the arbitrary complex thermal boundary conditions. Moreover, the design volume to be cooled can not be changed arbitrarily and be defined in advance when the constructal theory is adopted because of the definite time arrow of the construction from small to large. While the design volume can be defined in advance and the natural branch-like conductive cooling channel can be designed to fill up the volume with any shape by the suggested generation method. Therefore, it can be said that the suggested generation method is more powerful for designing the conductive cooling channel and the designed cooling channel can remove the heat generated in the matrix effectively.

4. Conclusions

By studying the growth mechanisms of branch systems in nature, an innovative layout design methodology of heat transfer systems is suggested in this paper. Having the similar optimality of branch systems in nature, the heat transfer system for cooling can be designed flexibly under any complex thermal boundary conditions within any specified shapes of perfusion volumes to be cooled, and can remove heat generated in the volume effectively. The effectiveness of the suggested design method has been validated by carrying out the FEM analysis and by comparing with other conventional design methods. It is expected that the suggested method can be applied to some more practical engineering applications, such as cooling channels in injection moulds, heat sinks in electronic packages, and so on.

References

[1] H. Honda, Description of form of trees by the parameters of tree-Like body: effects of the branching angle and the branch length on the shape of the tree-like body, *J. Theor. Bio.* , 31, pp.331-338, 1971.

[2] A. Takenaka, A simulation model of tree architecture development based on growth response to local light environment, *J. Plant Research* , 107, pp.321-330, 1994.

[3] R. Takaki, H. Kitaoka, Virtual construction of human lung, *Forma* , 14, pp.309-313, 1999.

[4] W. Schreiner, R. Karch, F. Neumann, M. Neumann, Constrained constructive optimization of arterial tree models, In: Scaling in Biology, Edited by Brown, J.H. & West G.B., Oxford university press, Inc., pp.145-165, 2000.

[5] X. Ding, K. Yamazaki, Stiffener layout design for plate structures by growing and branching tree model (application to vibration-proof design), *Struct. Multidisc. Optim.* , 26, pp.99-110, 2004.

[6] Bejan, A., *Shape and structure, from engineering to nature, Cambridge* , UK: Cambridge University Press, 2000.

IMPLEMENTATION OF MULTIOBJECTIVE OPTIMIZATION PROCEDURES AT THE PRODUCT DESIGN PLANNING STAGE

M. Yoshimura[1]

[1] *Department of Aeronautics and Astronautics, Mechanical Engineering Division, Graduate School of Engineering, Kyoto University, yoshimura@prec.kyoto-u.ac.jp*

Abstract In order to obtain maximally innovative and successful product designs, the utilization of optimization strategies at the product design planning stage is of prime importance, and the methods proposed in this paper enable multiobjective optimization technologies to be effectively applied. The necessity and effectiveness of utilizing optimization techniques at the product design planning stage are first explained, and the features that this requires are then clarified. Optimization solutions provided at the product design planning stage, while far from final, can nevertheless be used to obtain guidelines for more preferable product designs. For this purpose, even if characteristics evaluated at the product design planning stages are simplified and/or idealized, the interrelationships among all related characteristics should be simultaneously and thoroughly explored. The successful application of optimization techniques at the product design planning stage requires the rapid presentation and evaluation of a variety of alternative designs, a deeper understanding of the reasons for the optimized designs that are developed, and breakthrough of the initial optimized design solutions, so that the most effective design can ultimately be implemented in a manufactured product. This paper proposes multiobjective design optimization methodologies and procedures, utilized at the product design planning stage, to achieve these goals.

keywords: Product design planning stage, Multiobjective optimization, Pareto optimum solutions, Comparison of alternative designs, Hierarchical optimization problem, Rapid evaluation, Deeper insight into design solutions

1. Introduction

Today's rigorous manufacturing environments require the application of optimization techniques from wider points of view. To accomplish this, strategic utilization of optimization techniques at the product design planning stage, a process far upstream of product manufacturing, is essential. In this paper, the significance of conducting optimization at the product design planning stage

Please use the following format when citing this chapter:
Author(s) [insert Last name, First-name initial(s)], 2006, in IFIP International Federation for Information Processing, Volume 199, System Modeling and Optimization, eds. Ceragioli F., Dontchev A., Furuta H., Marti K., Pandolfi L., (Boston: Springer), pp. [insert page numbers].

is first clarified. Next, problem areas concerning the use of optimization techniques at the product design planning stage, and desirable features that such techniques should offer, are described.

At the product planning stage, principal product performances for the product should be considered and evaluated, and the conflicting relationships among the characteristics should be quickly but roughly evaluated. Many alternative designs are usually generated and compared at this stage and, for the most part, multiobjective optimization methods are applied. In multiobjective optimization problems, a Pareto optimum solution set, namely a set of feasible solutions for each of which there exists no other feasible solution that yields an improvement in one objective without causing degradation in at least one other objective, is obtained to evaluate conflicting objectives of the design optimization problem at hand [1] [2]. In order to effectively apply multiobjective optimization methods to the product planning stage, new methods need to be developed, to incorporate fundamental improvements in the multiobjective optimization. This paper presents methodologies for executing optimization at the product design planning stage, and several applied examples are given. Several methods developed by the author and his colleagues are organized and presented so that they can be effectively applied at the product design planning stage.

2. Significance of Optimization at the Product Design Planning Stage

2.1 Features of the product design planning stage in manufacturing processes

Figure 1 shows the sequence of manufacturing processes, where the product design planning is the first step. The product design planning, located furthest upstream, determines practically all of the downstream manufacturing process details. Current design environments require careful consideration not only of increasingly demanding requirements for product performances, qualities, and product cost, but also many other factors such as the product's environmental impact, lifecycle and recycling, and safety. Aggressive and relentless competition among companies developing new products under such circumstances makes the application of optimization strategies throughout product manufacturing processes a practical requirement. Particularly important to successful manufacturing is the application of optimization methods that start from the initial planning stages of product design.

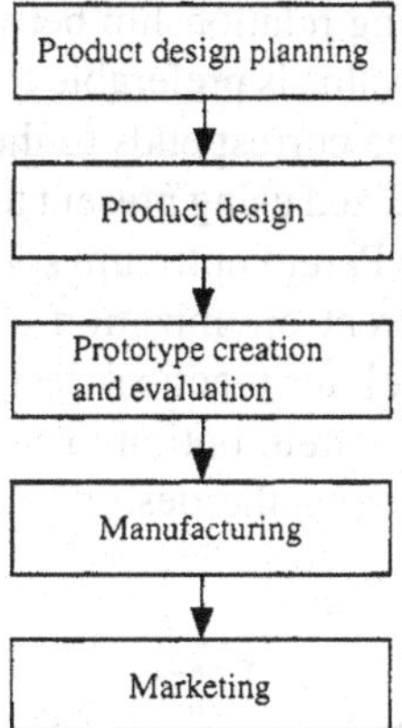

Figure 1. The sequence of manufacturing processes

2.2 Features of design optimization applied at the product design planning stage

The most effective application of optimization methods is based on careful inquiry and consideration of the features of the product design planning stage. This stage is more or less equivalent to the conceptual design, or fundamental design stage, where details of the product design have not yet been determined but various conceptual designs are considered, compared, and evaluated. This is when the design specifications for the product and its requirements and characteristics are usually given. Also, the principal characteristics used for evaluating the product performances can be defined. At the product design planning stage, the product performance and the product manufacturing cost for the entire product should be roughly evaluated even if the estimated values of the characteristics are imprecise.

At the product design planning stage, all characteristics should be systematically evaluated, and selection of the design candidates from among many design alternatives should be conducted, using an optimization procedure. Multiobjective optimization techniques can be effectively applied to systematical evaluations of the characteristics being regarded.

Solutions of optimization at the product design planning stages can be used for obtaining guidelines for product designs. For the purposes, the facts that characteristics considered at the product design planning stages may be simple and /or idealized can be acceptable. But, the relationships among the all related characteristics should be totally inquired.

When existing design solutions are available, searching methods that can achieve breakthrough or improved solutions are needed.

Figure 2 shows the conflicting relationship between two performance characteristics, f_1 and f_2. A larger value is preferable for each of these performance characteristics. The shaded area corresponds to the feasible region formed by design solutions that can be realized using present technologies and knowledge. The line PQ corresponds to the Pareto optimum solution set of global optimum solutions, achieved by concurrent optimization of all related characteristics. Product designers generally look for practical design solutions on a Pareto optimum solution line. From the Pareto optimum solution set, the most suitable solution is selected by considering the design requirements and the product environments.

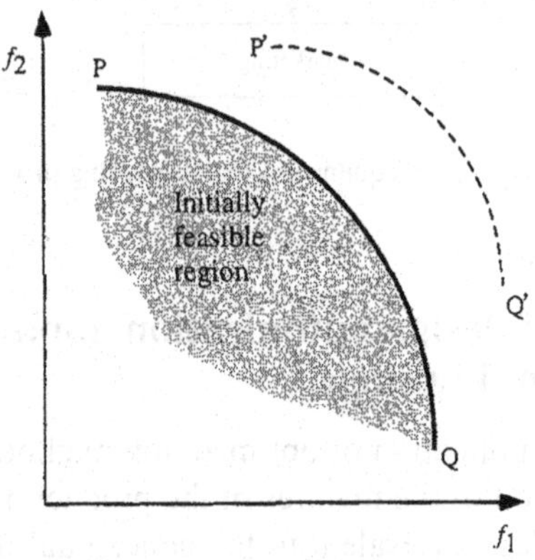

Figure 2. Concept of the proposed methods

If an alternative design results in a new optimum solution line, such as P'Q' shown in Fig.2, it can represent an improved new Pareto optimum solution set line. The display of such Pareto optimum solution sets is effective since designers can quickly understand the features of entire solutions.

2.3 Requirements for applying optimization techniques at the Product Design Planning Stage

In order to apply optimization techniques at the product design planning stage, the following points should be realized:

1 Since designers usually have a wide range of initial ideas, numerous alternative designs should be quickly formulated and effectively compared with each other, so that the most suitable small number of designs can be selected.

2 The many performance characteristics of the designs at hand must be concurrently evaluated and optimized.

Multiobjective optimization methods that incorporate new and improved advanced techniques can be applied to achieve the foregoing points. The details of such methods are explained below, along with some applied examples that illustrate their application.

3. Methodologies of Design Optimization at the Product Design Planning Stage

3.1 Comparison of many alternative designs

In the first method, many design alternatives in the multiobjective functions space are displayed in detail, and a relational tree diagram of design alternatives is shown to aid a deeper understanding of the optimized solutions [3]. To explain the process, the design of industrial multi-link manipulators that are used to transport an object in a workspace within a given operational time is used.

Alternative designs are constructed from a group of modules. Basic module is one link mechanism with a motor corresponding to a minimum unit of linked mechanisms. To create systems of practical complexity, we add design variables and increase the number of degrees of freedom. Here, two kinds of operators, operators 1 and 2, that control modules during the process of constructing a more complex system are introduced. Operator 1 adds a new module to the group of modules that make up the system. The degrees of freedom of the system are increased, and higher functionality can be realized. Operator 2 is an operator that alters the properties of a given module. Concerning operator 2, modification of link shapes by operator 2-1 and of the number of joints by operator 2-2 can be used any number of times.

Examples of system modifications by these operator actions are shown in Fig.3. The change from system f_1 to system f_2 is an example where operator 1 is applied to add a module. The linked mechanism is changed from having one degree of freedom to two degrees of freedom for the serial drive type mechanism (manipulator). The change from system f_2 to f_3 corresponds to the modifications of the internal variables of links. The action of operator 2-1 where a joint is added to a link produces a parallel drive type manipulator having two motors on a pedestal. By using these two kinds of operators, a variety of systems can be expressed using combinations of modules.

The requirements set by the user are: 1) minimization of consumed energy and 2) maximization of the operational simplicity of the link mechanism. The user's requirements concerning the amount of energy consumed and the dynamic manipulability are used as the criteria for this product design.

The consumed energy is calculated from the magnitudes of torques applied to the joints. The requirement concerning the consumed energy is expressed by minimizing the summation f_1 of the consumed energy of each motor over all motors, while the manipulability requirement is expressed by minimizing

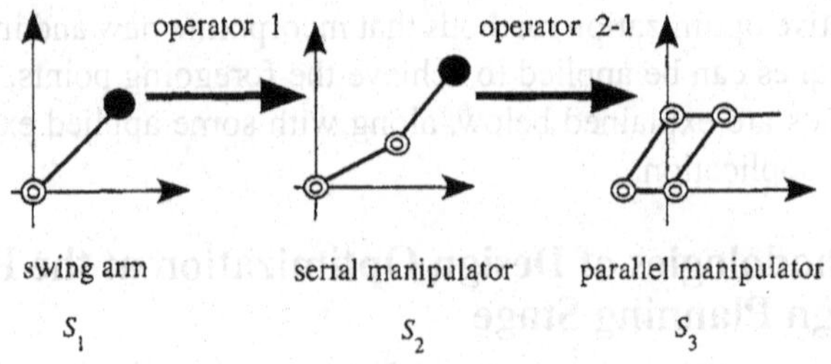

Figure 3. Examples of operator actions

the summation f_2 of the reciprocal of the dynamic manipulability measure ω_{d_j} over the all measuring points of the system. The objective functions f_1 and f_2 concerning the consumed energy and the manipulability are respectively formulated [3].

Fig.4 shows an example of a tree diagram for a design that was constructed and later stored in a database. Fig.5 shows changes in Pareto optimum solutions for the example. The history of the tree formation shown in Fig.4 is explained as follows. Operator i (i=1, 2-1, and 2-2) between various nodes represents operators active in the generation of subsequent systems. Operator type 1 alters the combination of modules, while type 2 operators change internal variables. Here, operator 2-1 changes the number of joints of the link, while operator 2-2 changes the position of the center of mass. Both operators 2-1 and 2-2 change internal variables of a given module.

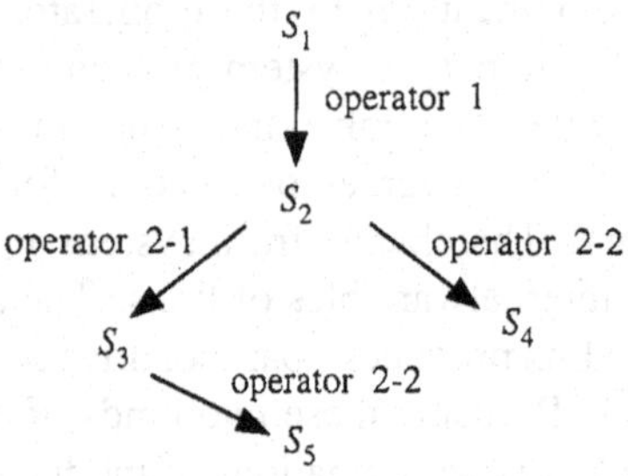

Figure 4. Tree form data of design solutions

First of all, from a unit link mechanism S_1, a serial drive type manipulator S_2 was obtained by the addition of a module. Next, the action of operator 2-1 upon S_2 generated S_3 , a parallel type drive manipulator, while applying operator 2-2 to S_2 yielded S_3, in which the center of mass was changed by changing the cross-sectional shapes of the links. After evaluation of S_3 and S_4 , S_5 was

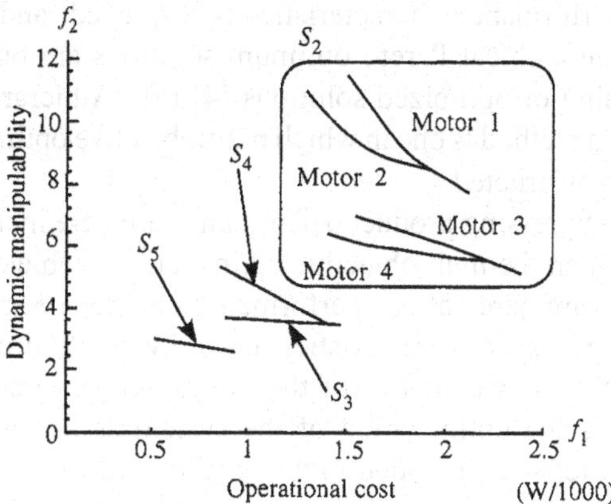

Figure 5. Changes in Pareto optimum solutions

obtained by applying operator 2-2 to S_3. But S_4 became a dead-end, since the action of any operator upon S_4 generated an identical system. For the same reason, S_5 became a dead-end. Ultimately, the tree-form data shown in Fig.4 was obtained.

Next, the changes in Pareto optimum solutions during the design generation processes are shown in Fig.5. Fig.5 is a space showing two criteria, f_1 and f_2, where the Pareto optimum solution set for each generated system is displayed. Since smaller magnitudes are preferable for each of the criteria, the design solution near the origin of the coordinates in the criteria space shown in Fig.5 is more preferable. For S_2, features of optimized results including discrete design variables of four kinds of motors are displayed for each kind of motor. For each of the other systems, the best Pareto optimum solution is shown. It can be understood from Fig.5 that the order of preferable solution lines is S_5, S_3, S_4, and S_2.

At this time, examining the generational history represented by the tree form diagram shown in Fig.4 will aid understanding the solution sequence, since the origin of the obtained preferable solution can be seen. From Fig.5, the optimum solution set for the design problem being regarded is S_5. The design system is a parallel drive type manipulator having two links.

3.2 Concurrent evaluations and optimization of related performance characteristics

Next, product performances that are related to the product design are concurrently optimized at the product design planning stage. For this purpose, a hierarchically decomposed structure of multiobjective optimization problems having multiple performance characteristics is displayed, and corresponding relations among hierarchical Pareto optimum solutions are obtained to aid a deeper understanding of optimized solutions [4] [5]. A hierarchical multiobjective optimization method is one in which multiobjective optimization models are hierarchically constructed.

Characteristics expressing product performance are here included in the objective functions when the multiobjective optimization problem is formulated. The characteristics are here called "performance characteristics". When each characteristic in a group of characteristics has individually different optimum design solutions, the characteristics of the group will have conflicting interrelationships during the optimization of the system as a whole. Generally, the group of characteristics included in the objective functions has conflicting interrelationships.

(i) Hierarchical construction of optimization problems

In the first stage of the proposed product design optimization method, each performance characteristic in the group of product performances is decomposed into simpler basic characteristics according to its structure. Alternatively, simpler characteristics are extracted from performance characteristics, to accommodate the specific features, or difficulties, of the particular design problem. Decomposition and extraction techniques are sequentially applied until the characteristics become sufficiently simple to use in the next stage of the procedure. The decomposed and extracted characteristics are placed in hierarchical levels that are below those of the original characteristics. The decomposed or extracted characteristics are here simply called "characteristics" to distinguish them from performance characteristics.

In this research, the decomposed or extracted characteristics and design variables are ordered in a hierarchical structure, creating a hierarchical display of system components, based on the clarification of input and output relationships among the components comprising the system. This ultimately provides an easily understandable global view of the system as a whole, such as is shown in Fig.6. The construction of optimization strategies is then based on this global structural model.

Characteristics on the same hierarchical level have different input variables. The set of characteristics sharing common input variables is denoted as a basic optimal unit group.

In Fig.6, characteristics f_7 and f_8 have common design variables, vector $\mathbf{d}_1$, while characteristics f_9 and f_{10} have common design variables, vector $\mathbf{d}_2$. In such cases, f_7 and f_8 , and f_9 and f_{10} are respectively unified as basic optimum unit groups. f_3 and f_4 have common input variables, namely f_7 , f_8 , f_9 and f_{10} . In such cases, f_3 and f_4 are unified as a basic optimum unit group at a higher hierarchical level. Characteristics existing in the same basic optimal unit group are essentially simultaneously optimized as a multiobjective optimization

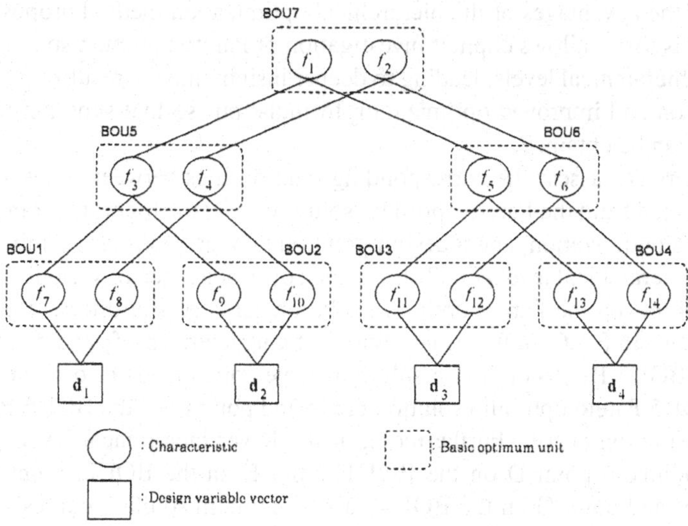

Figure 6. Hierarchical construction of the optimization problem

problem. Optimizations start at the bottom level of the basic optimal unit groups, for example, f_7 and f_8 , and then proceed to the higher levels. Basic optimum units existing at the same hierarchical level can be optimized separately or concurrently as needed, reducing the required computation time.

The Pareto optimum design solutions obtained in a basic optimal unit group are included in the input variables for the optimization of basic optimal unit groups located at higher levels along the decomposition path. Here, design solutions at discrete points on the Pareto optimum solution set are transferred for use in upper level optimizations. The Pareto optimum solutions obtained by each optimization are added one after another, to obtain Pareto optimum solutions for the whole basic optimal unit group. Finally, the Pareto optimum solutions at the top hierarchical level are achieved.

(ii) Deeper insight into the results of design optimization

The results derived from the design optimization are only solutions obtained based on the initially given formulations. Even if multidisciplinary optimizations are applied, it is impossible to include all the product design factors in the initial formulations. Optimization methods should not simply be used just to obtain final design solutions to the problem at hand, but also for effectively and rationally obtaining candidate design solutions for further design investigations and improvement. The information and knowledge obtained by the design optimization should, ideally, be used as investigational data for further design improvements.

One of the advantages of the hierarchical optimization method proposed in this paper is that it allows explicit investigation of Pareto optimum solutions at the lower hierarchical levels, leading to deeper insight into the results of design optimization and improved optimization formulations so that superior design solutions can be obtained.

Designers can assess the corresponding relationships between a design solution selected from the Pareto optimum solution set at the highest hierarchical level of the optimization, and a design point on a Pareto optimum solution set at a lower hierarchical level. Fig.7 shows the correspondence of various design solution points on the Pareto optimum solution set curves at different hierarchical levels. In Fig.6, BOU7 is composed of characteristics f_1 and f_2. Point A on the BOU7 Pareto optimum solution curve corresponds to both point B on the BOU5 Pareto optimum solution curve and point C on the BOU6 Pareto optimum solution curve. Furthermore, at the lower hierarchical level, point A corresponds to point D on the BOU1, point E on the BOU2, point F on the BOU3, and point G on the BOU4 Pareto optimum solution curves. Such detailed clarification of corresponding design points is a useful and important feature of the proposed method.

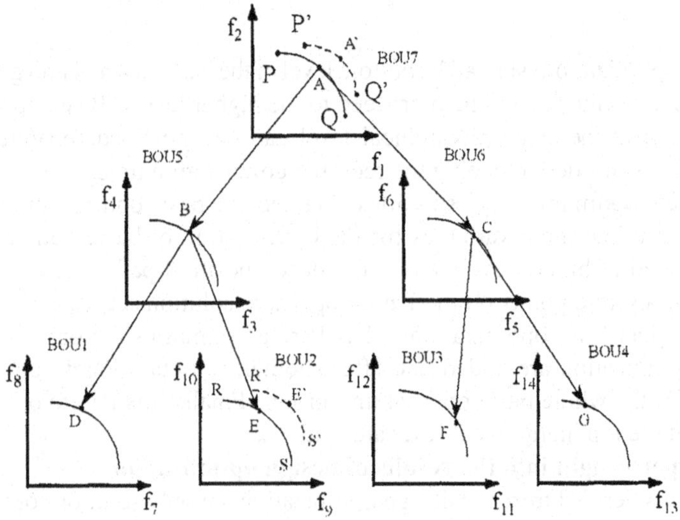

Figure 7. Correspondence of design solution points

In Fig.7, given the conflicting relationship of the essential characteristics f_9 and f_{10} at the lowest hierarchical level of the optimization problem, the breakthrough design alternative that yields the improved Pareto optimum solution shown by the dashed line R'S' can now be considered. When the new

f_9/f_{10} Pareto optimum solution is applied during further optimization, a new, enhanced Pareto optimum solution line for performance characteristics f_1 and f_2 at the highest hierarchical level is obtained, indicated by the dashed line P'Q' [4] [5].

4. Concluding Remarks

In current product design scenarios, rapidly changing customer preferences make reductions in product development/design time a practical necessity for many companies. Furthermore, competition among companies for products that can be offered at lower cost, while providing better performance and quality, higher reliability, and so on, is relentless and ongoing. Since product cost and product performances/characteristics are essentially determined at the product planning stage, the success or failure of product development depends on how appropriately optimization methods can be applied at this stage. As explained in this paper, many alternative designs should be generated and quickly compared at the product planning stage, and the relationships among associated characteristics should be effectively evaluated. The application of multiobjective optimization methods where Pareto optimum solution sets are obtained and displayed on the characteristics space can be useful and effective.

The methods proposed in this paper, based on the use of particularly sophisticated multiobjective optimization methods, can be effectively used for comparison of many alternative designs and concurrent evaluations of related performance characteristics.

References

[1] Edited by Stadler, W., 1988, Multicriteria Optimization in Engineering and in the Sciences, Plenum Press.

[2] Edited by Eschenauer, H, Koski, J, and Osyczka, A., 1990, Multicriteria Design Optimization, Springer-Verlag.

[3] M. Yoshimura and S. Horie: Concurrent Design of Mechanical Systems Using Operator-Acting Modules, JSME International Journal, Vol.43, No.2, Series C, 2000, pp.478-485.

[4] M. Yoshimura, K. Izui, S. Komori, and S. Nishiwaki: Breakthrough of Design Solutions Enabled by Extraction of Core Factors in Product Design Optimization, Proceedings of the fifth World Congress of Structural and Multidisciplinary Optimization, Paper No.82, 2003.

[5] M. Yoshimura, M. Taniguchi, K. Izui, and S. Nishiwaki: Hierarchical Arrangement of Characteristics in Product Design Optimization, ASME Journal of Mechanical Design (in process of publication).

ON THE NUMERICAL SOLUTION OF STOCHASTIC OPTIMIZATION PROBLEMS

J. Mayer[1]

[1] *Institute for Operations Research, University of Zurich, Zurich, Switzerland, mayer@ior.unizh.ch*

Abstract We introduce the stochastic linear programming (SLP) model classes, which will be considered in this paper, on the basis of a small–scale linear programming problem. The solutions for the various problem formulations are discussed in a comparative fashion. We point out the need for model and solution analysis. Subsequently, we outline the basic ideas of selected major algorithms for two classes of SLP problems: two–stage recourse problems and problems with chance constraints. Finally, we illustrate the computational behavior of two algorithms for large–scale SLP problems.

keywords: stochastic linear programming, numerical methods.

1. SLP problem formulations

Our starting point is a simple deterministic linear programming (LP) production problem which serves for illustrating various model formulations in stochastic linear programming (SLP). Two kinds of raw materials are used for producing a single good, and we consider a single planning period. The LP–formulation for minimizing costs reads as

$$\begin{array}{llrcrl} \textit{Costs:} & z = & 2\,x_1 & + & 3\,x_2 & \rightarrow \min \\ \textit{Capacity:} & & x_1 & + & x_2 & \le 100 \\ \textit{Demand:} & & a_1\,x_1 & + & a_2\,x_2 & \ge b \\ & & x_1, & & x_2 & \ge 0 \end{array} \tag{1}$$

where x_1 and x_2 denote the amounts of raw materials to be used for the production; these are our decision variables. The overall storage capacity for the raw materials is 100, and the prices are 2 and 3 in some monetary unit, respectively. b denotes the demand for the product whereas a_1 and a_2 stand for the productivity factors of the two raw materials, respectively.

a_1, a_2, and b will be considered as parameters. Choosing $a_1 = 5$, $a_2 = 8$, and $b = 640$, we get the solution $x_1^* = 0$, $x_2^* = 80$, $z^* = 240$. Note that in this solution the storage capacity is not fully utilized and only the second raw material is used for the production.

Please use the following format when citing this chapter:
Author(s) [insert Last name, First-name initial(s)], 2006, in IFIP International Federation for Information Processing, Volume 199, System Modeling and Optimization, eds. Ceragioli F., Dontchev A., Furuta H., Marti K., Pandolfi L., (Boston: Springer), pp. [insert page numbers].

In the sequel, we will assume that the parameters a_1, a_2, and b are stochastically independent, normally distributed random variables with the probability distribution not depending on x_1 and x_2: $a_1 \sim \mathcal{N}(5, 0.2)$, $a_2 \sim \mathcal{N}(8, 0.6)$, and $b \sim \mathcal{N}(640, 14)$. The question arises, how to interpret (1) under such circumstances.

1.1 First idea: the expected value problem

The simplest idea is to replace the random parameters by their expected values $\mathbf{E}[a_1] = 5$, $\mathbf{E}[a_2] = 8$, $\mathbf{E}[b] = 625$ and solve the resulting deterministic LP. In our case, this yields the solution discussed in the previous section.

A clear drawback of this approach is that we get the same solution for all probability distributions having the same expected value. Unfortunately, due to its simplicity, the expected value problem is widely used in practice as a substitute of the stochastic problem. As we will see later, the expected value solution behaves in our case extremely badly, when taking the true stochastic nature of the problem into the account.

1.2 A robust interpretation: "fat" solutions

The next idea is to take problem (1) as it stands, with each of the realizations generating a constraint. This idea is due to Madansky who termed the solution obtained this way as "fat solution". Having continuous distributions with an unbounded support, we arrive at a problem with infinitely many constraints, and have no chance to get a feasible solution. Thus, as a next step, let us replace the original distribution with an empirical one. Discretizing the distribution with $(a_1, a_2, b) \sim (9 \times 9 \times 9) = 729$ and with $\sim (5 \times 5 \times 5) = 125$ realizations, the problem turns out to be still infeasible. Finally, taking the rather crude discretization with $(a_1, a_2, b) \sim (3 \times 3 \times 3) = 27$ realizations we get an optimal solution. This illustrates the main drawback of the approach: typically we have no feasible solutions for the reformulated problems. Another drawback is that instead of the probability distribution only the support of the distribution enters the model; we obtain the same solution for any two probability distributions having the same support.

1.3 Chance constraints

Regarding the stochastic demand constraint in (1), the next idea is to evaluate the quality of a decision by computing the probability of the event that the constraint inequality holds. Prescribing the probability on a high level leads to chance constrained problems (or probabilistic constrained problems). In our case, we get a chance constrained problem by replacing the demand constraint in (1) by the probability constraint

$$\mathbf{P}(a_1\, x_1 + a_2\, x_2 \geq b\,) \geq \alpha \tag{2}$$

with α being a high probability level. We have solved our example with probability levels $\alpha = 0.95$ and $\alpha = 0.99$. Note that positive values of the quantity $\zeta(x_1, x_2) := b - a_1 x_1 - a_2 x_2$ represent unfulfilled demands. We interpret these as *losses*. Thus our chance constrained model provides a solution, for which the probability of a loss is small $(1 - \alpha)$. Nevertheless, losses my occur, and for the case when losses occur, chance constrained models have no built–in facilities for controlling the size of a loss.

1.4 Integrated chance constraints and CVaR constraints

Constraining the size of the expected loss leads to models with integrated chance constraints. In our case we obtain a model of this type by replacing the demand constraint in (1) by the constraint

$$\mathbf{E}[(\,b\; - \;a_1\, x_1 + a_2\, x_2\,)^{+}\,] \;\leq\; \gamma_{icc}$$

with γ_{icc} being a maximum tolerable loss and $u^{+} = \max\{0, u\}$ for any real number u. In our computations, we have chosen $\gamma_{icc} = 5$ and have discretized the probability distribution with $(a_1, a_2, b) \sim (10 \times 10 \times 10) = 1000$ realizations.

A related idea gaining increasing importance in financial applications, is based on conditional value–at–risk (CVaR). In our continuously distributed case, the idea can be interpreted as constraining the expected loss, given that it exceeds the α–quantile of the loss, $VaR_{\alpha}(\zeta(x_1, x_2))$. In our example, the demand constraint in (1) is substituted by the constraint

$$\mathbf{E}[\zeta(x_1, x_2) \mid \zeta(x_1, x_2) \geq VaR_{\alpha}(\zeta(x_1, x_2))] \leq \gamma_{cvar}$$

with γ_{cvar} being a maximum tolerable CVaR value. Although in our normally distributed case the problem can equivalently be formulated as a nonlinear programming problem, we have discretized the probability distribution as before and took $\gamma_{cvar} = 5$ in our computations. We have chosen the probability level as $\alpha = 0.95$.

1.5 Two–stage recourse model

We introduce penalty costs both for $\zeta(x_1, x_2) = b - a_1 x_1 - a_2 x_2 < 0$ and for $\zeta(x_1, x_2) > 0$ and consider the random variable
$Q(x_1, x_2; a_1, a_2, b) =$

$$= \min \left. \begin{array}{llll} 7y_1 & + & 2y_2 & \\ y_1 & - & y_2 & = b - a_1 x_1 - a_2 x_2 \\ y_1, & & y_2 & \geq 0 \end{array} \right\} \tag{3}$$

where the penalty costs of 7 arise if the demand is not fulfilled and the costs of 2 stand for overproduction. The idea is to evaluate solutions via the expected overall costs. The two–stage recourse model arises from (1) by augmenting the objective function by the expected costs, leading to

$$2x_1 + 3x_2 + \mathbf{E}[\,Q(x_1, x_2; a_1, a_2, b)],$$

and by dropping the demand constraint. Note that we still have a single time period, say $[0, T]$, but a two–stage decision. At time $t = 0$ we have to decide on x_1 and x_2, taking into account the expected costs of the recourse actions at $t = T$ (represented by the variables y_1 and y_2). The latter clearly depend on x_1, x_2, and also on the distribution of the random entries.

1.6 Wait–and–See solution

This means solving

$$\left.\begin{array}{llllll} \mathbf{E}[& \min & 2\,x_1 & + & 3\,x_2 & +Q(x_1, x_2; a_1, a_2, b)\ ] \\ & & x_1 & + & x_2 & \leq 100 \\ & & x_1, & & x_2 & \geq 0 \end{array}\right\} \qquad (4)$$

and amounts in computing the optimal objective values separately for the realizations and computing subsequently the expected value. In our computations, we took a discrete approximation with $10 \times 10 \times 10 = 1000$ realizations. In general, for the different realizations different solution vectors are obtained. One might get the idea to construct a solution by taking the expected value of the solutions for the separate realizations. As we will see, our example indicates that this is usually not a good idea.

1.7 Computational results, outlook on algorithms

Table 1.7 displays the results obtained by solving the SLP–variants of the production problem. The rows correspond to the expected value problem, to the fat formulation, to the chance constrained problem (with probability levels 0.95 and 0.99), to integrated chance constraint, to CVaR constraint, to the two–stage recourse problem, and to the wait–and–see problem, respectively. The second and third columns display the components of the optimal solution; the fourth column shows the optimal objective value of the corresponding SLP problem.

The column headed by $\mathbf{P}$ shows the probabilities (2) computed for the optimal solutions obtained from the various SLP models. The last column displays the overall expected costs in the two–stage recourse problem, when fixing the first–stage variables according to the optimal solutions from the second and third column.

Comparing the solutions obtained from the various approaches, we observe

	x_1^*	x_2^*	z^*	P	R–cost
Exp	0.00	80.00	240.00	0.49	378.71
Fat	0.00	94.05	282.14	0.98	285.82
CC95	28.24	71.76	271.76	0.95	277.63
CC99	9.63	90.37	290.38	0.99	291.37
ICC	44.26	55.74	255.74	0.77	290.73
CVar	21.66	78.34	278.33	0.97	281.49
RS	32.59	67.41	277.08	0.93	277.08
WSS	7.50	75.23	241.10	0.69	375.19

Table 1. Computational results for the example

a great diversity. The expected value solution and the fat solution, for instance, suggest a production plan, solely based on the second raw material. In addition, for these solutions the storage capacity is not fully utilized. Contrary to this, the ICC solution proposes a balanced usage of the two raw materials. The question arises: Which of these is the "true" solution of our stochastic problem? Clearly none of them can be identified as ultimately best; the proper choice depends on the modeling attitude and also on available solvers (implementations of solution algorithms).

According to Richard W. Hamming, "the purpose of computing is insight, not numbers." In our case, we have built and solved several SLP problems corresponding to different modeling paradigms and based on the same initial deterministic LP model and the same probability distribution. The last two columns in Table 1.7 display an evaluation of the solutions obtained, based on two quality measures: the probability that the demand will be fulfilled and the overall expected costs. According to this, the expected value solution is by far the worst, having the lowest probability and highest costs. Almost as worse is the solution obtained form the naïve application of the wait–and–see approach, with averaged solutions. The proper choice clearly depends on the risk–cost attitude of the modeler. Assuming a modeler who places approximately equal weights on risk and costs, a good solution appears to be the C95 solution.

When working with a single modeling paradigm, analysis of the model instance and the solution should be part of the modeling process. As an example for model instance analysis, let us consider the two–stage recourse formulation of our example. We may compute the expected value of perfect information (EVPI) and the value of stochastic solution (VSS) according to

$$EVPI := z^R - z^W = 35.96 \quad \text{and} \quad VSS := z^V - z^R = 101.24$$

where z^W and z^R are the optimal objective values of the wait-and-see problem and the two–stage recourse problem, respectively. z^V is the objective value of the two–stage problem with x fixed as a solution of the expected value problem. These quantities are interpreted as valuing the effort of building a stochastic model, instead of taking the expected value problem, for instance. Loosely speaking, $EVPI$ and VSS indicate a "degree of stochasticity" of the model instances. For details see [2] or [11]. According to these measures, our example counts as highly stochastic.

For SLP problems, the main numerical difficulties have their roots in the expected values and probabilities involved in the model formulations. In general, computing them amounts in computing multivariate integrals. Regarding expectations, the main solution approaches are based on approximating the probability distribution by finite discrete distributions. Thus, the integrals reduce to sums, leading to (typically large–scale) LP problems. For chance constraints the integrals are evaluated by Monte–Carlo methods, which is time–consuming and provides results with a relatively low accuracy.

In the next sections we will outline the basic ideas of the algorithms used in our computations. We will not discuss algorithms for integrated chance constraints and for CVaR constraints. For these methods see [15], [16], as well as [11]. Introductory textbooks for SLP algorithms are [2] and [14]. For algorithms discussed in a detailed fashion see the books [5], [7], [11], [18], [19], and [22]. For comparative computational results involving several algorithms see [13] and the references therein.

2. Algorithms: chance constraints

The general problem formulation is

$$\left.\begin{array}{lrcl} \min & c^{\mathrm{T}}x & & \\ & \mathbf{P}(\,T(\xi)x \geq h(\xi)\,) & \geq & \alpha \\ & x & \in & \mathcal{B} \end{array}\right\} \tag{5}$$

with α being a high probability level and $\mathcal{B} = \{x \mid Ax = b, x \geq 0\}$. The two basic classes of chance constraints are:

Separate chance constraints: The probability applies to a single inequality ($T(\xi)$ has a single row). For some distributions, including the normal, and sufficiently high probability levels, reformulations into convex NLP problems in algebraic terms exist, see [11] or [22].

Joint chance constraints: The probability applies to a vector inequality ($T(\xi)$ may involve several rows). If only the right–hand–side (RHS) is stochastic, the above problem is a convex programming problem for some distributions, including the normal, see [11] or [22].

In the special case, when only the RHS is stochastic, by taking $T(\xi) \equiv T$

and $h(\xi) \equiv \xi$, (5) can be written as

$$\left.\begin{array}{rcl} \min & c^{\mathrm{T}}x & \\ F(Tx) & \geq & \alpha \\ x & \in & \mathcal{B} \end{array}\right\} \quad (6)$$

where we utilized $\mathbf{P}(Tx \geq \xi) = F(Tx)$, with F being the probability distribution function of ξ. In the sequel, we assume that ξ has a multivariate normal distribution. In this case F turns out to be a logconcave function (see [22]) and (6) becomes a convex programming problem. Nevertheless, the problem turns out to be difficult to solve numerically.

On the one hand, the computation of F and its gradient ∇F is a numerically difficult problem, which can only be carried out via Monte–Carlo integration methods in higher dimensions. Therefore, as far as possible, algorithms are utilizing cheaply computable Boole–Bonferroni–type bounds. On the other hand, the graph of F, except for a relatively small non–convex region, consists of extremely flat regions with practically vanishing gradients. Therefore, algorithms utilize Slater–points (feasible points x with $F(Tx) > \alpha$) as navigation aids in the iteration process.

A detailed discussion of the numerical issues can be found in [20]. For the Monte–Carlo techniques applied for the multivariate normal distribution function and for the techniques for computing Boole–Bonferroni bounds see [11], [22], and the references therein.

The algorithms for jointly chance–constrained problems are constructed in the following way: a general nonlinear programming algorithm is taken and subsequently specialized to the problem structure. As an example let us consider the central cutting plane method of Elzinga and Moore [3], endowed with a moving Slater–point by Mayer [19]. Figure 2 displays two iterations of the method.

On the left–hand–side of the figure, the feasible domain of (6) is indicated by

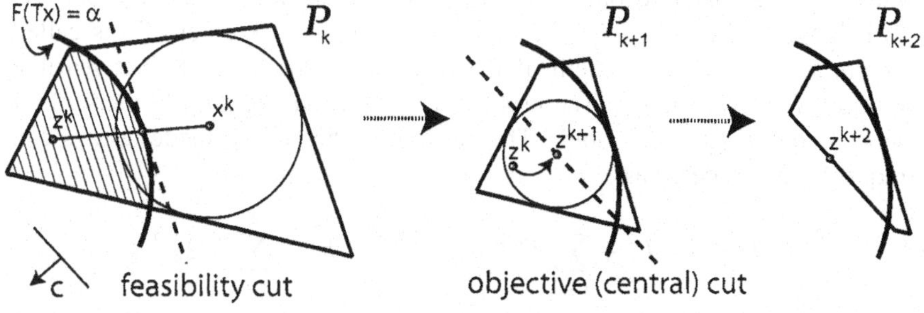

Figure 1. The central cutting plane method

the shaded region. P_k is a convex polyhedron, containing the feasible domain and z^k is the current Slater–point. First the center x^k of the largest hypersphere, inscribed into P_k is computed, which can be carried out by solving an LP problem. The center x^k lies in this case outside of the feasible region. Subsequently the intersection of the boundary of the feasible domain and of the straight line segment joining x^k and the Slater–point z^k is computed. For this computation Boole–Bonferroni bounds are utilized. Applying a *feasibility cut* via a supporting hyperplane leads to the convex polyhedron P_{k+1}, still containing the feasible domain, as it can be seen in the central part of the figure.

In P_{k+1}, the center of the largest inscribed hypersphere belongs to the feasible domain. In this case, the center becomes the new Slater–point z^{k+1} and an *objective cut* is carried out, where the cutting plane passes through z^{k+1} and is parallel to the contour–hyperplanes of the objective function. The objective cut cuts off a portion of the feasible domain, nevertheless, the optimal solution still belongs to the reduced convex polyhedron P_{k+2}, shown in the right–hand–side of the figure.

For details concerning this method, including a theoretical discussion, see [19].

3. Algorithms: two–stage recourse problems, empirical distribution

The general formulation of two–stage fixed recourse problems is

$$\left.\begin{array}{rl} \min & c^{\mathrm{T}}x + \mathbf{E}[\,Q(x,\xi)\,] \\ & x \;\in\; \mathcal{B} \end{array}\right\} \tag{7}$$

where $\mathcal{B} = \{x \mid Ax = b, x \geq 0\}$ and the *recourse subproblem* is

$$\left.\begin{array}{rcl} Q(x,\xi) & = & \min \;\; q^{\mathrm{T}}y \\ & & \;\; Wy \;\geq\; h(\xi) - T(\xi)x \\ & & \;\;\;\; y \;\geq\; 0. \end{array}\right\} \tag{8}$$

where W is called the recourse matrix. Due to the fact that W is not stochastic, (7) belongs to the class of *fixed recourse problems*. Problem (7) is called a *complete recourse* problem, if the recourse subproblem (8) has feasible solutions for any x and ξ. The problem counts as having *simple recourse* if $W = I$ and $T(\xi) \equiv T^0$ hold. The random entries in the model arrays can frequently be modeled as

$$T(\xi) = T^0 + \sum_{k=1}^{r} T^k \xi_k, \quad h(\xi) = h^0 + \sum_{k=1}^{r} h^k \xi_k,$$

for instance, via principal component analysis, where T^k, h^k are deterministic arrays. $\xi_1, \ldots, \xi_r$ are in many cases stochastically independent.

Now we assume that ξ has an empirical distribution with L realizations (scenarios) $\hat{\xi}^k$ and corresponding probabilities p_k, $k = 1, \ldots, L$. Let, furthermore, $T^k = T(\hat{\xi}^k)$, $h^k = h(\hat{\xi}^k)$, $k = 1, \ldots, L$.

In this case the problem can equivalently be formulated as a deterministic LP problem, having the structure as displayed in Figure 3.

Naïve view: the discretely distributed case is easy to handle numerically;

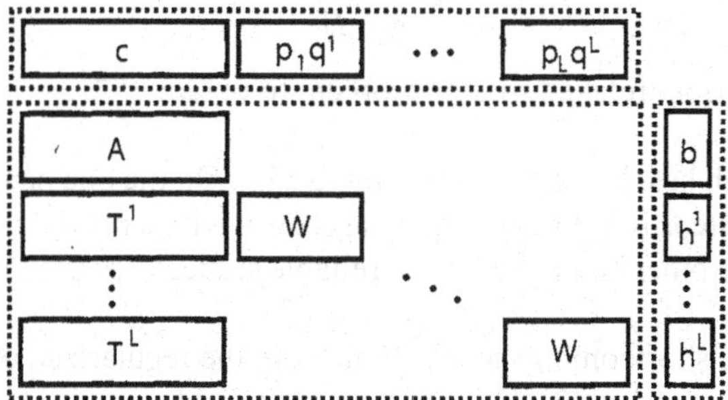

Figure 2. Dual block–angular structure of the equivalent LP

just solve this LP by readily available general–purpose LP solvers. To see the difficulty, just take 10 independent random variables, each with 10 realizations. The number of diagonal blocks will be $L = 10^{10}$. Thus, also in the discretely distributed case, ideas are needed.

In fact, a first idea is to utilize the special structure of the LP. There are two main classes of algorithm in this category.

The first class consists of decomposition methods, the most widely used algorithms will be discussed in the next section.

Interior point methods belong to the second class, where the algorithm of Mészáros [21] turned out to be one of the best in our numerical experiments.

3.1 Decomposition methods

These methods are based on the following basic observation: the expected value of the recourse function

$$f(x) := \mathbf{E}[\, Q(x, \xi)\,] = p_1\, Q(x, \xi^1) + \ldots, p_L\, Q(x, \xi^L)$$

is a piecewise linear convex function.

The basic decomposition method is due to Benders [1]. Its specialized version to SLP–problems, called L–shaped method, has been developed by Van Slyke and Wets [25]. The main idea is to apply the cutting plane method to the epigraph of f. Having x^ν as the current iterate, proceed as follows:

$\mapsto$ Compute $f(x^\nu)$ by solving L recourse problems (8) via the simplex method. Fortunately, utilizing the dual solutions, this also provides a supporting cutting

hyperplane.
$\mapsto$ Check the optimality criterion $f(x^\nu) \leq \theta^\nu + \varepsilon$.
$\mapsto$ If the algorithm does not stop, apply a cut. Technically, the cuts are collected as constraints in the *relaxed master problem*

$$\left.\begin{array}{rrll} \theta^\nu := \min & c^{\mathrm{T}} x & +z & \\ & D_k x & -z & \leq d_k, \quad k = 1, \ldots, \nu \\ & x & \in & \mathcal{B} \end{array}\right\}$$

$\mapsto$ Solve the current relaxed master problem to obtain $x^{\nu+1}$.

This approach has, however, some drawbacks. On the one hand, the method produces large steps in the beginning phase, even with a nearly optimal starting solution. On the other hand, there is no reliable strategy for dropping redundant cuts.

Both of these shortcomings are eliminated in the regularized decomposition method of Ruszczyński [23]. The main idea is to add a regularizing term to the objective of the relaxed master problem:

$$\theta^\nu \;:=\; \min c^{\mathrm{T}} x + \lambda \|x - \hat{x}^\nu\|^2 + z$$

where $\hat{x}_\nu$ is the current *candidate solution* and $\lambda > 0$ holds. The candidate solution is changed, only if the solution $f(x^\nu)$ is sufficiently smaller than $f(\hat{x}^\nu)$. Additionally, it turns out that it is sufficient to keep at most $n + L$ cuts.

4. Algorithms: two–stage recourse problems, general distributions

Decomposition methods certainly help to solve problems with a large number of realizations. It is still open, however, what to do if in the discretely distributed case we have, for instance, $L = 10^{10}$ joint realizations. A further problem is, what to do if ξ has a continuous distribution?

We will consider the main ideas of three basic approaches in the subsequent sections. An additional general approach is based on stochastic quasi–gradients; for these methods see Marti [18].

4.1 Successive discrete approximation (SDA)

This algorithm is due to Kall [8], Kall and Stoyan [9], Frauendorfer and Kall [4], Frauendorfer [5]. See also [11] and [14]. The basic idea is to approximate the original distribution by discrete distributions in a successive manner, via partitions of Ξ, which is an interval covering the support of ξ.

Having the partition $\Xi = \Xi_1 \cup \ldots \cup \Xi_L$, the approximate discrete distribution will be

$$p_k = \mathbf{P}(\xi \mid \xi \in \Xi_k), \quad \hat{\xi}^k = \mathbf{E}(\xi \mid \xi \in \Xi_k)$$

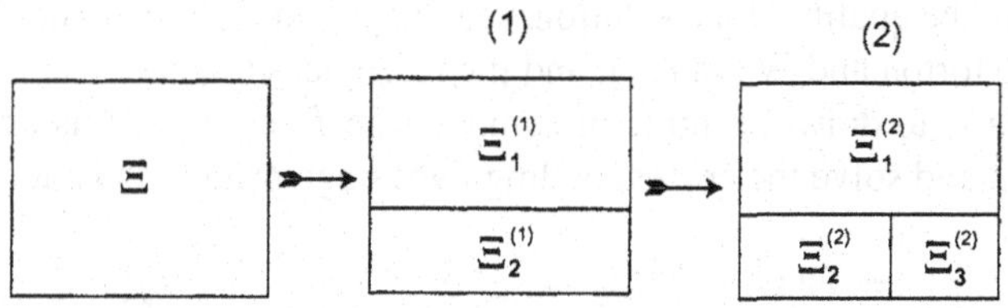

Figure 3. Successive subdivision of Ξ

for $k = 1, \ldots, L$. The key part of the method is the subdivision strategy.

The subdivision strategy is based on lower and upper bounds, for each of the cells in the partition

$$L_k(x^\nu, \xi) \leq \mathbf{E}[\, Q(x^\nu, \xi) \mid \xi \in \Xi_k \,] \leq U_k(x^\nu, \xi),$$

based on the Jensen and on the Edmundson–Madansky inequalities, respectively. For computing the upper bound, the recourse subproblem (8) has to be solved for each of the vertices of Ξ, with ξ taken as the vertex.

$\mapsto$ the cell to be subdivided next will have the maximal relative difference regarding the bounds.

$\mapsto$ the coordinate for the subdivision is selected by employing various heuristic measures of nonlinearity along the corresponding direction.

Great merit of the method: computable error bounds.

4.2 Stochastic decomposition (SD)

This algorithm is due to Higle and Sen [6], [7]. It can be considered as a stochastic, sampling–based version of Benders–decomposition. Let us denote by $\tilde{\xi}^1, \ldots, \tilde{\xi}^\nu, \ldots$ a sample according to the distribution of ξ.

The basic idea is the following: instead of $\mathbf{E}[Q(x, \xi)]$, build Benders–type cuts to the Monte–Carlo approximation

$$\mathbf{E}[Q(x, \xi)] \approx \frac{1}{\nu} \sum_{k=1}^{\nu} Q(x, \tilde{\xi}^k).$$

This is a moving target, therefore, besides adding new cuts, the existing cuts must also be updated. Sampling and adding cuts runs in a successive manner. The most efficient variants employ "incumbent solutions" and regularized master problems. New cuts are computed by taking into account all previous dual solutions, and the stopping rule is based on bootstrapping.

4.3 Sample average approximation (SAA)

The basic idea of this algorithm has been widely used by practitioners. It became increasing attention due to recent results concerning speed of convergence and judging the quality of the solution, see Shapiro and Homem–de–Mello [24] and Mak, Morton and Wood [17], and the references therein.

The idea is to draw a sample of sample–size L, consider this as a discrete distribution and solve the corresponding two–stage problem. Thus we have

$$f(x) := \mathbf{E}[\,Q(x,\xi)\,] = \frac{1}{L}\,[Q(x,\xi^1) + \ldots + Q(x,\xi^L)]$$

Subsequently the quality of the solution is to be judged, and if needed, the procedure repeated with a larger sample–size.

Crucial issue: judging solution quality. The best estimators are based on the optimality gap between statistical lower and upper bounds.

5. Illustrative computational results

We have randomly generated test problem batteries for two–stage recourse problems, with dimensions A (10×20), W (5×10). T and h are both stochastic and the random vector ξ is 5–dimensional. Each battery consists of 10 test problems.

The batteries were generated as follows: first we have generated a basis–battery with ξ having a normal distribution with stochastically independent components. This has been used to generate 5 further batteries by discretizing the distribution, resulting in test problem batteries with the following amounts of joint ralizations L: $2^{10} = 1'024$, $2^{15} = 32'768$, ; $2^{20} = 1'048'576$, $2^{25} = 33'554'432$, and $2^{30} \approx 1'056'964'608$.

The testing environment was SLP–IOR, our model management system for SLP, developed jointly with P. Kall, see [10], [12].

Computer: 2.6 GHz Pentium-III PC with 1 GB RAM, under the operating system Windows 2000.

Figure 4 displays the minimum, maximum, and average computing times

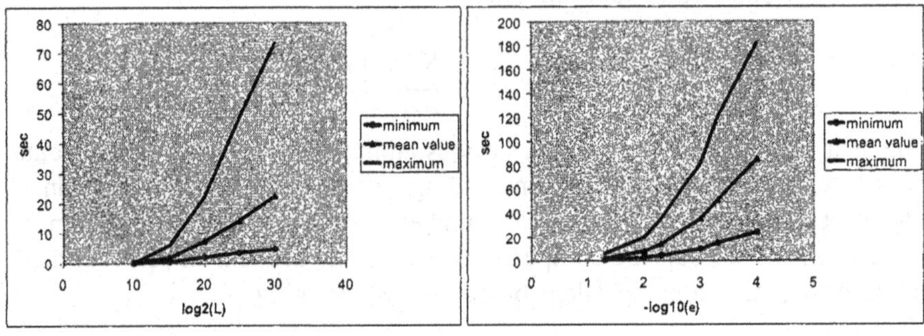

Figure 4. DAPPROX: computing time in seconds

by DAPPROX, our implementation of the successive discrete approximation

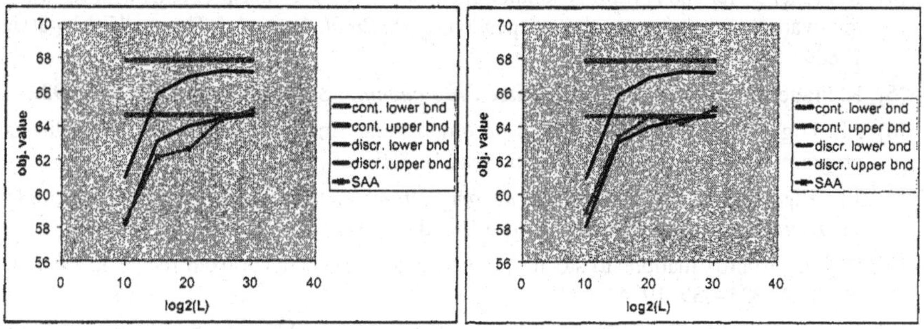

Figure 5. DAPPROX and SAA: objective values at termination

method (jointly developed with P. Kall). On the left–hand–side the dependence of the computing time on L is displayed, whereas the right–hand–side chart shows the dependence on the relative accuracy of the solution. For the latter we took accuracies $\varepsilon = 5 \cdot 10^{-2}$, 10^{-2}, $5 \cdot 10^{-3}$, 10^{-3}, $5 \cdot 10^{-4}$, and 10^{-4}. The computing times were quite acceptable, even for SLP problems with $\sim$ one billion realizations.

In Figure 5 the objective values at termination are displayed, for DAPPROX and SAA, the latter being our implementation of the SAA algorithm. For the computations we took test problem #1. The two horizontal lines correspond to the lower and upper bounds, obtained by DAPPROX for the basis problem with the normal distribution. The approximately parallel increasing curves labeled as "discr. lower bnd" and "discr. upper bnd" correspond to the results obtained by DAPPROX (5% relative accuracy).

For SAA (lowest curve in the left–hand–side chart) we took $L = 500$ and generated 5 samples. After solving the corresponding 5 problems, the objective values of the solutions have been estimated using a sample–size M, which has been chosen for the two charts as $M = 1000$ and $M = 5000$, respectively. The solution with the best estimated objective value was returned by the solver as solution. Observe that the quality of the SAA solution improves dramatically by a relatively moderate change of the run–time parameter M.

References

[1] J.F. Benders. Partitioning procedures for solving mixed-variables programming problems. *Numer. Math.*, 4:238–252, 1962.

[2] J.R. Birge and F. Louveaux. *Introduction to Stochastic Programming.* Springer–Verlag, Berlin/Heidelberg, 1997.

[3] J. Elzinga and T.G. Moore. A central cutting plane method for the convex programming problem. *Math. Progr.*, 8:134–145, 1975.

[4] K. Frauendorfer and P. Kall. A solution method for SLP recourse problems with arbitrary multivariate distributions – the independent case. *Probl. Contr. Inf. Theory*, 17:177–205, 1988.

[5] K. Frauendorfer. *Stochastic Two-Stage Programming*. Springer–Verlag, Berlin, 1992.

[6] J.L. Higle and S. Sen. Stochastic decomposition: An algorithm for two stage linear programs with recourse. *Math. Oper. Res.*, 16:650–669, 1991.

[7] J.L. Higle and S. Sen. *Stochastic decomposition. A statistical method for large scale stochastic linear programming*. Kluwer Academic Publ., 1996.

[8] P. Kall. Approximations to stochastic programs with complete fixed recourse. *Numer. Math.*, 22:333–339, 1974.

[9] P. Kall and D. Stoyan. Solving stochastic programming problems with recourse including error bounds. *Math. Operationsforsch. Statist., Ser. Opt.*, 13:431–447, 1982.

[10] P. Kall and J. Mayer. SLP-IOR: An interactive model management system for stochastic linear programs. *Math. Prog. B*, 75:221–240, 1996.

[11] P. Kall and J. Mayer. *Stochastic linear programming. Models, theory and computation.* Springer-Verlag, 2005.

[12] P. Kall and J. Mayer. Building and solving stochastic linear programming models with SLP-IOR. In S.W. Wallace and W.T. Ziemba, editors, *Applications of Stochastic Programming*, pages 77–90. MPS SIAM, SIAM, Philadelphia, 2005.

[13] P. Kall and J. Mayer. Some insights into the solution algorithms for SLP problems. *Ann. Oper. Res.*, 2005. To appear.

[14] P. Kall and S.W. Wallace. *Stochastic programming*. John Wiley & Sons, Chichester, 1994.

[15] W.K. Klein Haneveld and M.H. van der Vlerk. Integrated chance constraints: reduced forms and an algorithm. *Comp. Management Sci.*, 2005. To appear.

[16] A. Künzi-Bay and J. Mayer. Computational aspects of minimizing conditional value-at-risk. *Comp. Management Sci.*, 2005. To appear.

[17] W-K. Mak, D.P. Morton, and R.K. Wood. Monte Carlo bounding techniques for determining solution quality in stochastic programs. *Oper. Res. Lett.*, 24:47–56, 1999.

[18] K. Marti. *Stochastic optimization*. Springer-Verlag, 2005.

[19] J. Mayer. *Stochastic Linear Programming Algorithms: A Comparison Based on a Model Management System*. Gordon and Breach Science Publishers, 1998.

[20] J. Mayer. On the numerical solution of jointly chance constrained problems. In S. Uryasev, editor, *Probabilistic constrained optimization: Methodology and applications*, pages 220–233. Kluwer Academic Publ., 2000.

[21] Cs. Mészáros. The augmented system variant of IPMs in two–stage stochastic linear programming computation. *Eur. J. Oper. Res.*, 101:317–327, 1997.

[22] A. Prékopa. *Stochastic programming*. Kluwer Academic Publ., 1995.

[23] A. Ruszczyński. A regularized decomposition method for minimizing a sum of polyhedral functions. *Math. Progr.*, 35:309–333, 1986.

[24] A. Shapiro and T. Homem–de–Mello. On the rate of convergence of optimal solutions of Monte Carlo approximations of stochastic programs. *SIAM J. Opt.*, 11:70–86, 2000.

[25] R. Van Slyke and R. J-B. Wets. *L*-shaped linear programs with applications to optimal control and stochastic linear programs. *SIAM J. Appl. Math.*, 17:638–663, 1969.

PARAMETER ESTIMATION OF PARABOLIC TYPE FACTOR MODEL AND EMPIRICAL STUDY OF US TREASURY BONDS

S.I. Aihara,[1] and A. Bagchi,[2]

[1] *Tokyo University of Science, Suwa Toyohira,5000-1 Chino, Nagano, Japan, aihara@rs.suwa.tus.ac.jp*

[2] *FELab and Department of Applied Mathematics, University of Twente, P.O.Box 217, 7500AE Enschede, The Netherlands, a.bagchi@ewi.utwente.nl*

Abstract

In this paper we study the parameter estimation problem for stochastic distributed parameter systems by using the modified maximum likelihood method. More specifically, by using the US treasury bond data, the parameter estimation is performed for the stochastic hyperbolic and parabolic models describing the behavior of the term-structure of the US bond. From the prediction results, we can show that the parabolic factor models work better than the hyperbolic ones.

Key words: Factor model, US bonds, MLE, Stochastic Parabolic Equation, Maximum likelihood estimate

1. Introduction

Parameter estimation problem for stochastic distributed parameters has a long history and there still exist many open problems. In this paper, we present a practical application of the parameter estimation to a financial engineering problem. Let $P(t,T)$ denote the bond price where t is a present time and T denotes the maturity. The bond price $P(t,T)$ changes randomly in value and at $t = T$ $P(T,T)$ takes the preassigned value.

From the relation that $P(t,T) = \exp\{-\int^{T-t} f(t,x)dx\}$, the forward rate process $f(t,x)$ may be directly modeled instead of P. In this paper, we check the feasibility of the model selection of forward rate process by using some real data.

Here we use the treasury bills data which are easily obtained from the website. In US government securities, we used the constant maturity bond data,i.e., 1 year (starting date 01/02/1962) 2 year (starting date 06/01/1976) 3 year (starting date 01/02/1962) 5 year (starting date 06/01/1962) 7 year (starting

Please use the following format when citing this chapter:
Author(s) [insert Last name, First-name initial(s)], 2006, in IFIP International Federation for Information Processing, Volume 199, System Modeling and Optimization, eds. Ceragioli F., Dontchev A., Furuta H., Marti K., Pandolfi L., (Boston: Springer), pp. [insert page numbers].

date 07/01/1964) 10 year (starting date 01/02/1962) 20 year (starting date 10/01/1993) .

Noting that 20-year bond only starts at 10/01/1993, we cut past date for other bonds and set all data which start from this date up to 05/28/2004. In Fed data, there are missing parts and so we adjust these data by using the method proposed by Cochrane .(See http://gbs.uchikago.edu/fac /john.cochrane/)

To derive the forward rate process, the obtained yield data are regarded as zero-coupon curve. Hence we have the following relation between forward rate $f(t, x)$ and the yield curve $Y(t, T)$ such that

$$\frac{1}{T-t}\int_0^{T-t} f(t,x)dx = \log(1 + Y(t,T)).$$

Theoretically speaking, if we differentiate the above equation with respect to $T - t$, we can get the forward rate process $f(t, x)$. However, we only obtain 7 different maturity bonds. Firstly, we use the usual curve fitting procedure as stated in [1] and next we differentiate this process with respect to $T - t$ and obtain the forward rate process. As was mentioned in [1], the obtained results strongly depend on the methods used. For example, if we use the cubic spline and differentiate the interpolated process, the obtained forward process is largely volatile at the long maturity part. To aviod this we use the interpolation with cubic-function which is found in MATLAB as 'interp1(..., 'cubic').m'.

In Fig.1, you can see the original T-bond yield curves. By using the cubic interpolation (interp1 with 'cubic' in MATLAB) we obtain the smooth yield curve and differentiate this process. In Fig.1, the derived forward rate process is demonstrated. Now from this process, we shall try to identify parameters contained in the dynamics. Here we use the classical procedure to identify the several parameter functions. The main aim of this paper is to show that the parabolic type dynamics is experimentally accepted as the forward rate dynamics.

2. Hyperbolic system modeling

The most popular dynamics of the forward rate processes is a hyperbolic type partial differential equation which was first introduced by Heath, Jarrow and Morton from the absence of arbitrage argument and developed further by Santa-Clara et. al [2] and Aihara and Bagchi [3].

The general hyperbolic model is given by

$$df(t,x) = \frac{\partial f(t,x)}{\partial x}dt + \mu(x)dt + dw(t,x) \quad (1)$$

$$f(0,x) = f_o(x), \quad (2)$$

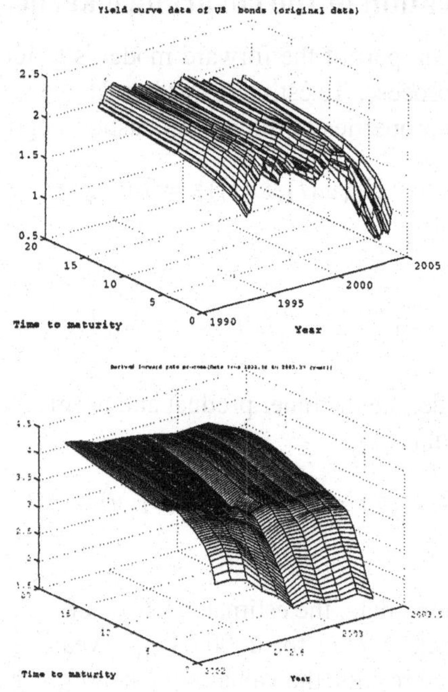

Figure 1. Original yield curve data (US bonds) and derived forward rate

where

$$E\{w(t,x)w(t,y)\} = q(x,y)t.$$

We need the following regularity property for $f(t,x)$ to perform the parameter identification.

THEOREM 1 *We assume that*

$$f_o \in L^2(\Omega; H^m(\tilde{G})),\ \mu \in H^m(\tilde{G}) \tag{3}$$

and

$$Tr\{\frac{\partial^m}{\partial x^m}(\frac{\partial^m}{\partial x^m}Q)^*\} < \infty, \tag{4}$$

where $\tilde{G} =]0, \hat{T} + t_f[$. *with* $G =]0, \hat{T}[$ H^m *denotes the m-th order Sobolev space and* $Q = \int_{\tilde{G}} q(x,y)(\cdot)dy$. *Then*

$$f \in L^2(\Omega; C(\bar{T}_f; H^m(G)), \tag{5}$$

where $T_f =]0, t_f[$.

The proof can be found in [3].

2.1 Identification of the covariance kernel

The most important part of the forward model is to identify the covariance kernel of the noise process. To estimate this kernel, we use the classical procedure by using some properties of the Ito stochastic integral. Noting that

$$E\{(w(t), w(t))\} = Tr\{Q\}t,$$

we have

$$Tr\{Q\}t = |f(t)|^2 - |f(0)|^2 - 2\int_0^t (f(s), df(s)), \tag{6}$$

where $(\cdot,\cdot)$ and $|\cdot|$ denote the inner product and norm in $L^2(\tilde{G})$. The discrete-version of the formula (6) is

$$\sum (f(t_{i+1}, x) - f(t_i, x))(f(t_{i+1}, y) - f(t_i, y)) \sim q(x, y)t. \tag{7}$$

Applying (6) to T-bond data, the estimated kernel of $q(x, y)$ is shown in Fig.2. Here we used the data $f(t, x)$ for $2000.64 \leq t(\text{year}) \leq 2002.183$ shown in Fig.1. In the obtained results, the value of the kernel at the long maturity parts

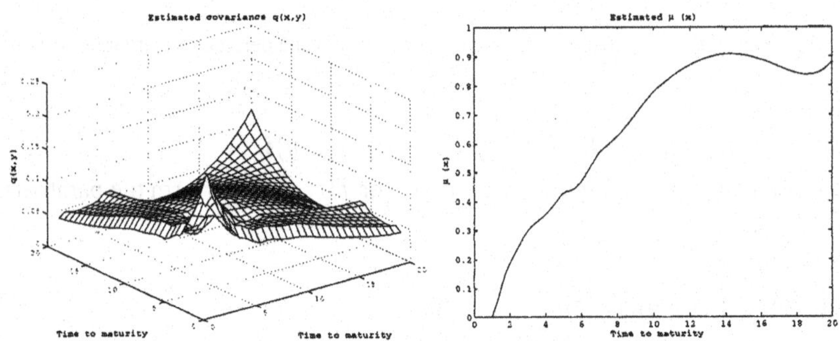

Figure 2. Estimated kernel $q(x, y)$ and $\int_0^x q(x, y)dy$

seems to be rather big. This phenomena may be caused by the interpolation method "cubic-function".

(i) Modeling in the risk neutral world

Hereafter we set the kernel $q(x, y)$ as the estimated one. In the risk neutral world, the function $\mu(x)$ is set as

$$\mu(x) = \int_0^x q(x, y)dy. \tag{8}$$

Hence the $\mu(x)$-process becomes as shown in Fig.2.

In order to check the feasibility of this model, we simulated the hyperbolic equation, setting the value of the initial condition as $f(2002.183,:)$. Without adding the noise $w(t,x)$, we obtain $E\{f(t,x)|f_o = f(2002.183,:)\}$. In Fig.3, the predicted value is shown.

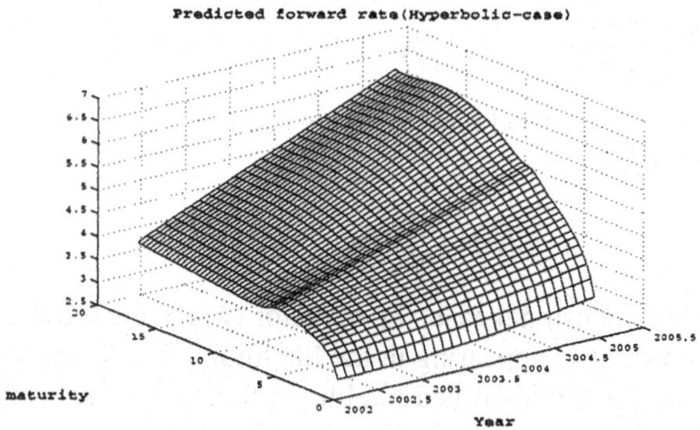

Figure 3. Predicted forward rate(Hyperbolic case)

From this result, we clearly see that our observed data is not in the risk neutral world. So we need to identify the market price of risk in the next subsection.

(ii) Identification of market price of risk:
Here we consider the following restriction: The market price of risk has a form;

$$\lambda\sqrt{q(x,x)}$$

i.e, we reset $\mu(x)$ as

$$\mu(x) = \lambda\sqrt{q(x,x)} + \int_0^x q(x,y)dy.$$

This λ has primarily been invented to price consistently interest rate derivatives rather than to fit the historical evolution of the forward rate process. However, this parameter λ is still needed to reproduce the forward rate process.

It is interesting that the parameter λ of market price of risk may be identified to maximize the modified log likelihood functional. The used data are the same as those used in the previous identification.

The exact likelihood function for an infinite-dimensional system is difficult to derive without any strict conditions. However, from Thorem 1, we can define the modified likelihood functional by setting $m = 1$ and get $f \in L^2(\Omega; C(\bar{T}_f; H^1))$. Hence

$$MLF = \int_{t_1}^{t_2} (\frac{\partial f(s,x)}{\partial x} + \lambda\sqrt{q(x,x)}$$

$$+\int_0^x q(x,y)dy, df(s,x))$$

$$-\frac{1}{2}\int_{t_1}^{t_2} |\frac{\partial f(s,x)}{\partial x} + \lambda\sqrt{q(x,x)} + \int_0^x q(x,y)dy|^2 ds, \tag{9}$$

where $t_2 = 2002.185, t_1 = 2000.646$ and

$$(\phi_1, \phi_2) = \int_0^{19} \phi_1(x)\phi_2(x)dx, \ \ |\phi|^2 = (\phi, \phi).$$

To derive the exact likelihood functional we need to support the invertibility of the covariance kernel

$$Q(\cdot) = \int_0^{19} q(x,y)(\cdot)dy.$$

In the infinite dimensional case, the operator Q is not invertible. Ultimately we replace the weight Q appearing in the likelihood functional by the identity operator. We call this the modified likelihood functional. To avoid this ambiguity, using the principal component analysis, we can pick up finite principal components. In such a case, we can derive the inverse of Q and the exact likelihood can be derived. However, the proposed modified likelihood is easily constructed without using principal component analysis and still contains the infinite number of random sources. The maximum MLE $\hat{\lambda}$ is given by

$$\hat{\lambda} = [\int_{t_1}^{t_2} (\sqrt{q(x,x)}, df(t,x))$$

$$-\int_{t_1}^{t_2} (\sqrt{q(x,x)}, \frac{\partial f(t,x)}{\partial x} + \mu(x))ds]/TrQ(t_2 - t_1). \tag{10}$$

The derived $\hat{\lambda}$ and the predicted forward rate process are respectively shown in Fig.4.

In Fig.5, we present the real forward rate and predicted processes, respectively.

3. Parabolic modeling

In this section, we introduce the parabolic type partial differential equation for the forward rate process instead of the hyperbolic type. This model was already proposed by Bouchaud et.al [4] and [5] to support the smoothness of the forward process with respect to time-to-maturity and that the information diffuses from one maturity to the next.

In the empirical studies, we find that the adjusting term is needed to fit the historical data. In addition to the $\lambda\sqrt{q(x,x)}$ term, we add the diffusion term in the model, because from Fig.5 it seems that the shape of the real forward rate process is a diffused shape rather than the predicted shape from the

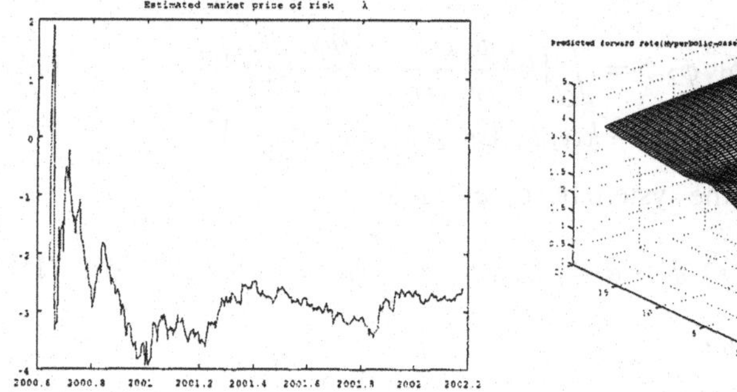

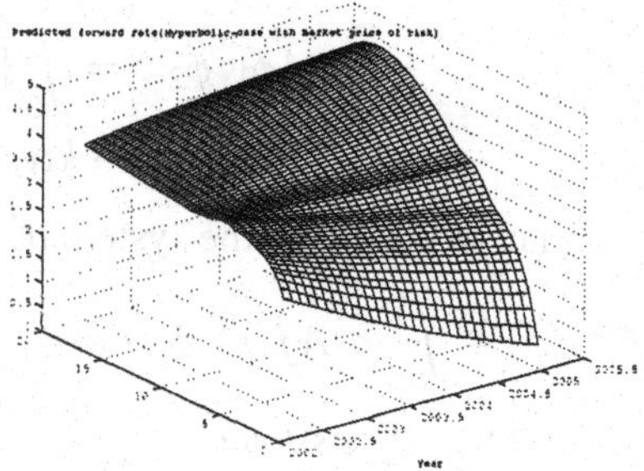

Figure 4. Estimated market price of risk and predicted forward rate

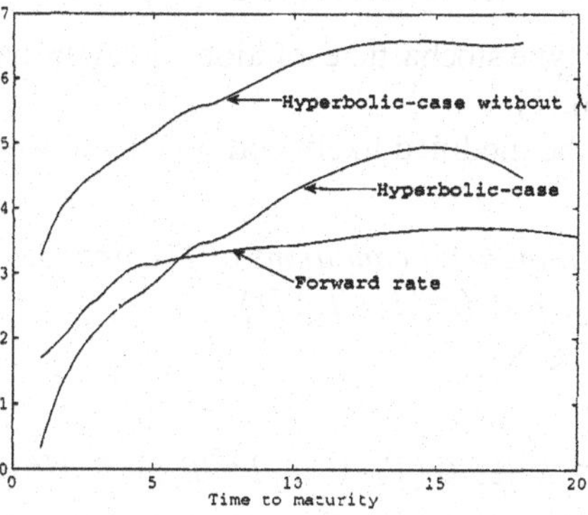

Figure 5. Predicted forward rates and real rate

hyperbolic model. Hence we set the simple parabolic type equation for the forward rate model and identify the systems parameters from the data used from $1999.5 \leq t \leq 2001.79$. From this experiment, we can conclude that the parabolic modeling is more efficient than the hyperbolic modeling.

We consider the following model: for $x \in G =]0, 19[$,

$$df(t,x) = k\frac{\partial^2 f(t,x)}{\partial x^2}dt + \frac{\partial f(t,x)}{\partial x} + \mu(x)dt + dw(t,x) \quad (11)$$

$$f(0,x) = f_o(x) \quad (12)$$

$$\frac{\partial f(t,0)}{\partial x} + \alpha_1 f(t,0) = \beta_1, \quad \frac{\partial f(t,19)}{\partial x} + \alpha_2 f(t,19) = \beta_2. \quad (13)$$

We work in the following Hilbert spaces:

$$V = H^1(G) \subset H = L^2(G) \subset V' = \text{dual of } V.$$

Define

$$< A\phi_1, \phi_2 >= \int_G \{k\frac{\partial \phi_1}{\partial x}\frac{\partial \phi_2}{\partial x} - \frac{\partial \phi_1}{\partial x}\phi_2\}dx \tag{14}$$
$$+k\alpha_2\phi_1(19)\phi_2(19) - k\alpha_1\phi_1(0)\phi_2(0), \forall \phi_1, \phi_2 \in V$$

The variational form of the system becomes $\forall \phi \in V$,

$$(f(t), \phi) + \int_0^t < Af(s), \phi > ds = (f_o, \phi) + \int_0^t (\mu, \phi)ds + (w(t), \phi). \tag{15}$$

THEOREM 2 *We assume*
(C-1) $k > 0$
(C-2) $f_o \in L^2(\Omega; H), \mu \in L^2(\Omega \times T_f; V')$
(C-3) $Tr\{Q\} < \infty$. *(15) has a unique solution in*

$$L^2(\Omega; C(\bar{T}_f; H) \cap L^2(T_f; V)).$$

Proof. The parabolic type stochastic evolution equations have been studied by many authors,e.g., [6],[7].

In order to define the modified likelihood functional we need the following theorem:

THEOREM 3 *In addition to all conditions of Theorem 3.1, we set*
(C-4) $f_o \in L^2(\Omega; V)$, $\mu \in L^2(\Omega \times T_f; H)$
(C-5) $Tr\{\frac{\partial}{\partial x}(\frac{\partial Q}{\partial x})^*\} < \infty$
Hence we have

$$f \in L^2(\Omega; C(\bar{T}_f; V) \cap L^2(T_f; V \cap H^2)).$$

By using the similar method used in [3] we can prove this theorem.

We can use the same technique for identifying $q(x, y)$ in section 2. Furthermore we also set

$$\mu(x) = \lambda\sqrt{q(x,x)} + \tilde{\mu}(x)$$
$$\tilde{\mu}(x) = \int_0^x q(x,y)dy.$$

We need to identify unknown parameters $k, \alpha_1, \alpha_2, \beta_1, \beta_2$ and λ.

(i) Identification of boundary parameters:
Noting that from Theorem 3, $\frac{\partial f(t,19)}{\partial x}$ and $\frac{\partial f(t,0)}{\partial x}$ belong to $L^2(\Omega \times T_f; R^1)$, respectively, we can apply the usual least square method and obtain the following algorithm:

$$\hat{\alpha}_1 = \frac{\int_{t_1}^{t_2} (\frac{\partial f(s,0)}{\partial x} - \frac{\partial f(t_1,0)}{\partial x})(f(s,0) - f(t_1,0))ds}{\int_{t_1}^{t_2} (f(s,0) - f(t_1,0))^2 ds} \tag{16}$$

$$\hat{\beta}_1 = \frac{\int_{t_1}^{t_2} \frac{\partial f(s,0)}{\partial x}ds + \int_{t_1}^{t_2} \hat{\alpha}(s)f(s,0)ds}{t_2 - t_1}, \tag{17}$$

and for the boundary $x = 19$ we can set the similar algorithm.

The estimated results are shown in Fig.6 for $1999.5 \leq t_2 \leq 2002$.

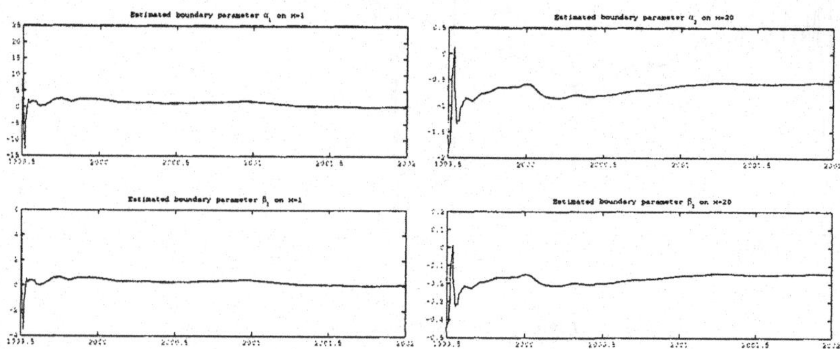

Figure 6. Sample runs of estimated boundary parameters

(ii) Identification of k and λ:

In order to identify the diffusion coefficient k and λ, we also introduced the modified likelihood functional:

$$MLF = \int_{t_1}^{t_2} (k\frac{\partial^2 f(t,x)}{\partial x^2} + \frac{\partial f(t,x)}{\partial x} + \tilde{\mu}(x) + \lambda\sqrt{q(x,x)}, df(t,x)) - \frac{1}{2}\int_{t_1}^{t_2} |k\frac{\partial^2 f(t,x)}{\partial x^2} + \frac{\partial f(t,x)}{\partial x} + \tilde{\mu}(x) + \lambda\sqrt{q(x,x)}|^2 ds, \quad (18)$$

where we already find that $f \in L^2(\Omega \times T_f; H^2(G))$ from 2. The maximum MLF $\hat{k}$ and $\hat{\lambda}$ are given by

$$\begin{bmatrix} \hat{k} \\ \hat{\lambda} \end{bmatrix} = M^{-1}N, \quad (19)$$

where

$$M = \begin{bmatrix} \int_{t_1}^{t} |\frac{\partial^2 f(s,x)}{\partial x^2}|^2 ds & \int_{t_1}^{t} (\frac{\partial^2 f(s,x)}{\partial x^2}, \sqrt{q(x,x)})ds \\ \int_{t_1}^{t} (\frac{\partial^2 f(s,x)}{\partial x^2}, \sqrt{q(x,x)})ds & \int_{t_1}^{t} (\sqrt{q(x,x)}, \sqrt{q(x,x)})ds \end{bmatrix}$$

$$N = \begin{bmatrix} \int_{t_1}^{t} (\frac{\partial^2 f(s,x)}{\partial x^2}, df(s,x)) - \int_{t_1}^{t} (\frac{\partial f(t,x)}{\partial x} + \hat{\mu}(x), \frac{\partial^2 f(s,x)}{\partial x^2})ds \\ \int_{t_1}^{t} (\sqrt{q(x,x)}, df(s,x)) - \int_{t_1}^{t} (\frac{\partial f(t,x)}{\partial x} + \hat{\mu}(x), \sqrt{q(x,x)})ds \end{bmatrix}.$$

The sample runs of the estimated λ and k are shown in Fig.7. The predicted value of forward rate is given in Fig.8 with its true value.

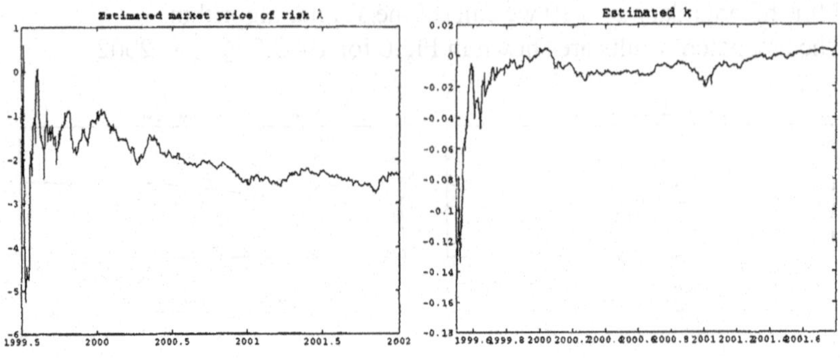

Figure 7. Estimated λ and k

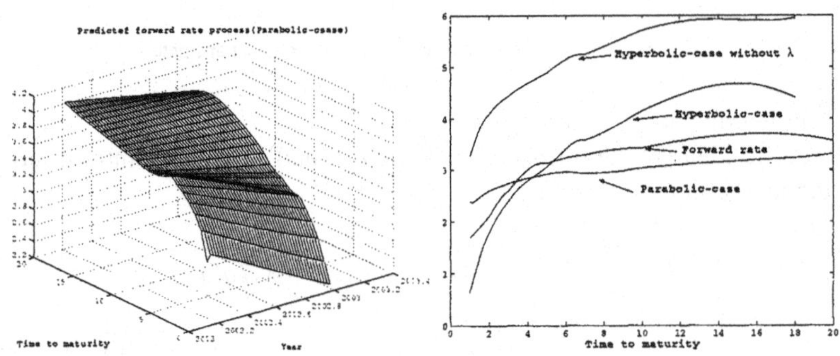

Figure 8. Predicted forward rate (Parabolic case) and true value

4. Concluding remarks

As shown in Fig.8, we could construct the parabolic type partial differential equation for the forward rate dynamics whose solution fits the future value of the forward rate better than the hyperbolic model. From the existence for the diffusion term, the shape of the predicted forward rate becomes flat and so the predicted forward rate for the parabolic case fits the real rate quite well. The calibration of the proposed model is very important to applying the mathematical algorithm to the practical situation and this should be done in the empirical probability rather than risk-neutral probability. For the identification problem in the parabolic case, we identified the term $k\frac{\partial^2 f(t,x)}{\partial x^2} + \lambda\sqrt{q(x,x)}$ for the unknown k and λ. From the obtained results, in the empirical probability world we need an extra term $k\frac{\partial^2 f(t,x)}{\partial x^2}$ or $k\frac{\partial^2 f(t,x)}{\partial x^2} + \lambda\sqrt{q(x,x)}$. The form of these terms are not theoretically derived and we only set the function form artificially. It should be noted that in the parabolic case, $\lambda = 0$ does not mean the risk neutral world because we still have an extra term $k\frac{\partial^2 f(t,x)}{\partial x^2}$.

Although Cont [5] proposed that the boundary value processes be stripped out of the original partial differential equation, we set the mixed boundary condition for the forward rate process . In our case, we can consider the more general type than Cont's model, e.g., the presence of boundary noises. If the covariance kernel is finite dimensional, we can transform the original measure to the risk neutral measure . If we do not consider the pricing of the future derivatives, it seems that this finite dimensionality condition is not required. However for the optimal portfolio problem with power utility we need this finite-dimensionality condition to support the optimal portfolio.

The most important part of this paper is how to fit the proposed model to the historical data. In the risk-neutral probability world, we only need to identify the kernel of the noise. However we can not reproduce the real future forward rate process from risk neutral case. The obtained empirical results strongly depend on the interpolation method which was used to convert the yield curve to forward rate process. In order to avoid this differential instability problem, we should reformulate the parameter identification problem as the filtering problem with discrete-time observation data. In such a reformulation, we need not to use the interpolation method and differentiation with respect to time-to-maturity variable.

References

[1] J. James and N. Webber. *Interest Rate Modeling*. John Willey & Sons, Ltd, Chicheter New York, 2002.

[2] P. Santa-Clara and D. Sornette. The dynamics of the forward rate curve with stochastic string shocks. *Review of Financial Studies*, 14:149–185, 2001.

[3] S. Aihara and A. Bagchi. Stochastic hyperbolic model for infinite-dimensional forward rates and option pricing. *Mathematical Finance*, 15:27–47, 2005.

[4] J-P Bouchaud, N. Sagna, R. Cont, N.EL-Karoui, and M. Potters. Phenomenology of the interest rate curve. *Applied Mathathematical Finance*, 6:209–232, 1999.

[5] R. Cont. Modeling term structure dynamics: An infinite dimensional approach. *Int. J. Theoretical and Applied Finace*, 8:357–380, 2005.

[6] G. Da Prato and J. Zabczyk. *Stochastic equations in infinite dimensions*. Cambridge University Press, 1992.

[7] E. Pardoux. Stochastic partial differential equations and filtering of diffusion processes. *Stochastics*, 3:127–167, 1979.

MULTI-STAGE STOCHASTIC ELECTRICITY PORTFOLIO OPTIMIZATION IN LIBERALIZED ENERGY MARKETS

R. Hochreiter,[1] G. Ch. Pflug,[1] and D. Wozabal[1]

[1] *Department of Statistics and Decision Support Systems, University of Vienna; Universitätsstraße 5/9, 1010 Vienna, Austria, {ronald.hochreiter,georg.pflug,david.wozabal}@univie.ac.at*

Abstract In this paper we analyze the electricity portfolio problem of a big consumer in a multi-stage stochastic programming framework. Stochasticity enters the model via the uncertain spot price process and is represented by a scenario tree. The decision that has to be taken is how much energy should be bought in advance, and how large the exposition to the uncertain spot market, as well as the relatively expensive production with an own power plant should be. The risk is modeled using an Average Value-at-Risk (AVaR) term in the objective function. The results of the stochastic programming model are compared with classical fix mix strategies, which are outperformed. Furthermore, the influence of risk parameters is shown.

keywords: Stochastic Optimization, Scenario Generation, Energy Markets, Optimal Electricity Portfolios, Average Value-at-Risk

1. Introduction

In this paper, we present a multi-stage stochastic optimization model for calculating optimal electricity portfolios. We refer to [10] for an overview of stochastic programming and to [1] for applications to the energy market. The general formulation of a multistage stochastic optimization program is

$$\begin{array}{ll} \text{minimize} & \mathbf{E}(f(x(\xi),\xi)) \\ \text{subject to} & (x(\xi),\xi) \in \mathcal{X} \\ & x \in \mathcal{N} \end{array}$$

where ξ denotes a multi-dimensional stochastic process describing the future uncertainty. The constraint-set $\mathcal{X}$ contains feasible solutions (x, ξ) and the (non-anticipativity) set $\mathcal{N}$ of functions $\xi \mapsto x$ is necessary to ensure, that the decisions x_t are only based on realizations up to stage t $(\xi_0, \ldots, \xi_t)$. $f(x(\xi), \xi)$ is some cost function.

Please use the following format when citing this chapter:
Author(s) [insert Last name, First-name initial(s)], 2006, in IFIP International Federation for Information Processing, Volume 199, System Modeling and Optimization, eds. Ceragioli F., Dontchev A., Furuta H., Marti K., Pandolfi L., (Boston: Springer), pp. [insert page numbers].

We apply this framework to the optimization of electricity portfolios. Additionally, an Average Value-at-Risk functional is included, enabling modern risk management, which is necessary to survive in liberalized energy markets economically.

This paper is organized as follows. Section 2 describes the estimation and simulation of the (uncertain) electricity spot market, which is the most import input for the stochastic program. Section 3 provides a detailed overview of the underlying model. Section 4 summarizes numerical results, while Section 5 concludes the paper.

2. Scenario Generation

2.1 Estimation and simulation of the spot market

The generation of scenarios for the possible development of spot prices is based on an econometric model which is designed to capture the past movements of the spot price as good as possible. This model will be capable of giving good estimates for the expected price at every hour of the period under consideration. To generate realistic scenarios we simulate the residuals of the model and thereby distort the prediction to get a possible trajectory for the spot price of energy. At the end, we compute the mean for 4 consecutive values and therefore reduce the price movement in one day to six data points, representing the average price from 0-4, 4-8, 8-12, 12-16, 16-20, 20-24 o'clock respectively.

The modeling of the spot prices is done using linear regression where the main explanatory variables are: the hour of the day, the day of the week, and the season. The regressors related to time are modeled in such a way that there are initially $24 \times 7 \times 3 = 504$ dummy variables indicating which hour of which day in what season a specific data point belongs to. Obviously, this yields to an unnecessary huge model, which can be reduced in a further step. The reduction is based on the observation that the coefficient of a dummy variable will be the mean of the data points that it points to. A feasible way to reduce the number of regressors would be to compare the means of the different hours on the different days in the different seasons and club two regressors if the means are only insignificantly different. To determine whether two means are different, we use the Kruskal-Wallis test (see for example [5]), since it is based on rank order and does not assume the data to be distributed according to any specific distribution. The necessity of such a non-parametric approach will become clear, when we discuss the residuals of the model. With this procedure, we are able to significantly reduce the number of regressors without sacrificing much of the accuracy in predicting the expected (mean) price in the respective hours.

As already mentioned, we also use temperatures as explanatory variables. This proved to be beneficial (in terms of explanatory power) and also supports the intuition of modeling temperatures not as a single variable, but to split it up into six variables measuring the effect of temperature on power prices - not for the whole day, but for the six 4-hour blocks described above.

This model of the stock prices focuses heavily on the demand side of the market, which is reasonable, since the average prices and the daily patterns of price movements can be explained with these factors pretty well (adjusted R^2=0.6). To explain long term changes in the market and to understand the peaks in the hourly spot price, it would be a valuable idea to include the supply side too. This would allow for a better understanding of the energy market and probably boost the insample accuracy too. However, when it comes to simulation these gains would probably be lost, because the supply factors are hard to forecast.

To capture some of these effects we add an AR(1) and AR(25) term to our Regression model and obtain a R^2 of 0.84.

An inspection of the residuals shows, that those are clearly not normal. Since we need to simulate from the residual distribution, we have to fit some parametric models to the empirical residuals. It turns out that the distributions are extremely wide stretched (peaks in the prices) and are therefore heavy tailed. We chose to use a stable distribution to fit the data, since the family of stable distributions contains heavy tailed distributions too, and it is known that sums of independent identically distributed random variables follow some stable law. The family of stable distributions can be characterized through their characteristic function

$$\mathbf{E}\left(e^{iuX}\right) = \begin{cases} e^{-\gamma^\alpha |u|^\alpha \left(1+i\beta(\tan\frac{\pi\alpha}{2}) sign(u)(|\gamma u|^{1-\alpha}-1)\right)+i\delta u} & ,\alpha \neq 1 \\ e^{-\gamma|u|\left(1+i\beta\frac{2}{\pi} sign(u) log(\gamma|u|)\right)+i\delta u} & ,\alpha = 1 \end{cases}$$

where $0 < \alpha \leq 2$ can be interpreted as an index of stability (everything below 2 is heavy trailed), β is a skewness parameter, γ a scale parameter, and δ a location parameter. For a in depth description of heavy tailed distributions, their properties and how to fit them see [8]. To fit the stable distribution to our residuals, we split the residuals into 6 groups according to which time slot they belong to and separately estimate stable distributions for these time slots. In a next step we generate a sufficient number of random draws from these distributions and use those together with the predicted prices to obtain spot price trajectories. For fitting the stable distribution and the generation of the random variates we use the software *stable.exe* (see [6]).

For fitting the model we use one year of hourly data (01.06.2005-31.05.2006) from the European Energy Exchange (EEX). Using the fitted model and distribution of the residuals, we simulate price trajectories corresponding to a price development of half a year. Figure 1 dipicts the means of the simulated values in each of the 4 hour slots (left) and 20 days of a typical simulation trajectory (right).

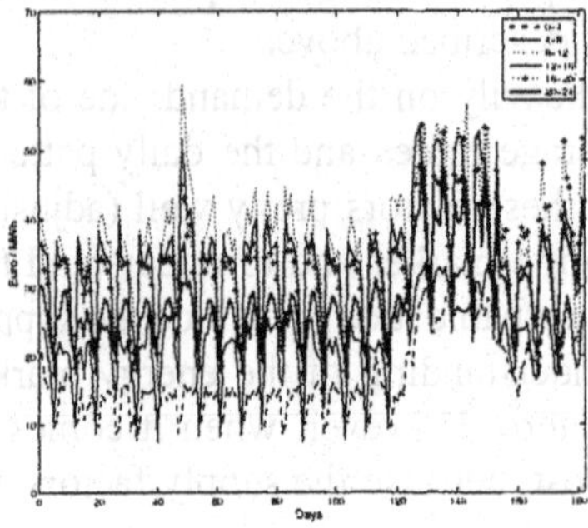

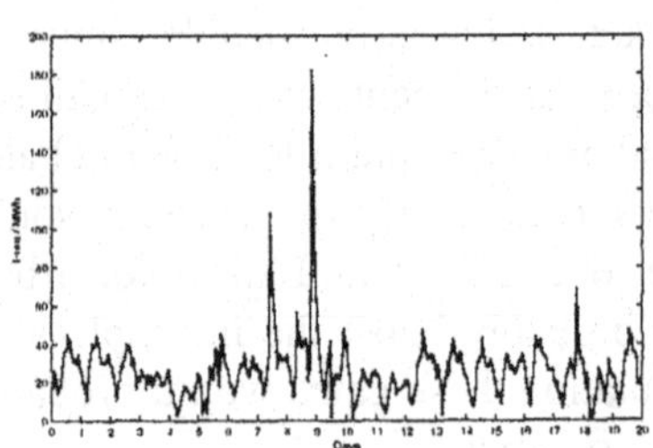

Figure 1. Averages of simulated price trajectories (left), Sample simulated spot price trajectory (right).

2.2 Generation of scenario trees

We apply a scenario generation method based on [7]. This method is optimization-problem related, and aims at minimizing a probability metric with a ζ-structure, i.e. the uniform distances of expectations of functions taken from a class $\mathcal{H}$ of measurable functions. The Wasserstein distance (from the class $\mathcal{H}$ of Lipschitz continuous functions), which plays an important role for stability results and approximation of stochastic programming models, has been used.

The implemented method generates scenario trees with a stagewise-fixed structure, which differs from other methods, e.g. [2] or e.g. an explicit method for the energy market in [4].

The size of the generated underlying spot market scenario tree with the stagewise-fixed structure, which was used for numerical experiments, is

$$n_t = [100, 200, 200 + (t-2) \cdot 2],$$

where n_t denotes the number of nodes in stage t, resulting in 560 scenarios.

3. Optimization Model

The goal of the optimization is to determine the amount of energy that should be bought in advance for a time period of half a year. The driving factor of the optimization is the expected demand for energy at all points of the considered time period. This demand is assumed to be non-stochastic and can be met in three ways. One may

1 buy electricity on the spot market,

2 produce electricity, and/or

3 buy supply contracts for future delivery of energy.

The supply contracts are designed as follows: every month the producer can consume a certain amount of energy u_i in every of the six periods of the day. This amount is part of the contract and can be specified by the consumer. On every day the energy consumption is bounded by a fraction of the overall monthly consumption, of course with the additional restriction, that the monthly consumption adds up to the pre-specified amount.

On the tree described in section 2.2 the program is shown in equation (1), which minimizes the expected total cost and the (terminal) Average Value-at-Risk (AVaR). Let AV@R$_\alpha$ be defined as the solution of the optimization problem

$$\inf \{a + \frac{1}{1-\alpha}\mathbf{E}[Y - a]^+ : a \in \mathbf{R}\},$$

where Y is a random cost variable. With a finite set of scenarios this optimization problem can be reformulated as a linear program (see [9]).

$$\begin{array}{llcll}
\text{minimize} & \sum_{n\in\mathcal{N}^{(T)}} P_{s(n)} c_n^{(t)} + \kappa(\gamma + \sum_{n\in\mathcal{N}^{(T)}} \frac{P_{s(n)} z_n}{1-\alpha}) & & & \\
\text{subject to} & g_{n,h} + s_{n,h} + m_{n,h} & \geq & D_{T(n),h} & (a) \\
 & g_{n,h} & \leq & G^{max} & (a) \\
 & m_{n,h}\xi_{n,h} + g_{n,h} C^{prod} + s_{n,h}\phi F^p_{T(n)} & = & c_{n,h} & (b) \\
 & m_{n,h}\xi_{n,h} + g_{n,h} C^{prod} + s_{n,h}\phi F^o_{T(n)} & = & c_{n,h} & (c) \\
 & s_{n,h} & \leq & \beta u_{M(n),h} & (a) \\
 & \sum_{n\in\mathcal{N}_i} s_{n,h} & \leq & u_{m,h} & (d) \\
 & \sum_{t\in\mathcal{T}} c_{R(n,t),h} & = & c_n^{(t)} & (e)
\end{array} \quad (1)$$

$M(n)$ is the month of the node n, $g_{n,h}$, $s_{n,h}$, $m_{n,h}$ electricity coming from own production (generation), the supply contract and the spot market in node n and hour-block h respectively, while $D_{n,h}$ denotes the (deterministic) demand. β represent the daily upper constraint on consumption, defined as fraction of the overall monthly consumption.

$R(n,t)$ returns the scenario predecessor node of terminal node n in stage t, $P_{s(n)}$ is the scenario probability of the scenario terminating in node n. $T(n)$ returns the stage of node n, F_t^p, F_t^o are the costs of the peak and offpeak future at stage t, and ϕ is a factor by which the contract is cheaper than the future price. $u_{m,h}$ is the optimal contract volume for month n and hour-block h. $\xi_{n,h}$ is the stochastic spot price.

The constraints the model (1) are defined for different parameter sets. The letter in parenthesis at the right indicates, which group of sets the constraint is defined for. Let $\mathcal{H}$ be the set of all hour-blocks, and $\mathcal{H}_p$ and $\mathcal{H}_o$ peak and off-peak hour-blocks. $\mathcal{N}$ the set of all nodes and $\mathcal{N}_T$ the set of all terminal

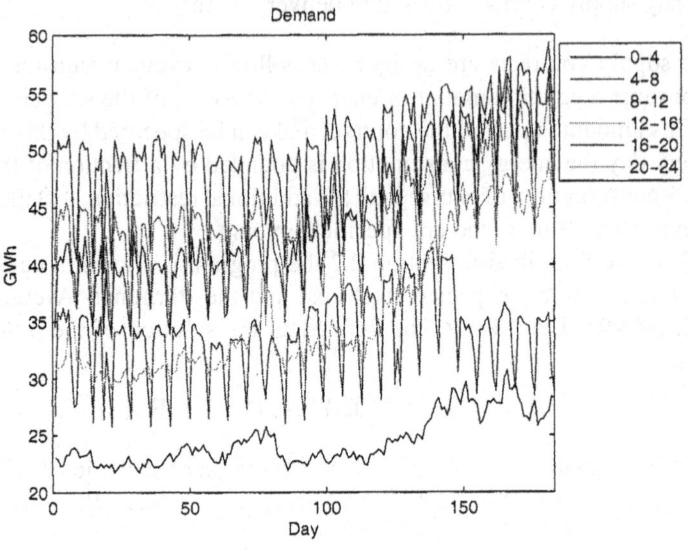

Figure 2. Demand of a large local energy distributor.

nodes. $\mathcal{M}_m$ is the set of all stages/nodes within month m. The following group of sets are applied:

(a) $\forall n \in \mathcal{N}, \forall h \in \mathcal{H}$,

(b) $\forall n \in \mathcal{N}, \forall h \in \mathcal{H}_p$,

(c) $\forall n \in \mathcal{N}, \forall h \in \mathcal{H}_o$,

(d) $\forall i \in \mathcal{M}_m, \forall m \in \{June, \ldots, November\}, \forall h \in \mathcal{H}$.

(e) $\forall n \in \mathcal{N}_T, \forall h \in \mathcal{H}$.

4. Numerical Results

A (deterministic) demand forecast of a large local energy distributor has been used. This demand is shown in Figure 2 for each of the six 4-hour blocks described in Section 2.1.

To enable a numerical comparison, the following parameters have been fixed: The cost of producing energy C^{prod} is Euro 70 and the maximum production G^{max} per 4-hour block is 5000MW. Factor β has been set to 0.1 and ϕ to 0.9.

Spot Market	Supply Contract	Production	Expected Total Cost (Euro)
20%	60%	20%	1.838.258.100
20%	70%	10%	1.706.414.546
20%	80%	0%	1.574.570.991

Table 1. Expected Total Cost for Fix Mix Portfolios

	June	July	August	September	October	November
0-4	229153	546926	510129	449503	407692	802547
4-4	208374	705748	423220	249749	197142	872110
8-12	169746	624004	374384	133423	155421	869426
12-16	253450	600427	287414	131299	187684	835807
16-20	721441	1113900	923807	832003	595822	1262270
20-24	582319	693024	216780	190338	386261	1376000

Table 2. Example contract for $\alpha = 0.9$ (MWh)

α	Expected Total Cost (Euro)	$\sum$ Contracted Volume (GWh)
0.7	1.221.079.145	19990
0.8	1.220.568.060	19637
0.9	1.220.026.120	19118
0.95	1.219.763.475	18747

Table 3. Aggregated results of the stochastic optimization for different risk parameters

4.1 A Fix Mix Solution

To compare the stochastic solution, we calculated some fix mix strategies. The results are shown in Table 1.

4.2 The Stochastic Solution

The stochastic optimization models have been implemented in AMPL (see [3]). The workflow has been developed in MatLab and some parsing scripts have been implemented in Python. The optimization problems have been solved with the MOSEK interior point solver.

The optimization problems were solved with a Pentium 4 (2GHz) with 1GB RAM running Debian GNU/Linux. The average solution time of the underlying problems is half an hour.

A typical optimal contract volume sheet is shown in Table 2 for $\alpha = 0.9$. To see the influence of the risk parameter α, the expected total cost of the portfolio in Euro and the sum of contracted volume in GWh is shown in Table 3. Additionally, these results show that the stochastic solution clearly outperforms fix mix strategies.

5. Conclusion

In this paper we proposed a model to solve the electricity portfolio problem of a big consumer in a multi-stage stochastic programming framework. The decision that has to be taken is how much energy should be bought in advance and how large the exposition to the uncertain spot market, and the relatively expensive production by an own power plant should be. It has been shown that the underlying spot price can be realistically estimated and simulated with a regression model. The underlying scenario trees representing the uncertain future spot prices are used to calculate optimal electricity portfolios. Different supply contract details have been included, such that the model is ready to be applied for practical usage.

The results show that the solution of the multi-stage stochastic program clearly outperforms classical fix mix strategies. Furthermore, by varying the risk parameter α, the consumer can fine-tune his optimal decision.

References

[1] W. Römisch, A. Eichhorn and I. Wegner. Mean-risk optimization of electricity portfolios using multiperiod polyhedral risk measures. IEEE St. Petersburg Power Tech 2005 (to appear), 2005.

[2] J. Dupačová, N. Gröwe-Kuska, and W. Römisch. *Scenario reduction in stochastic programming. An approach using probability metrics*. Mathematical Programming, 95(3, Ser. A):493–511, 2003.

[3] R. Fourer, D. M. Gay, and B. W. Kernighan. *AMPL: A Modeling Language for Mathematical Programming*. Duxbury Press / Brooks/Cole Publishing Company, 2002.

[4] H. Heitsch and W. Römisch. *Generation of multivariate scenario trees to model stochasticity in power management*. IEEE St. Petersburg Power Tech 2005 (to appear), 2005.

[5] M. Hollander and A. Wolfe. *Nonparametric statistical inference*. John Wiley & Sons, New York, 1973.

[6] J. Nolan. *stable.exe*, 2004. A program to fit and simulate stable laws (http://academic2.american.edu/jpnolan/stable/stable.html for further details).

[7] G. Ch. Pflug. *Scenario tree generation for multiperiod financial optimization by optimal discretization*. Mathematical Programming, 89(2, Ser. B):251–271, 2001.

[8] S. Rachev and S. Mittnik. *Stable Paretian Models in Finance*. John Wiley & Sons, New York, 2000.

[9] R. T. Rockafellar and S. Uryasev. *Optimization of Conditional Value-at-Risk*. The Journal of Risk, 2(3):21–41, 2000.

[10] A. Ruszczynski and A. Shapiro, editors. *Stochastic programming*, volume 10 of Handbooks in Operations Research and Management Science. Elsevier Science B.V., Amsterdam, 2003.

SSD CONSISTENT CRITERIA AND COHERENT RISK MEASURES

W. Ogryczak,[1] and M. Opolska-Rutkowska,[2]
[1]*Warsaw University of Technology, Institute of Control & Computation Engineering, 00-665 Warsaw, Poland, ogryczak@ia.pw.edu.pl* *, [2]*Warsaw University of Technology, Institute of Mathematics, Warsaw, Poland*

Abstract The mean-risk approach quantifies the problem of choice among uncertain prospects in a lucid form of only two criteria: the mean, representing the expected outcome, and the risk: a scalar measure of the variability of outcomes. The model is appealing to decision makers but it may lead to inferior conclusions. Several risk measures, however, can be combined with the mean itself into the robust optimization criteria thus generating SSD consistent performances (safety) measures. In this paper we introduce general conditions for risk measures sufficient to provide the SSD consistency of the corresponding safety measures.

keywords: decisions under risk, stochastic dominance, mean-risk.

1. Introduction

We consider the general problem of comparing real-valued random variables (distributions), assuming that larger outcomes are preferred. Two methods are frequently used for modeling choice among uncertain prospects: stochastic dominance, and mean-risk analysis. The former is based on an axiomatic model of risk-averse preferences but it does not provide us with a simple computational recipe. It is, actually, a multiple criteria model with a continuum of criteria. The mean-risk approach quantifies the problem in a lucid form of only two criteria: the mean, representing the expected outcome, and the risk: a scalar measure of the variability of outcomes. The mean-risk model is appealing to decision makers but it is not capable of modeling the entire gamut of risk-averse preferences. Moreover, for typical dispersion statistics used as risk measures, the mean-risk approach may lead to inferior conclusions.

*Paper written with financial support of grant 3T11C 005 27 from The State Committee for Scientific Research.

Please use the following format when citing this chapter:
Author(s) [insert Last name, First-name initial(s)], 2006, in IFIP International Federation for Information Processing, Volume 199, System Modeling and Optimization, eds. Ceragioli F., Dontchev A., Furuta H., Marti K., Pandolfi L., (Boston: Springer), pp. [insert page numbers].

In this paper we analyze conditions that are necessary and sufficient for risk measures to provide the SSD consistency of the corresponding mean-risk models. Actually, we show that under simple and natural conditions on the risk measures they can be combined with the mean itself into the robust optimization criteria thus generating SSD consistent performance (safety) measures. The analysis is performed for general distributions but we also pay attention to special cases such as discrete or symmetric distributions. We demonstrate that, while considering risk measures depending only on the distributions, the conditions similar to those for the coherency are, essentially, sufficient for SSD consistency.

2. Stochastic dominance and mean-risk models

In the stochastic dominance approach random variables are compared by pointwise comparison of some performance functions constructed from their distribution functions. Let X be a random variable representing some returns. The first performance function $F_1(X,r)$ is defined as the right-continuous cumulative distribution function itself: $F_1(X,r) = \mathbf{P}[X \leq r]$ for $r \in R$. We say that X weakly dominates Y under the FSD rules ($X \succeq_{FSD} Y$), if $F_1(X,r) \leq F_1(Y,r)$ for all $r \in R$, and X FSD dominates Y ($X \succ_{FSD} Y$), if at least one strict inequality holds. Actually, the stochastic dominance is a stochastic order thus defined on distributions rather than on random variables themselves. Nevertheless, it is a common convention, that in the case of random variables X and Y having distributions P_X and P_Y, the stochastic order relation $P_X \succeq P_Y$ might be viewed as a relation on random variables $X \succeq Y$ [11]. It must be emphasized, however, that the dominance relation on random variables is no longer an order as it is not antisymmetric.

The second degree stochastic dominance relation is defined with the second performance function $F_2(X,r)$ given by areas below the cumulative distribution function itself, i.e.: $F_2(X,r) = \int_{-\infty}^{r} F_1(X,t)dt$ for $r \in R$. Similarly to FSD, we say that X weakly dominates Y under the SSD rules ($X \succeq_{SSD} Y$), if $F_2(X,r) \leq F_2(Y,r)$ for all $r \in R$, while X SSD dominates Y ($X \succ_{SSD} Y$), when at least one inequality is strict. Certainly, $X \succ_{FSD} Y$ implies $X \succ_{SSD} Y$. Function $F_2(X,r)$, used to define the SSD relation can also be presented as follows [12]: $F_2(X,r) = \mathbf{E}[\max\{r - X, 0\}]$, thus representing the mean below-target deviations from real targets.

If $X \succ_{SSD} Y$, then X is preferred to Y within all risk-averse preference models that prefer larger outcomes. In terms of the expected utility theory the SSD relation represent all the preferences modeled with increasing and concave utility functions. It is therefore a matter of primary importance that an approach to the comparison of random outcomes be consistent with the second degree stochastic dominance relation. Our paper focuses on the consistency of mean-risk approaches with the SSD.

Alternatively, the stochastic dominance order can be expressed on the inverse cumulative functions (quantile functions). Namely, for random variable X, one may consider the performance function $F_{-1}(X,p)$ defined as is the left-continuous inverse of the cumulative distribution function $F_1(X,r)$, i.e., $F_{-1}(X,p) = \inf \{\eta : F_1(X,\eta) \geq p\}$. Obviously, X dominates Y under the FSD rules ($X \succ_{FSD} Y$), if $F_{-1}(X,p) \geq F_{-1}(Y,p)$ for all $p \in [0,1]$, where at least one strict inequality holds. Further, the second quantile function (or the so-called *Absolute Lorenz Curve* ALC) is defined by integrating F_{-1}, which provides an alternative characterization of the SSD relation,

Mean-risk approaches are based on comparing two scalar characteristics (summary statistics), the first, denoted $\mu(X)$, represents the expected outcome (reward), and the second, denoted $\varrho(X)$, is some measure of risk. The original Markowitz portfolio optimization model [9] uses the variance or the standard deviation. Several other risk measures have been later considered thus creating the entire family of mean-risk models. Risk measures in Markowitz-type mean-risk models, similar to the standard deviation, are translation invariant and risk relevant deviation type measures (dispersion parameters). Thus, they are not affected by any shift of the outcome scale $\varrho(X + a) = \varrho(X)$ for any real number a and they are equal to 0 in the case of a risk-free portfolio while taking positive values for any risky portfolio. Unfortunately, such risk measures are not consistent with the stochastic dominance order [11]. Indeed, in the Markowitz model its efficient set may contain SSD inferior portfolios characterized by a small risk but also very low return [15]. Unfortunately, it is a common flaw of all Markowitz-type mean-risk models where risk is measured with some dispersion measures. In order to overcome this flaw of the Markowitz model, already Baumol [2] suggested to consider a performance measure, he called the expected gain-confidence limit criterion, $\mu(X) - \lambda\sigma(X)$ to be maximized instead of the minimization of $\sigma(X)$ itself. Similarly, Yitzhaki [18] considered maximization of the criterion $\mu(X)-\varrho(X)$ for the Gini's mean difference and he demonstrated its SSD consistency. Recently, similar consistency results have been introduced [12, 13] for measures corresponding to the standard semideviation and to the mean semideviation (half of the mean absolute deviation).

Hereafter, for any dispersion type risk measure $\varrho(X)$, the performance function $S(X) = \mu(X) - \varrho(X)$ will be referred to as the corresponding safety measure. Note that risk measures, we consider, are defined as translation invariant and risk relevant dispersion parameters. Hence, the corresponding safety measures are translation equivariant in the sense that any shift of the outcome scale results in an equivalent change of the safety measure value (with opposite sign as safety measures are maximized), or in other words, the safety measures distinguish (and order) various risk-free portfolios (outcomes) according to their values. The safety measures, we consider, are risk relevant but in the sense that the value of a safety measure for any risky portfolio is less than the value for the risk-free portfolio with the same expected returns. Moreover, when risk measure $\varrho(X)$ is a convex function of X, then the corresponding safety measure $S(X)$ is concave.

Relation of the SSD consistency of the safety measures directly involves criterion $\mu(X)-\varrho(X)$. However, the SSD dominance always implies the means inequality. Hence, in the case of $X \succeq_{SSD} Y$ we have both $\mu(X) \geq \mu(Y)$ and $\mu(X) - \varrho(X) \geq \mu(Y) - \varrho(Y)$. Thus, by combining inequalities, one may easily notice that $X \succ_{SSD} Y$ implies $\mu(X) - \lambda\varrho(X) \geq \mu(Y) - \lambda\varrho(Y)$ for all $0 \leq \lambda \leq 1$. On the other hand, one may just consider $\varrho_\beta(X) = \beta\varrho(X)$ as a basic risk measure, like the mean absolute semideviation equal to the half of the mean absolute deviation itself. In such a case one may gets another (possibly higher) upper bound for the trade-off coefficient guaranteeing the SSD consistency. Therefore, following [12], in this paper we say that the (deviation) risk measure is SSD α-safety consistent if *there exists a positive constant* α *such that for all* X *and* Y:

$$X \succeq_{SSD} Y \quad \Rightarrow \quad \mu(X) - \alpha\,\varrho(X) \geq \mu(Y) - \alpha\,\varrho(Y). \qquad (1)$$

For the sake of simplicity, the SSD 1-safety consistency of a risk measure we will usually call simply SSD safety consistency. The relation of SSD (safety) consistency is called *strong* if, in addition to (1), the following holds

$$X \succ_{SSD} Y \quad \Rightarrow \quad \mu(X) - \alpha\,\varrho(X) > \mu(Y) - \alpha\,\varrho(Y). \qquad (2)$$

An important advantage of mean-risk approaches is that having assumed a trade-off coefficient λ between the risk and the mean, one may directly compare real values of $\mu(X) - \lambda\varrho(X)$. If the risk measure $\varrho(X)$ is SSD α-safety consistent, then except for random variables with identical $\mu(X)$ and $\varrho(X)$, every random variable that is maximal by $\mu(X) - \lambda\varrho(X)$ with $0 < \lambda < \alpha$ is efficient under he SSD rules. In the case of strong SSD safety consistency, every such maximal random variable is, unconditionally, SSD efficient. Therefore, the strong SSD safety consistency is an important property of a risk measure.

The stochastic dominance partial orders are defined on distributions. The risk measures are commonly considered as functions of random variables. One may focus on a linear space of random variables $\mathcal{L} = L^k(\Omega, \mathcal{F}, \mathbf{P})$ with some $k \geq 1$ (assuming $k \geq 2$ whenever variance or any related measure is considered). Although defined for random variables, typical risk measures depend only on the corresponding distributions themselves and we focus on such measures. In other words, we assume that $\varrho(X) = \varrho(\hat{X})$ whenever random variables X and $\hat{X}$ have the same distribution, i.e. $F_1(X, r) = F_1(\hat{X}, r)$ for all $r \in R$ or equivalently $F_{-1}(X, p) = F_{-1}(\hat{X}, p)$ for all $p \in [0, 1]$.

Table 1. SSD consistency limits for general distributions

Risk Measure		Consistency	
Standard semideviation	$\bar{\sigma}(X)$	1 [12]	
Mean absolute semideviation	$\bar{\delta}(X)$	1 [12]	
Mean absolute deviation	$\delta(X)$	1/2 [12]	
Conditional β-semideviation	$\Delta_\beta(X)$	1 [14]	
Mean abs. dev. from median	$\Delta_{0.5}(X)$	1 [14]	
Maximum semideviation	$\Delta(X)$	1 [14]	
Gini's mean difference	$\Gamma(X)$	1 [18]	(strong [14])
Tail Gini's mean difference	$\Gamma_\beta(X)$	1 [14]	

Table 2. SSD consistency limits for symmetric distributions

Risk Measure		Consistency	
Standard semideviation	$\bar{\sigma}(X)$	$\sqrt{2}$ [12]	(strong)
Standard deviation	$\sigma(X)$	1 [12]	(strong)
Mean absolute semideviation	$\bar{\delta}(X)$	2 [12]	
Mean absolute deviation	$\delta(X)$	1 [12]	
Gini's mean difference	$\Gamma(X)$	2 [14]	(strong)

Within the class of arbitrary uncertain prospects allowing to consider stochastic dominance (the class of random variables with finite expectations $\mathbf{E}[|X|] < \infty$, or $\mathbf{E}[X^2] < \infty$ while for standard deviation), several consistency results have been shown, as summarized in Table 1 (where the maximum value of $alpha$ is presented). Obviously, any convex combination of measures preserves their SSD safety consistency which justifies several combined measures [8]. It turns out that when limiting the analysis to outcomes described with the symmetric distribution some consistency levels α increase and one gets additionally SSD 1-safety consistency of the standard deviation (see Table 2).

3. SSD consistency conditions

The risk measures we consider from the perspective of the stochastic dominance are defined as (real valued) functions of distributions rather than random variables themselves. Nevertheless, in many various applications it might be more convenient to analyze their properties as functions of random variables. Recently, a class of coherent risk measures [1] have been defined by means of several axioms. The axioms depicts the most important issues in the risk comparison for economic decisions. therefore, they have been quite commonly recognized as the standard requirements for risk measures. Let us consider a linear space of random variables $\mathcal{L} = L^k(\Omega, \mathcal{F}, \mathbf{P})$ with some $k \geq 1$ (recall, we assume $k \geq 2$ whenever variance or any related measure is considered). A real valued performance function $C : \mathcal{L} \to \mathbf{R}$ is called a coherent risk measure on

$\mathcal{L}$ if for any $X, Y \in \mathcal{L}$ it is monotonous ($X \geq Y$ implies $C(X) \leq C(Y)$), positively homogeneous ($C(hX) = hC(X)$ for real number $h > 0$), subadditive ($C(X+Y) \leq C(X)+C(Y)$), translation equivariant ($C(X+a) = C(X)-a$, for real number a), risk relevant ($X \leq 0$ and $X \neq 0$ implies $C(X) > 0$), where or inequalities on random variables are understood in terms 'a.s.'. If $\varrho(X) \geq 0$ is a convex, positively homogeneous and translation invariant (dispersion type) risk measure, then the performance function $C(X) = \varrho(X) - \mu(X)$ does satisfy the axioms of translation equivariance, positive homogeneity, and subadditivity. Further, if $X \geq Y$, then $X = Y + (X - Y)$ and $X - Y \geq 0$. Hence, the convexity together with the expectation boundedness

$$X \geq 0 \Rightarrow \varrho(X) \leq \mu(X) \tag{3}$$

of the risk measure imply that the performance function $C(X)$ satisfies also the axioms of monotonicity and relevance [8].

In order to derive similar conditions for the SSD consistency we will use the SSD separation results. Namely, the following result [11, Th. 1.5.14] allows us to split the SSD dominance into two simpler stochastic orders: the FSD dominance and the Rotschild-Stiglitz (RS) dominance (or concave stochastic order), where the latter is the SSD dominance restricted to the case of equal means.

THEOREM 1 *Let X and Y be random variables with $X \succeq_{SSD} Y$. Then there is a random variable Z such that*

$$X \succeq_{FSD} Z \succeq_{RS} Y$$

The above theorem allows us to separate two important properties of the SSD dominance and the corresponding requirements for the risk measures.

COROLLARY 2 *Let $\varrho(X) \geq 0$ be a (dispersion type) risk measure. The measure is SSD 1-safety consistent if and only if it satisfies both the following conditions:*

$$X \succeq_{FSD} Y \quad \Rightarrow \quad \mu(X) - \varrho(X) \geq \mu(Y) - \varrho(Y), \tag{4}$$

$$X \succeq_{RS} Y \quad \Rightarrow \quad \varrho(X) \leq \varrho(Y). \tag{5}$$

Proof. If $X \succeq_{SSD} Y$, then according to separation theorem $X \succeq_{FSD} Z \succeq_{RS} Y$ where $\mathbf{E}[Z] = \mathbf{E}[Y]$. Hence, applying (4) and (5) one gets

$$X \succeq_{SSD} Y \quad \Rightarrow \quad \mu(X) - \varrho(X) \geq \mu(Z) - \varrho(Z) \geq \mu(Y) - \varrho(Y).$$

On the other hand, both the requirements are obviously necessary. □

For strict relation $X \succ_{SSD} Y$, the separating Z satisfies $X \succ_{FSD} Z$ or $Z \succ_{RS} Y$. Hence, the corresponding strong forms of both (4) and (5) are necessary and sufficient for the strong SSD 1-safety consistency.

Condition (4) represents the stochastic monotonicity and it may be replaced with more standard monotonicity requirement

$$X \geq Y \quad \Rightarrow \quad \mu(X) - \varrho(X) \geq \mu(Y) - \varrho(Y), \tag{6}$$

where the inequality $X \geq Y$ is to be viewed in the sense o holding almost surely (a.s.). Essentially, $X \geq Y$ implies $X \succeq_{FSD} Y$, but not opposite. However, the relation $X \succeq_{FSD} Y$ is equivalent [11, Th. 1.2.4] to the existence of a probability space and random variables $\hat{X}$ and $\hat{Y}$ on it with the distribution functions the same as X and Y, respectively, such that $\hat{X} \geq \hat{Y}$. Hence, for risk measures depending only on distributions, we consider, one gets requirements (4) and (6) equivalent. Note that for any $X \geq 0$ and $a \in R$ one gets $X + a \geq a$ while $\varrho(X + a) = \varrho(X)$ and, therefore, the monotonicity (6) implies $\varrho(X) \leq \mu(X)$. This justifies the expectation boundedness (3) as a necessary for monotonicity (6) or (4).

Condition (5) represents the required convexity properties to model diversification advantages. Note that the second cumulative distribution functions $F_2(X, r)$ are convex with respect to random variables X [13]. Hence, taking two random variables Y' and Y'' both with the same distribution as X one gets $F_2(\alpha Y' + (1 - \alpha)Y'', r) \leq F_2(X, r)$ for any $0 \leq \alpha \leq 1$ and any $r \in \mathbf{R}$. Thus, $\alpha Y' + (1 - \alpha)Y'' \succeq_{RS} X$ and convexity of $\varrho(X)$ is necessary to meet the requirement (5).

The concept of separation risk measures properties following Theorem 1 is applicable while considering general (arbitrary) distributions. It may be, however, adjusted to some specific classes of distribution. In particular, we will show that it remains valid for a class of symmetric distributions. Indeed, a more subtle construction

$$F_1(Z, r) = \begin{cases} F_1(X, r), & r < \mu(Y) \\ F_1(X, r + 2(\mu(X) - \mu(Y))), & r \geq \mu(Y) \end{cases} \tag{7}$$

preserves symmetry of the distribution thus leading us to the following assertion. For any symmetric random variables with $X \succeq_{SSD} Y$, there is a symmetric random variable Z such that $X \succeq_{FSD} Z \succeq_{RS} Y$. One may also notice that the 'a.s.' characteristic of the FSD relation may be, respectively, enhanced for symmetric distributions.

It follows from the majorization theory [6, 10] that in the case of simple lotteries constructed as random variables corresponding n-dimensional real vectors (probability $1/n$ is assigned to each coordinate if they are different, while probability k/n is assigned to the value of k coinciding coordinates) a convex, positively homogeneous and translation invariant (dispersion type) risk measure is SSD 1-safety consistent if and only if it is additionally expectation bounded (3). We will demonstrate this for more general space of lotteries. Hereafter, a *lottery* is a discrete random variable with a finite number of steps.

LEMMA 3 *Lotteries X and Y satisfies $X \succeq_{RS} Y$ if and only if $F_{-2}(X,p) \geq F_{-2}(Y,p)$ for all p - cumulative probability of a step of $F_1(X,\alpha)$ or $F_1(Y,\alpha)$ and $F_{-2}(X,1) = F_{-2}(Y,1)$.*

Proof. From quantile characterization of SSD we have inequality for all $p \in (0,1)$. On the other hand, we can see that p - cumulative probability of steps are sufficient. Let $p_1, p_2, \ldots, p_m$ - cumulative probability of the steps of $F_1(X,\alpha)$ or $F_1(Y,\alpha)$, and let $c \in (p_i, p_{i+1})$. Then, $F_{-2}(X,c) = F_{-2}(X,p_i)+(c-p_i)F_{-1}(X,p_{i+1}) = F_{-2}(X,p_{i+1})-(p_{i+1}-c)F_{-1}(X,p_{i+1})$. Hence, $F_{-2}(X,c) \geq F_{-2}(Y,c)$ whenever $F_{-2}(X,p_i) \geq F_{-2}(Y,p_i)$ and $F_{-2}(X,p_{i+1}) \geq F_{-2}(Y,p_{i+1})$. Furthermore, $F_{-2}(X,1) = \mathrm{E}[X] = \mathrm{E}[Y] = F_{-2}(Y,1)$ is necessary for RS-dominance. □

Let $X = Y_n \succ_{RS} Y_{n-1} \succ_{RS} \ldots \succ_{RS} Y_1 = Y$ and for all k: $Y_k = \lambda_{k-1}Y'_{k-1} + (1-\lambda_{k-1})Y''_{k-1}$, where $\lambda_{k-1} \in (0,1)$ and $Y'_{k-1} \neq Y''_{k-1}$ are the same distributed as Y_{k-1}, then it is obvious to say that Y is more risky that X for all ρ - convex risk measures. Rothschild and Stiglitz have formulated that RS-dominance between two random variables is equivalent to existing a sequence of mean preserving spreads (MPS) that transform one variable to the other. Two variables X and Y differ by MPS if there exists some interval that the distribution of X one gets from the distribution of Y by removing some of the mass from inside the interval and moving it to some place outside this interval. Gaining by MPS we will show that for lotteries with rational probability X and Y if only $X \succeq_{RS} Y$ then Y is always more risky than X for all convex risk measures.

THEOREM 4 **Proof.** *Let X, Y - lotteries with rational probability of steps. If $X \succ_{RS} Y$ then there exists a sequence of lotteries $Y_1, Y_2, \ldots, Y_n$ satisfying the following conditions:*

1. $X = Y_n \succ_{RS} Y_{n-1} \succ_{RS} \ldots \succ_{RS} Y_1 = Y$
2. $Y^{i+1} = (1-\lambda^i)Y_1^i + \lambda^i Y_2^i$, *for* $i = 1, \ldots, n-1$, *where* $0 < \lambda^i < 1$ *and* $Y_1^i \neq Y_2^i$ *are identically distributed as* Y^i.

We will construct a sequence of MPS – $Y_1, Y_2, \ldots, Y_n$ using quantile characterization of RS-dominance. From Rothschild and Stiglitz theorem [7] one knows that the sequence $(Y_k)_{k=1,\ldots,n}$ exists. We will build it, however, as a convex combination of two identically distributed random variables.

Let $c_1, c_2, \ldots, c_m$ – cumulative probability of steps of $F_1(X,\alpha)$ or $F_1(Y,\alpha)$. Due to Lemma 3, we may focus on the steps of distributions.
Let $p_i = c_i - c_{i-1}$, for $i = 2, \ldots, m$ and $p_1 = c_1$, $\vec{p} = (p_1, \ldots, p_m)$,
x_i – c_i-quantile of X for $i = 1, \ldots, m$, $\vec{X} = (x_1, \ldots, x_m)$,
y_i – c_i-quantile of Y for $i = 1, \ldots, m$, $\vec{Y} = (y_1, \ldots, y_m)$.
There exists the first index i such that $x_i \neq y_i$ (actually $x_i > y_i$, due to the

dominance) as well as there exists the last index i for which $x_i \neq y_i$ (actually $x_i < y_i$, due to the equality: $F_{-2}(X,1) = \mathbf{E}[X] = \mathbf{E}[Y] = F_{-2}(Y,1)$). With no loss of generality we can assume that the first index is 1 and the last one is m. By definition, we get $F_{-2}(Y, c_i) = \sum_{j=1}^{i} p_j y_j$.
Define: $\Delta_j^i := x_j - y_j^i \; \forall_{i=1,\dots,n,\, j=1,\dots,m}$,

First step. $\Delta_1^1 > 0$ and $\Delta_k^1 < 0$, where $k = \min\{i : \Delta_i^1 < 0\}$, let $\Delta^1 = \min\{\Delta_1^1, -\Delta_k^1\}$ and $p^1 = \min\{p_1, p_k\}$

$$\begin{array}{llllllll}
\vec{p}^{\,1}: & p_1, & & p_2 & \dots & p_k, & & \dots & p_m \\
\vec{Y}^{\,1}: & y_1, & & y_2 & \dots & y_k, & & \dots & y_m \\
\vec{X}^{\,1}: & x_1, & & x_2 & \dots & x_k, & & \dots & x_m
\end{array}$$

$$\begin{array}{llllllll}
\vec{p}^{\,2}: & p^1, & p_1 - p^1, & p_2 & \dots & p^1, & p_k - p^1 & \dots & p_m \\
\vec{Y}^{\,2}: & y_1, & y_1, & y_2 & \dots & y_k, & y_k & \dots & y_m \\
\vec{X}^{\,2}: & x_1, & x_1, & x_2 & \dots & x_k, & x_k & \dots & x_m
\end{array}$$

Note that $\vec{p}^{\,2}$ has at least one coordinate equal 0: $p_1 - p^1 = 0$ or $p_k - p^1 = 0$ while $\vec{Y}^{\,2}$ and $\vec{X}^{\,2}$ have at least one new the same coordinate: $x_1 = y_1^1 + \Delta^1$ or $x_k = y_k^1 - \Delta^1$.

With a finite number of steps we can transform y_1^1 to x_1 or y_k^1 to x_k. Thus, m coordinates of Y can be transformed with a finite number of steps to m coordinates of X.

The i-th step has the same idea: we choose first index where Δ_j^i is positive and it can be treated as Δ_1^1 in first step, because for all indexes before $\Delta_j^i = 0$. Then, we choose first index when Δ_j^i is negative as Δ_k^1 in first step. Δ^i, p^i are formed in the same way as Δ^1, p^1.

$\vec{Y}^{\,i}$ is built from $\vec{Y}^{\,i-1}$ by moving the same mass – Δ^i from one coordinate to another with the same probability – p^i. Let j, k be these coordinates, the rest of them are the same in the vectors.

$$\begin{array}{llllll}
\vec{p}^{\,i}: & \dots & p^i, & \dots & p^i, & \dots \\
\vec{Y}^{\,i}: & \dots & y_j^i, & \dots & y_k^i, & \dots \\
\vec{Y}'^{\,i}: & \dots & y_k^i, & \dots & y_j^i, & \dots \\
\vec{Y}^{\,i+1}: & \dots & y_j^i + \Delta^i, & \dots & y_k^i - \Delta^i, & \dots
\end{array}$$

$\vec{Y}_1^{\,i}$ and $\vec{Y}^{\,i}$ are the same distributed and $(1 - \lambda^i)\vec{Y}_1^{\,i} + \lambda^i \vec{Y}^{\,i} = \vec{Y}^{\,i+1}$ where

$\lambda^i = \Delta^i / (y_k^i - y_j^i)$ □

$X, Y', Y'' \in (\Omega, \mathcal{F}, \mathbf{P})$; Y', Y'' – the same distributed lotteries with rational probability of steps. If $X = \lambda Y' + (1 - \lambda)Y''$, then for all convex positive functions ϱ (where $\varrho(Y') < \infty$), one gets $\varrho(X) = \varrho(\lambda Y' + (1 - \lambda)Y'') \leq \lambda\varrho(Y') + (1 - \lambda)\varrho(Y'')$. Moreover, if ϱ depends only on distributions, then $\varrho(Y') = \varrho(Y'')$. Hence, $X \succeq_{RS} Y'$ implies $\varrho(X) \leq \varrho(Y')$, and $X \succ_{RS} Y'$ implies $\varrho(X) < \varrho(Y')$ if ϱ is strictly convex on identically distributed random variables. Recall that expectation boundedness together with convexity guarantee the corresponding monotonicity with strict monotonicity properties for strictly expectation bounded risk measures. This leads to the following assertion.

THEOREM 5 *Let us consider a linear space $\mathcal{L} \subset L^k(\Omega, \mathcal{F}, \mathbf{P})$ of lotteries with rational probability of steps. If risk measure $\varrho(X) \geq 0$ depending only on distributions is convex, positively homogeneous, translation invariant and expectation bounded, then the measure is SSD 1-safety consistent on $\mathcal{L}$. If $\varrho(X)$ is also strictly convex on identically distributed random variables and strictly expectation bounded (on risky r.v.), then it is strongly SSD 1-safety consistent on $\mathcal{L}$.*

Note that Theorem 5 applies to the important class of distributions where one may take advantages of the LP computable risk measures [8]. It justifies then the sufficient conditions for the coherency as simultaneously sufficient for SSD (safety) consistency. The basic consistency results could be also derived for continuous distributions from the relation [3]

$$X \succeq_{RS} Y \quad \Leftrightarrow \quad X \in \overline{\mathrm{conv}}\{\hat{Y} : F_{\hat{Y}} = F_Y\}.$$

However, the strong consistency results cannot be achieved in this way.

4. Concluding remarks

One may specify risk dependent performance functions to transform several risk measures into SSD consistent and coherent safety measures. We have introduced convexity and expectation boundedness as necessary and sufficient conditions which allow us to justify various risk measures with respect to such coherent transformation. While focusing on the space of finite lotteries, where one may take advantages of the LP computable risk measures, it turns out that these sufficient conditions for the coherency are also sufficient for SSD consistency. Moreover, when enhanced to strict convexity (on identically distributed random variables) and strict expectation boundedness (on risky random variables) they are also sufficient for strong SSD consistency. The latter is crucial to guarantee the SSD efficiency of the corresponding safety maximization optimal solutions.

References

[1] Artzner, P., Delbaen, F., Eber, J.-M., Heath, D. (1999), Coherent measures of risk. *Math. Finance*, 9, 203-228.

[2] Baumol, W.J. (1964), An expected gain-confidence limit criterion for portfolio selection. *Manag. Sci.*, 10, 174-182.

[3] Dentcheva, D., Ruszczyński, A. (2004), Convexification of stochastic ordering constraints, *Compt. Res. l'Acad. Bulgare Sci.*, 57, 11-16.

[4] Fishburn, P.C. (1964), *Decision and Value Theory*, Wiley, New York.

[5] Hanoch, G., Levy, H. (1969), The efficiency analysis of choices involving risk, *Revue Econ. Studies*, 36, 335-346.

[6] Hardy, G.H., Littlewood, J.E., Polya, G. (1934), *Inequalities*, Cambridge University Press, Cambridge, MA.

[7] Leshno, M., Levy, H., Spector, Y. (1997), A Comment on Rothschild and Stiglitz's "Increasing risk: I. A definition", *J. Econ. Theory*, 77, 223-228.

[8] Mansini, R., Ogryczak, W., Speranza, M.G. (2003), LP solvable models for portfolio optimization: A classification and computational comparison, *IMA J. Manag. Math.*, 14, 187-220.

[9] Markowitz, H.M. (1952), Portfolio selection, *J. Finance*, 7, 77-91.

[10] Marshall, A.W., Olkin, I. (1979), *Inequalities: Theory of Majorization and Its Applications*, Academic Press, NY, 1979.

[11] Müller, A., Stoyan, D. (2002), *Comparison Methods for Stochastic Models and Risks*, Wiley, New York.

[12] Ogryczak, W., Ruszczyński, A. (1999), From stochastic dominance to mean-risk models: Semideviations as risk measures, *European J. Opnl. Res.*, 116, 33-50.

[13] Ogryczak, W., Ruszczyński, A. (2001), On stochastic dominance and mean-semideviation models. *Mathematical Programming*, 89, 217-232.

[14] Ogryczak, W., Ruszczyński, A. (2002), Dual stochastic dominance and related mean-risk models, *SIAM J. Optimization*, 13, 60-78.

[15] Porter, R.B. (1974), Semivariance and stochastic dominance: A comparison, *American Econ. Review*, 64, 200-204.

[16] Rockafellar, R.T., Uryasev, S. (2000), Optimization of conditional value-at-risk. *J. Risk*, 2, 21-41.

[17] Rothschild, M., Stiglitz, J.E. (1970), Increasing risk: I. A definition, *J. Econ. Theory*, 2, 225-243.

[18] Yitzhaki, S. (1982), Stochastic dominance, mean variance, and Gini's mean difference, *American Econ. Review*, 72, 178-185.

OPTIMAL POLICIES UNDER DIFFERENT PRICING STRATEGIES IN A PRODUCTION SYSTEM WITH MARKOV-MODULATED DEMAND

E. L. Örmeci,[1] J. P. Gayon,[2] I. Talay-Değirmenci,[3] and F. Karaesmen[1]
[1]*Koç University, İstanbul, TURKEY, lormeci@ku.edu.tr* [2] *INPG, Grenoble, FRANCE,* [3] *Duke University, Durham, USA*

Abstract We study the effects of different pricing strategies available to a continuous review inventory system with capacitated supply, which operates in a fluctuating environment. The system has a single server with exponential processing time. The inventory holding cost is nondecreasing and convex in the inventory level, the production cost is linear with no set-up cost. The potential customer demand is generated by a Markov-Modulated (environment-dependent) Poisson process, while the actual demand rate depends on the offerred price. For such systems, there are three possible pricing strategies: Static pricing, where only one price is used at all times, environment-dependent pricing, where the price changes with the environment, and dynamic pricing, where price depends on both the current environment and the stock level. The objective is to find an optimal replenishment policy under each of these strategies. This paper presents some structural properties of optimal replenishment policies, and a numerical study which compares the performances of these three pricing strategies.

Keywords: Inventory control, pricing, Markov Decision processes

1. Introduction

During the last few decades, it is realized that the joint optimization of pricing and replenishment decisions results in significant improvements on the firm's profit (see e.g., [3]). The inspiring results obtained on this topic so far encouraged us to analyse an inventory pricing and replenishment problem. On the other hand, the environmental factors affect the density of the demand distribution unpredictably, and the focus in the recent studies of inventory control has been shifting to model the impact of fluctuating demand on the optimal replenishment policy. Hence, we consider an inventory system operating in a fluctuating demand environment, which controls the prices as well as the replenishment. As a result, our work stands at the junction of three main-stream

Please use the following format when citing this chapter:
Author(s) [insert Last name, First-name initial(s)], 2006, in IFIP International Federation for Information Processing, Volume 199, System Modeling and Optimization, eds. Ceragioli F., Dontchev A., Furuta H., Marti K., Pandolfi L., (Boston: Springer), pp. [insert page numbers].

research topics, inventory control, price control and the effects of environmental changes on the control policies.

We study a continuous review, infinite horizon inventory pricing and replenishment problem with capacitated supply. The system has a single server with exponential processing time. There is no set up cost, and the production cost is linear. The inventory holding cost, on the other hand, is nondecreasing and convex in the inventory level. In order to model a fluctuating environment, we assume that the potential customer demand is generated by a Markov-Modulated (environment-dependent) Poisson process. Moreover, the actual demand depends on the price offered at the time of the transaction, such that the actual demand rate decreases as the price increases. For a system operating in this environment, there are three possible pricing strategies: Static pricing, where only one price is used at all times, environment-dependent pricing, where the price is allowed to change with the environment, and dynamic pricing, where price depends on both the current environment and the stock level. In this paper, we use a Markov Decision Process framework to model this system as a make-to-stock queue operating under each of these strategies. Using this framework, we show that optimal replenishment policies are of environment-dependent base-stock level policies for these pricing strategies. We also compare the performances of these three strategies by an extensive numerical study.

The objective of inventory management is to reduce the losses caused by the mismatches that arise between supply and demand processes. With the advances in computers and communication technology, the role of inventory management has changed from cost control to value creation. Therefore, the issues inventory management studies now include both the traditional decisions such as inventory replenishment and the strategic decisions made by the firm such as pricing. In fact, there has been an increasing amount of research on pricing with inventory/production considerations, see the excellent review papers [4], [9], and [1].

The widely known results in inventory control model the randomness of demand by using a random component with a well-known density in the definition of the demand process. However, the focus in the recent studies of inventory control has been shifting to model the impact of fluctuating demand on the optimal replenishment policy (see [7] and [2] among others). In particular, changes in the demand distribution might be caused by economic factors such as interest rates, or they might be caused by the changes in business environment conditions such as progress in the product-life-cycle or the consequences of rivals' actions on the market. The model we present below considers the effect of external factors on the demand distribution.

This paper is organized as follows: In the next section we introduce the models for the pricing strategies described above. Section 3 will present structural results for an optimal replenishment policy for each of the pricing strategies.

In section 4, we will present our numerical results, which compare the performances of the three policies and provide insights, and point out possible directions of future research.

2. Model formulation

In this section we present a make-to-stock production system with three different pricing strategies: (1) the static pricing problem where a unique price has to be chosen for the whole time horizon regardless of the environment and the inventory level, (2) the environment-dependent pricing where the price can be changed over time depending on the environment, but not on the inventory level (3) the dynamic pricing where the price can be changed over time depending on both the inventory level and the environment. The production system should also decide on the replenishment of the items.

Consider a supplier who produces a single part at a single facility. The processing time is exponentially distributed with mean $1/\mu$ and the completed items are placed in a finished goods inventory. The unit variable production cost is c and the stock level is $X(t)$ at time t, where $X(t) \in \mathbb{N} = \{0, 1, ...\}$. We denote by h the induced inventory holding cost per unit time and h is assumed to be a convex function of the stock level.

The environment state evolves according to a continuous-time Markov Chain with state space $E = \{1, \cdots, n\}$ and transition rates q_{ej} from state e to state $j \neq e$. We assume that this Markov chain is recurrent to avoid technicalities. For all environment states, the set of allowable prices $\mathcal{P}$ is identical. The customers arrive according to a Markov Modulated Poisson process (MMPP) with rate Λ_e when the state of the exogenous environment is e. We assume that the potential demand rates are bounded, i.e., $\max\{\Lambda_e\} < \infty$; a reasonable assumption which will be necessary to uniformize the Markov decision process. The customers decide to buy an item according to the posted price p, so that the actual demand rate in environment e is $\lambda_e(p)$ when a price of p is offerred. Obviously, the actual demand rate is bounded by the potential demand rate so that $\lambda_e(p) \leq \Lambda_e$ for all e and for all p. We note that the domain of the prices, $\mathcal{P}$, may be either discrete or continuous. When $\mathcal{P}$ is continuous, it is assumed to be a compact subset of the set of non-negative real numbers $\mathbb{R}^+$.

For a fixed environment state e, we impose several mild assumptions on the demand function. First, we assume that $\lambda_e(p)$ is decreasing in p and we denote by $p_e(\lambda)$ its inverse. One can then alternatively view the rate λ as the decision variable, which is more convenient to work with from an analytical perspective. Thus the set of allowable demand rates is $\mathcal{L}_e = \lambda_e(\mathcal{P})$ in environment state e. Second, the revenue rate $r_e(\lambda) = \lambda p_e(\lambda)$ is bounded. Finally we assume that p_e is a continuous function of λ when the set of prices $\mathcal{P}$ is continuous.

At any time, the decision maker has to decide whether to produce or not. The decision maker may also choose a price $p \in \mathcal{P}$, or equivalently a demand rate $\lambda \in \mathcal{L}_e$ at certain times specified by the pricing strategies described above. If we are in search of optimal replenishment policies for the pricing strategies described above, then the optimal policy is known to belong to the class of stationary Markovian policies, see [8]. Therefore the current state of the system is exhaustively described by the state variable (x, e) with x the stock level and e the environment state and (x, e) belongs to the state space $\mathbb{N} \times E$. Then, for dynamic pricing strategy $p(x, e)$ is the price of the item when the system operates in environment e with x units of item on inventory, for environment-dependent pricing policy $p(e)$ is the price chosen a priori for environment e so that $p(e)$ is charged whenever the system enters environment e regardless of the current inventory level, and p_s is the static price to be always offerred regardless of the environment and the inventory level.

2.1 Optimal static pricing strategy

In static pricing, the decision maker has to choose a unique price in $\mathcal{P}$ for the whole horizon. The static pricing problem can be viewed in two steps. First, we determine the optimal production policy, which depends on both the environment and inventory level, for a given static price, p. Hence, let $v_s^\pi(x, e; p)$ be the expected total discounted reward when the replenishment control policy π is followed with $p_s = p$ over an infinite horizon starting from the state (x, e). If we denote by α the discount rate, by $N(t)$ the number of demands accepted up to time t, and by $W(t)$ the number of items produced up to time t when the posted price is always p and the replenishment policy π is followed, then:

$$\begin{aligned} v_s^\pi(x, e; p) &= E_{x,e}^\pi \left[\int_0^{+\infty} e^{-\alpha t} p \, dN(t) \right. \\ &\quad \left. - \int_0^{+\infty} e^{-\alpha t} h(X(t)) \, dt - \int_0^{+\infty} e^{-\alpha t} c \, dW(t) \right], \end{aligned}$$

where $X(t)$ is the inventory level at time t, as defined previously. We seek to find the policy π^* which maximizes $v_s^\pi(x, e; p)$ for a given price p. Let v_s^* be the optimal value function associated to π^*, so that:

$$v_s^*(x, e; p) = \max_\pi \{v_s^\pi(x, e; p)\}.$$

Now we can formulate this problem as a Markov Decision Process (MDP): Without loss of generality, we can rescale the time by taking $\mu + \sum \Lambda_e + \sum_e \sum_{j \neq e} q_{ej} + \alpha = 1$. After a uniformization, v_s^* satisfies the following optimality equations:

$$v_s^*(x, e; p) \;=\; -h(x) + \mu T_0 v_s^*(x, e; p) + \lambda_e(p) v_s^*(x, e; p) + \sum_{j \neq e} q_{ej} v_s^*(x, j; p)$$

$$+(\sum_i \Lambda_i - \lambda_e(p) + \sum_{i \neq e} \sum_{j \neq i} q_{ij}) v_s^*(x, e; p),$$

where the operator T_0 for any function $f(x, e)$ is defined as

$$T_0 f(x, e) = \max\{f(x, e), f(x+1, e) - c\}. \tag{1}$$

Hence, the operator T_0 corresponds to the production decision. We define $a_s(x, e)$ as the optimal replenishment decision in state (x, e) such that $a_s(x, e) = 1$ if it is optimal to produce the item, and $a_s(x, e) = 0$ otherwise. We also define also the operator T_s such that $v_s^* = T_s v_s^*$. Therefore, whenever a price p is given, we can find an optimal replenishment policy by solving an MDP.

The second step is to find the optimal price p_s^* in the set of prices $\mathcal{P}$, where there might exist potentially several optimal prices. Since we assume that the exogeneous environment state follows a recurrent Markov chain, we choose the price p_s^* such that $p_s^* = \text{argmax}\{v_s^*(0, 1; p) : p\}$ without loss of generality.

2.2 Optimal environment-dependent pricing strategy

The problem of environment-dependent pricing strategy is similar to the static pricing as it is also solved in two steps.

In the first step, the optimal production policy, π^*, is identified for a given set of prices $\bar{p}_{ed} = (p(1), ..., p(N))$. Let v_{ed}^* be the optimal value function associated to π^*. Then:

$$v_{ed}^*(x, e; \bar{p}_{ed}) = \max_{\pi} \left\{ E_{x,e}^{\pi} \left[\int_0^{+\infty} e^{-\alpha t} p(E(t))\, dN(t) - \int_0^{+\infty} e^{-\alpha t} h(X(t))\, dt - \int_0^{+\infty} e^{-\alpha t} c\, dW(t), \right] \right\}$$

where $E(t)$ is the state of the exogeneous environment at time t, $p(E(t))$ is the posted price when the current environment is $E(t)$, and α, $X(t)$, $N(t)$ and $W(t)$ are defined as above. Optimal replenishment policy π^* can be determined by using uniformization as in the static pricing problem. Hence:

$$\begin{aligned} v_{ed}^*(x, e; \bar{p}_{ed}) = & -h(x) + \mu T_0 v_{ed}^*(x, e; \bar{p}_{ed}) \\ & + \lambda_e(p) v_{ed}^*(x, e; \bar{p}_{ed}) + \sum_{j \neq e} q_{ej} v_{ed}^*(x, j; \bar{p}_{ed}) \\ & + (\sum_i \Lambda_i - \lambda_e(p) + \sum_{i \neq e} \sum_{j \neq i} q_{ij}) v_{ed}^*(x, e; \bar{p}_{ed}), \end{aligned} \tag{2}$$

where the operator T_0 is defined as in (1). Now $a_{ed}(x, e)$ is the optimal replenishment decision in state (x, e), so that $a_{ed}(x, e) = 1$ if it is optimal to produce the item, and $a_{ed}(x,e) = 0$ otherwise. We also define the operator T_{ed} such that $v^*_{ed} = T_{ed}v^*_{ed}$.

In the second step, an optimal price vector $\bar{p}^*_{ed} = (p^*(1), ..., p^*(n))$ is chosen such that $\bar{p}^*_{ed} = \text{argmax}\{v^*_{ed}(0,1;\bar{p}_{ed}) : \bar{p}_{ed}\}$, without loss of generality due to the recurrent Markov chain governing the environment process.

2.3 Optimal dynamic pricing strategy

The system with dynamic pricing is an extension of Li (1988), who analyzes the same system operating in a stationary enviornment, to the one operating in a fluctuating environment. This problem is different from the static and environment-dependent pricing in the following way: Since both optimal replenishment and optimal pricing policies depend on the current inventory level as well as the environment, both policies are determined as a result of an MDP. We let $v^*_d(x,e)$ be the maximal expected total discounted reward when an optimal dynamic control policy π^*, which controls both the replenishment decisions and prices, is followed over an infinite-horizon with initial state (x,e). Then we have:

$$\begin{aligned} v^*_d(x,e) &= \max_{\pi}\left\{E^{\pi}_{x,e}\left[\int_0^{+\infty} e^{-\alpha t}p(X(t),E(t))\,dN(t)\right.\right. \\ &\qquad \left.\left. -\int_0^{+\infty} e^{-\alpha t}h(X(t))\,dt - \int_0^{+\infty} e^{-\alpha t}c\,dW(t)\right]\right\}, \end{aligned}$$

where $E(t)$, α, $X(t)$, $N(t)$ and $W(t)$ are defined as above. We can still use uniformization, so that v^*_d should satisfy the following optimality equations:

$$\begin{aligned} v^*_d(x,e) &= -h(x) + \mu T_0 v^*_d(x,e) + T_e v^*_d(x,e) \\ &\quad + \sum_{j\neq e} q_{ej}v^*_d(x,j) + \left(\sum_{i\neq e}\Lambda_i + \sum_{i\neq e}\sum_{j\neq i} q_{ij}\right)v^*_d(x,e), \end{aligned}$$

where the operator T_0 is defined as in (1), and T_e is given by:

$$T_e v_d(x,e) = \max_{\lambda\in\mathcal{L}_e} g_{x,e}(\lambda),$$

and the function $g_{x,e}$ is defined for any λ in $\mathcal{L}_e$ by:

$$g_{x,e}(\lambda) = \begin{cases} r_e(\lambda) + \lambda v_d(x-1,e) + (\Lambda_e - \lambda)v_d(x,e) & : \ if\ x > 0 \\ \Lambda_e v_d(x,e) & : \ if\ x = 0. \end{cases}$$

Therefore, the operator T_e corresponds to the arrival rate decision, or equivalently the price decision in environment e. Optimal replenishment decision in state (x,e) is denoted by $a_d(x,e)$, where $a_d(x,e) = 1$ if it is optimal to produce the item, and $a_d(x,e) = 0$ otherwise. Finally, we define the operator T_d such that $v^*_d = T_d v^*_d$.

2.4 Discussion on different pricing strategies

Before describing our results, we want to discuss the advantages and disadvantages of these three pricing strategies. Obviously, optimal dynamic pricing policies always generate more profit than optimal environment-dependent policies, which in turn generate more than optimal static policies. Now we turn to the "qualitative" effects of these policies: Static pricing represents the traditional pricing since the price remains fixed over time, regardless of the changes in the environment and in the stock level. This type of policies is easy to implement. In addition, consumers may prefer the transparency of a known price that is not subject to any changes. At the other extreme, we have dynamic pricing that leads to frequent price changes, since even a change in the stock level may trigger a change in price. Therefore, dynamic pricing may create negative consumer reactions. Moreover, its implementation requires sophisticated information systems that can accurately track sales and inventory data in real time, and can be extremely difficult especially if price changes require a physical operation such as a label change. Environment-depending pricing, on the other hand, allows the price to change only with the environmental state. Hence, the associated system changes the prices, but not as frequently as the one with the dynamic pricing does. As a result, this policy is in between static and dynamic policies regarding to the practical problems and difficulties they bring.

3. Structural results

The MDP formulations of the replenishment problems given in Section 2 provide not only a tool to numerically solve the corresponding problem but also an effective methodology to establish certain structural properties of optimal policies. In particular, we will use these formulations to prove that there exists an optimal environment-dependent base-stock policy under each of the pricing strategies. We first present the definition of an environment-dependent base-stock policy:

DEFINITION 1 *A replenishment policy which operates in a fluctuating demand environment, as described in Section 2, is an environment-dependent base-stock policy, if it always produces the item in environment e whenever the current inventory level is below a fixed number* $S(e)$*, i.e.,* $x < S(e)$*, and it never produces in environment e whenever* $x \geq S(e)$*, where the numbers* $\{S(1), ..., S(N)\}$ *are called the base stock levels with* $S(e) \in \mathbb{N}$*.*

Now we argue that each of the pricing strategies yields to an optimal environment-dependent base-stock policy, if the corresponding value function is concave. Hence assume that $v_\pi^*(x, e)$ is concave with respect to x for each environment e, i.e.:

$$v_\pi^*(x+1,e) - v_\pi^*(x,e) \leq v_\pi^*(x,e) - v_\pi^*(x-1,e).$$

If it is optimal to replenish in a state (x,e), from equation (1) we have:

$$v_\pi^*(x,e) \leq v_\pi^*(x+1,e) - c, \iff c \leq v_\pi^*(x+1,e) - v_\pi^*(x,e).$$

Then, by concavity, we have:

$$c \leq v_\pi^*(x+1,e) - v_\pi^*(x,e) \leq v_\pi^*(x,e) - v_\pi^*(x-1,e),$$

implying that it has to be optimal to replenish in state $(x-1,e)$ as well. Therefore, whenever an optimal policy replenishes in a state (x,e), it replenishes in all states (k,e) with $k \leq x$. We can, similarly, show that if an optimal policy does not replenish in a state (x,e), it continues not to replenish in all states (k,e) with $k \geq x$. These two statements together imply the existence of an optimal base-stock level in each environment e, $S_\pi^*(e)$:

$$S_\pi^*(e) = \min\{x : a_\pi(x,e) = 0\},$$

where $a_\pi(x,e)$ is the optimal replenishment decision in state (x,e) with policy π. Now we show that the corresponding value functions are concave for all pricing strategies we describe above:

LEMMA 2 *For a fixed environment e, for all $\pi = s, ed, d$: If $v_\pi^*(x,e)$ is concave with respect to x, then $T_\pi v_\pi^*$ is also concave with respect to x.*

Proof. $\pi = s$ is a special case of $\pi = ed$ if we set $p(e) = p$ for all e, and we refer to [5] for the proof of $\pi = d$. Hence, we show the statement for $\pi = ed$. In this proof we denote $v_{ed}^*(x,e;\bar{p}_e)$ by $v_{ed}^*(x,e)$. Assume that v_{ed}^* is concave in x for each environment e.

Now we consider each term in equation (2) separately. By assumption $-h$ is concave. To prove that T_0 preserves concavity, we need to show:

$$\delta = T_0 v_{ed}^*(x+1,e) - 2T_0 v_{ed}^*(x,e) + T_0 v_{ed}^*(x-1,e) \leq 0$$

Now let $a' = a_{ed}(x+1,e)$ and $a'' = a_{ed}(x-1,e)$. By our observation above, there exists an optimal environment-dependent base-stock policy, so that $a' \leq a''$. Since $v_{ed}^*(x+a',e) \leq T_0 v_{ed}^*(x,e)$ and $v_{ed}^*(x+a'',e) \leq T_0 v_{ed}^*(x,e)$:

$$\delta \leq v_{ed}^*(x+1+a',e) - ca' - v_{ed}^*(x+a',e) + ca' -$$
$$v_{ed}^*(x+a'',e) + ca'' + v_{ed}^*(x-1+a'',e) - ca'' \leq 0.$$

If $a' = a''$, then the statement is true by the concavity of v_{ed}^*. If $a' = 0$ and $a'' = 1$, then the term in the second inequality is exactly 0. All other terms

ϵ	$\max\{PG_{d,s}\}$	$\max\{PG_{ed,s}\}$	$\max\{PG_{d,ed}\}$
0.3	5.68%	2.25%	3.36%
0.6	10.19%	7.58%	2.87%
0.8	13.04%	11.25%	3.23%

Table 1. Maximum profit gain for different demand variability.

μ	0.11	0.21	0.31	0.41	0.51	0.61	0.71
$PG_{d,s}$	12.50%	8.67%	6.80%	5.72%	4.75%	3.80%	3.12%
$PG_{ed,s}$	10.90%	6.70%	4.16%	2.60%	1.47%	0.7%	0.5%
$PG_{d,ed}$	1.45%	1.85%	2.53%	3.04%	3.23%	3.07%	2.66%

Table 2. Profit gain of pricing policies for different service rates with $\epsilon = 0.8$.

in (2) are concave by concavity of v^*_{ed}. Thus, $T_{ed}v^*_{ed}$ is concave in x for an environment e, whenever v^*_{ed} is concave.

Now the above argument immediately implies the existence of optimal environment-dependent base-stock policies:

THEOREM 3 *For all pricing strategies $\pi = s, ed, d$: The optimal replenishment policy is an environment-dependent base stock policy.*

Optimality of environment-dependent base stock policies shows that information about the environment in which a firm operates is crucial.

4. Numerical results

In our model formulation, the system is controlled directly by the demand rate, defined as a function of the offered price. In this section we explicitly refer to the prices. We consider a linear demand rate function, which is frequently used in the pricing literature. Let p be the price offered. Then we define the linear demand function, and its associated revenue rate by:

$$\lambda_e(p) = \Lambda_e(1 - ap)\,,\, p \in [0, 1/a],$$

where a is a positive real number.

For a given problem, let g^*_π be the optimal average profit using policy π, where discount rate is set to 0, i.e., $\alpha = 0$. We define the relative Profit Gain for using policy π instead of policy π', $PG_{\pi,\pi'}$, by

$$PG_{\pi,\pi'} = \frac{g^*_\pi - g^*_{\pi'}}{g^*_{\pi'}}.$$

ϵ	$S_s^*(L)$	$S_s^*(H)$	$S_{ed}^*(L)$	$S_{ed}^*(H)$	$S_d^*(L)$	$S_d^*(H)$
0.3	6	11	7	9	12	20
0.6	4	14	5	10	7	22
0.8	2	13	3	10	3	23

Table 3. The optimal base stock levels for different ϵ with $\mu = 0.11$.

ϵ	p_s^*	$p_{ed}^*(L)$	$p_{ed}^*(H)$	$\bar{p}_d^*(L)\}$	$\underline{p}_d^*(L)$	$\bar{p}_d^*(H)$	$\underline{p}_d^*(H)$
0.3	0.78	0.74	0.82	0.82	0.42	0.87	0.51
0.6	0.75	0.65	0.84	0.75	0.33	0.88	0.51
0.8	0.78	0.57	0.84	0.65	0.19	0.88	0.51

Table 4. The optimal prices for different ϵ with $\mu = 0.11$, where $\bar{p}_d^*(e) = \max\{p_d^*(x, e)\}$, and $\underline{p}_d^*(e) = \min\{p_d^*(x, e)\}$.

As we know that $g_s \leq g_{ed} \leq g_d$, we will consider $PG_{d,ed}$, $PG_{d,s}$ and $PG_{ed,s}$.

We consider a system which operates in two environments, with low demand rate (L) and with high demand rate (H). The demand rates in these environments are $\Lambda_L = 1 - \epsilon$ and $\Lambda_H = 1 + \epsilon$. The factors that affect optimal policies are the ratios λ/μ and h/p, so we vary the service rate μ and the holding cost h, where we set $a = 1$, $c = 0$, and the average demand rate as 1. Moreover, here we only report $h = 0.01$ and $q_{LH} = q_{HL} = q = 0.01$, although we experimented with different h and q as well as asymmetric transitions rates. In the whole numerical study, we restrict our attention to the recurrent states of the Markov chain generated by an optimal policy.

As ϵ increases, the demand variability increases. We observe that optimal gain for each pricing policy decreases with ϵ. The profit gain of π_{ed} and π_d with respect to π_s also increases with ϵ (see Table 1), which shows the ability of these policies to adjust the highly uncertain environments. For small ϵ, on the other hand, $PG_{d,s} < 6\%$, suggesting that optimal static policy performs good enough with mild uncertainty. From Table 2, we observe that policy π_s performs the worst with capacitated supply ($\mu < 0.4$) and volatile demand with respect to π_{ed} and π_d. Optimal static prices are closer to the optimal environment-dependent prices in environment H, rather than those in environment L (see Table 4). Hence, the demand fluctuation hurts not only the firm by decreasing its average gain, but also the customers due to high prices, when static pricing strategy is followed.

We see that policy π_{ed} performs very closely to policy π_d with $\max\{PG_{d,ed}\}$ $< 3.5\%$ (see Table 1). In fact, it brings most of the benefit of π_d, compare $PG_{d,s}$ with $PG_{ed,s}$ in Table 2. Moreover, policy π_{ed} has the advantage of lower inventory levels (see Table 3) and of smaller price differences (see Table 4). Hence, we can conclude that it is better to use π_{ed}, since it brings most of the benefit of π_d, while causing less reaction on the customer side with less variability in prices, and requiring a reasonable storage space with less variability in the stock levels.

Optimal pricing and replenishment policies may have further monotonicities under certain conditions: If we order the environment states with respect to the potential demand rates, i.e., $\Lambda_e \leq \Lambda_{e+1}$ for $e = 1, .., n-1$, then we expect to have monotone base stock levels, i.e., $S_\pi^*(e) \leq S_\pi^*(e+1)$ for all pricing strategies $\pi = s, ed, d$. The optimal environment-dependent prices as well as the effective demand rates should also be ordered with the potential demand rates. Our future work will focus on these monotonicities.

References

[1] L. M. A. Chan, Z. J. Max Shen, D. Simchi-Levi, J. L. Swann. Coordination of pricing and inventory decisions: a survey and classification. In *Handbook of quantitative supply chain analysis: modeling in the e-business era.* Kluwer, 2004.

[2] F. Chen, J. S. Song. Optimal Policies for Multiechelon Inventory Problems with Markov-Modulated Demand. *Oper. Research.* 49-2: 226-234.

[3] H. Chen, O. Wu, D.D. Yao. Optimal Pricing and Replenishment in a Single-Product Inventory System. Working Paper, Columbia University, Dept. of Industrial Engineering and Operations Research, 2004.

[4] W. Elmaghraby, P. Keskinocak. Dynamic pricing in the presence of inventory considerations: research overview, current practices and future directions. *Management Science.* 49-10:1287-1309, 2003.

[5] J.-P. Gayon, I. Talay-Değirmenci, F. Karaesmen, L. Ormeci. Dynamic Pricing and Replenishment in a Production/Inventory System with Markov-Modulated Demand. Working Paper, Koç University, Department of Industrial Engineering, 2004.

[6] L. Li. A stochastic theory of the firm. *Math. of Oper. Res..* 13:3, 1988.

[7] S. Özekici, M. Parlar. Inventory models with unreliable suppliersin a random environment. *Annals of Operations Research* . 91:123-136, 1999.

[8] M. Puterman. *Markov Decision Processes.* John Wiley and Sons Inc, New York, 1994.

[9] C.A Yano, S.M. Gilbert. Coordinated pricing and production/procurement decisions: a review. In *Managing Business Interfaces.* Kluwer Academic Publishers, 2003.

AN ADAPTATION OF BICGSTAB FOR NONLINEAR BIOLOGICAL SYSTEMS

E. Venturino,[1] P. R. Graves-Morris,[2] and A. De Rossi[1]
[1] *Università di Torino, Dipartimento di Matematica, via Carlo Alberto 10, I-10123, Torino, Italy, {ezio.venturino,alessandra.derossi}@unito.it**, [2] *University of Bradford, Department of Computing, Bradford BD7 1DP, UK, p.r.graves-morris@bradford.ac.uk*

Abstract Here we propose a new adaptation of Van der Vorst's BiCGStab to nonlinear systems, a method combining the iterative features of both sparse linear system solvers, such as BiCGStab, and of nonlinear systems, which in general are linearized by forming Jacobians, and whose resulting system usually involves the use of a linear solver. We consider the feasibility and efficiency of the proposed method in the context of a space-diffusive population model, the growth of which depends nonlinearly on the density itself.

keywords: BiCGStab, iterative methods, population models, sparse nonlinear systems.

1. Introduction

Many popular methods for the solution of a sparse system of linear equations such as BiCGStab are iterative. Methods for the solution of nonlinear systems are usually recursive and the recursion usually involves forming Jacobians for linearization of the nonlinear terms. The solution of a large sparse nonlinear system usually involves the use of a linear solver after the Jacobian has been formed.

The task of solving sparse systems of linear or nonlinear equations comes up in many large-scale problems of scientific computing. The iterative approach to the solution of large linear systems is preferable to the direct one in some situations, especially when we have to solve problems arising from applications in which the coefficient matrix is sparse. Among the many existing iterative methods, the Lanczos-Type Product Methods (LTPMs) are characterized by residual polynomials that are products of a Lanczos polynomial and another

*Paper written with financial support of the Dipartimento di Matematica, Progetto per la ricerca locale "Modelli matematici avanzati per le applicazioni: studio analitico e numerico II" and of the M.I.U.R. of Italy (P.R.I.N. 2003).

Please use the following format when citing this chapter:
Author(s) [insert Last name, First-name initial(s)], 2006, in IFIP International Federation for Information Processing, Volume 199, System Modeling and Optimization, eds. Ceragioli F., Dontchev A., Furuta H., Marti K., Pandolfi L., (Boston: Springer), pp. [insert page numbers].

polynomial of the same degree. LTPMs enjoy some remarkable properties: in fact, they incorporate stabilization, so as to reduce and smooth the residuals as much as possible; they are transpose-free and gain one dimension of the Krylov space per matrix-vector multiplication. For a good survey of LTPMs, see for example [10].

A subset of the LTPMs is given by the BiCGStab family. The biconjugate gradient method (BiCG) has the property that the estimates of the solution of (1) below have residuals that are orthogonal to dual (or shadow) Krylov subspaces which increase in dimension with n and this feature is retained *implicitly* by LTPMs. In 1992, Van der Vorst proposed the use of the BiCGStab method so as to get smoother convergence of the estimates of the solution of (1), see [13]. In the BiCGStab algorithm, the stabilising polynomials are built up in factored form, with a new linear factor being included at each step in a way such that the residual undergoes a one-dimensional minimization process. This basic stabilisation was soon advanced to BiCGStab$l$ and BiCGStab(l) to help to avoid possible breakdowns, see [10].

In 1974, Gragg observed a connection between BiCG and vector-Padé approximation [5]. Vector-Padé methods in general construct vector-valued rational functions which approximate functions specified by their vector-valued power series. The idea is usually to accelerate the convergence of the given vector-valued power series (see [1]). A linear system of equations is often denoted by

$$A\mathrm{x} = \mathrm{b}. \tag{1}$$

Many models arising in science and engineering require the solution of large sparse systems of non-linear equations. We consider the feasibility and efficiency of the methods proposed here in the context of models of a population whose growth depends non-linearly on the density of the population and we allow for diffusion of the population through space.

2. The method

A basic iterative method [11] for the solution of the nonlinear system of equations

$$F(x) = 0, \text{ with } F : \mathcal{R}^d \to \mathcal{R}^d. \tag{2}$$

may summarily be expressed by

$$z_{k+1} = S(z_k) = z_k + \Pi F(z_k), k = 0, 1, 2, \ldots, \tag{3}$$

where S is the successor functional, and Π is an approximation to the Jacobian of the system

$$\Pi \approx -J_F^{-1}, \text{ with } J_F \equiv \frac{\partial F(z_k)}{\partial z}. \tag{4}$$

We develop acceleration methods for (3), not requiring explicit evaluation of the sequence $\{z_k\}$. A sequence $\{x_k\}$ of estimates of the root will instead be formed using three-term inhomogeneous recursion relations and an optimal successor functional. To assess the accuracy of a successor functional, use its actual definition and let its residual be

$$R(z) = S(z) - z = \Pi F(z). \tag{5}$$

Thus $R(z)$ is a residual *preconditioned* by Π, while $F(z)$, given algebraically by (2), denotes the vector residual of the original system.

The core three-term inhomogeneous recursion relations, designed originally for convergence acceleration of linear systems, are taken directly from VPAStab and here adapted to a nonlinear system [7]. Define preconditioned residuals of the estimates x_i by

$$r_i = R(x_i) = S(x_i) - x_i, \text{ for } i = 0, 1, 2, \ldots \tag{6}$$

with initializations given by $x_0 = z_0$ and $x_1 = z_1 = S(x_0)$.

The VPAStab recursion formulas are [7]

$$\begin{aligned} x_{2k} &= (1+\alpha_k)x_{2k-1} - \alpha_k x_{2k-2} \\ &\quad + (1-\theta_k)[(1+\alpha_k)r_{2k-1} - \alpha_k r_{2k-2}], \end{aligned} \tag{7}$$

$$\begin{aligned} x_{2k+1} &= (1+\beta_k)x_{2k} - \beta_k x_{2k-1} + (1+\beta_k)r_{2k} \\ &\quad - \beta_k(1-\theta_k)r_{2k-1}, \end{aligned} \tag{8}$$

with coefficients α_k, β_k defined by

$$\alpha_k = e_{2k-1}(e_{2k-2} - e_{2k-1})^{-1}, k = 1, 2, 3, \ldots, \tag{9}$$

$$\beta_k = e_{2k}(e_{2k-1}(1-\theta_k) - e_{2k})^{-1}, k = 1, 2, 3, \ldots, \tag{10}$$

where

$$e_N := w^{\mathrm{T}} r_N, N = 0, 1, 2, \ldots. \tag{11}$$

For the vector w we can take $w = r_0$.

For $k = 1, 2, \ldots$, each parameter θ_k is chosen to minimise an estimate of r_{2k}. Grouping the terms in equn (7)

$$\bar{x}_{2k} := (1+\alpha_k)x_{2k-1} - \alpha_k x_{2k-2}, \tag{12}$$

$$\bar{r}_{2k} := (1+\alpha_k)r_{2k-1} - \alpha_k r_{2k-2}. \tag{13}$$

and starting with the definition (6), we find that

$$r_{2k} \approx R(\bar{x}_{2k} + (1-\theta_k)\bar{r}_{2k}) \tag{14}$$

We estimate r_{2k} using first order Taylor expansion of the right-hand side of (14), under the assumption that the residual terms in (7) are smaller in norm than the estimates of x. Thus

$$r_{2k} \approx R(\bar{x}_{2k}) + (1-\theta_k)\frac{\partial R(\bar{x}_{2k})}{\partial z}\bar{r}_{2k}. \tag{15}$$

The matrix $J_R(z) = \partial R/\partial z$ is the Jacobian of this preconditioned residual, and it is only the product $J_R(\bar{x}_{2k})\bar{r}_{2k}$ that is required in (15). The value of θ_k is determined by minimising the right-hand side of (15) in norm. A suitable choice of the preconditioning operator Π still has to be made. Equation (3) allows dynamical updating during iteration. Then equations (5) - (15) complete the specification of the nonlinear algorithm.

3. An application

We consider the feasibility and efficiency of the method proposed in the context of models of a population diffusing through space whose growth depends nonlinearly on the density itself ([8] and [9]).

Let $\rho(t, \mathbf{x})$ denote the density at time t of a population living in an environment, the spatial variable being represented by the vector $\mathbf{x}$. The density is assumed to reproduce and diffuse along the space direction. For simplicity here we first formulate the case of one spatial dimension and then outline the case of two spatial dimension $x \in \mathcal{R}^2$, although our framework is also designed for $x \in \mathcal{R}^3$.

3.1 The uni-dimensional case

We have

$$\frac{\partial \rho}{\partial t} = f(\rho) + \epsilon\frac{\partial^2 \rho}{\partial x^2}, \tag{16}$$

where ϵ denotes the diffusion coefficient and the function f expresses the reproduction mechanism. One of the most frequently used forms is the logistic law, which gives a nonlinear model:

$$f(\rho) \equiv r\rho(t,x)\left[1 - \frac{\rho(t,x)}{K}\right], \tag{17}$$

where K is the carrying capacity and r is the reproduction rate. Thus the complete one dimensional model reads

$$\frac{\partial \rho}{\partial t} = r\rho(t,x)\left[1 - \frac{\rho(t,x)}{K}\right] + \epsilon\frac{\partial^2 \rho}{\partial x^2}, \tag{18}$$

where $x \in [a, b]$. The discretization method used is the Crank-Nicholson formula, which evaluates the equation at a suitable point P (see [12]). Letting the grid $(t_n, x_i) \equiv (nk, ih)$ where k and h are respectively the time and space discretization stepsizes, $i = 1, 2, \ldots, N-1$, and the evaluation point P is taken as the time midpoint in the mesh, $\left(\frac{1}{2}(t_n + t_{n+1}), x_i\right)$. Letting $\rho_{n,i} \equiv \rho(t_n, x_i)$, the equation gets discretized as follows

$$\frac{1}{k}[\rho_{n+1,i} - \rho_{n,i}] = \frac{\epsilon}{2}\left[\frac{1}{h_x^2}(\rho_{n+1,i-1} - 2\rho_{n+1,i} + \rho_{n+1,i+1}) + \frac{1}{h_x^2}(\rho_{n,i-1} - 2\rho_{n,i} + \rho_{n,i+1})\right] + \frac{1}{2}[f(\rho_{n+1,i}) + f(\rho_{n,i})] \tag{19}$$

so that, introducing as a shorthand $q_x \equiv \frac{k}{h^2}$, the system reads finally

$$-\epsilon\frac{q_x}{2}\rho_{n+1,i-1} + \rho_{n+1,i}[1 + \epsilon q_x - \frac{rk}{2}] - \epsilon\frac{q_x}{2}\rho_{n+1,i+1} + \frac{rk}{2K}(\rho_{n+1,i})^2 \tag{20}$$

$$= \epsilon\frac{q_x}{2}\rho_{n,i-1} + \rho_{n,i}[1 - \epsilon q_x + \frac{rk}{2}] + \epsilon\frac{q_x}{2}\rho_{n,i+1} - \frac{rk}{2K}(\rho_{n,i})^2 \tag{21}$$

and in matrix form

$$A\rho_{(n+1)} + \lambda[\rho_{(n+1)}]^2 = B\rho_{(n)} + \lambda[\rho_{(n)}]^2, \quad \forall n = 0, 1, 2, \ldots, \tag{22}$$

where

$$\lambda = \frac{rk}{2K}, \quad \rho_{(n)} = (\rho_{n,1}, \rho_{n,2}, \ldots, \rho_{n,N-1})^t, \tag{23}$$

$$A = \begin{bmatrix} 1 + \epsilon q - \frac{rk}{2} & -\epsilon\frac{q}{2} & 0 & \ldots & 0 \\ -\epsilon\frac{q}{2} & \ddots & \ddots & \ddots & \vdots \\ 0 & \ddots & \ddots & \ddots & 0 \\ \vdots & \ddots & \ddots & \ddots & -\epsilon\frac{q}{2} \\ 0 & \ldots & 0 & -\epsilon\frac{q}{2} & 1 + \epsilon q - \frac{rk}{2} \end{bmatrix}, \tag{24}$$

and

$$B = \begin{bmatrix} 1 - \epsilon q + \frac{rk}{2} & \epsilon\frac{q}{2} & 0 & \ldots & 0 \\ \epsilon\frac{q}{2} & \ddots & \ddots & \ddots & \vdots \\ 0 & \ddots & \ddots & \ddots & 0 \\ \vdots & \ddots & \ddots & \ddots & \epsilon\frac{q}{2} \\ 0 & \ldots & 0 & \epsilon\frac{q}{2} & 1 - \epsilon q + \frac{rk}{2} \end{bmatrix}. \tag{25}$$

The matrices A and B have dimension $N \times N$. Suitable boundary conditions are used.

For simplicity, we can also write

$$A\rho + \lambda\rho^2 = b \tag{26}$$

Introducing an extra index m to count the iterations we then define the iteration scheme

$$A\rho_{(n+1)}^{(m+1)} = B\rho_{(n)} + \lambda[\rho_{(n)}]^2 - \lambda[\rho_{(n+1)}^{(m)}]^2, \quad m \geq 1. \tag{27}$$

Now, we consider the splitting $A = L + U$ and the iteration then becomes

$$L\rho_{(n+1)}^{(m+1)} = B\rho_{(n)} + \lambda[\rho_{(n)}]^2 - U[\rho_{(n+1)}^{(m)}]^2 - \lambda[\rho_{(n+1)}^{(m)}]^2, \quad m \geq 1. \tag{28}$$

The solution is obviously obtained by operating on the right hand side by L^{-1} in the usual sequential way.

Another iteration scheme which we consider is given by

$$(LU)\rho_{(n+1)}^{(m+1)} = (LU - A)\rho_{(n)}^{(m)} - \lambda\rho_{(n)}^{(m)}, \tag{29}$$

where one could also use an incomplete LU decomposition of the Jacobian (incomplete LU *preconditioning*)

$$(LU)_{\text{inc}} \approx A + 2\lambda \text{diag}\rho. \tag{30}$$

Then, we obviously obtain the solution as if by using $U_{\text{inc}}^{-1}L_{\text{inc}}^{-1}$, but without explicitly forming the inverse.

3.2 The two-dimensional case

The two-dimensional case is a simple extension of the former one. We consider the solution of the problem in the square $[0,1] \times [0,1]$. Letting the grid $(t_n, x_i, y_j) \equiv (nk, ih_x, jh_y)$, where k is the time stepsize and h_x, h_y are respectively the step-sizes along the x and y axes, the evaluation point P will be the average of the Laplacian, discretized via the five point formula at time n and $n+1$. Letting $\rho_{n,i,j} \equiv \rho(t_n, x_i, y_j)$, and then $q_x \equiv \frac{k}{h_x^2}$ and $q_y \equiv \frac{k}{h_y^2}$, the discretized equation gives the nonlinear system

$$\begin{aligned}
-\epsilon\frac{q_y}{2}\rho_{n+1,i,j-1} - \epsilon\frac{q_x}{2}\rho_{n+1,i-1,j} + \rho_{n+1,i,j}\left[1 + \epsilon q_x + \epsilon q_y - \frac{rk}{2}\right] \\
-\epsilon\frac{q_x}{2}\rho_{n+1,i+1,j} - \epsilon\frac{q_y}{2}\rho_{n+1,i,j+1} + \frac{rk}{2K}(\rho_{n+1,i,j})^2 \\
= \epsilon\frac{q_y}{2}\rho_{n,i,j-1} + \epsilon\frac{q_x}{2}\rho_{n,i-1,j} + \rho_{n,i,j}\left[1 - \epsilon q_x - \epsilon q_y + \frac{rk}{2}\right] \\
+\epsilon\frac{q_x}{2}\rho_{n,i+1,j} + \epsilon\frac{q_y}{2}\rho_{n,i,j+1} - \frac{rk}{2K}(\rho_{n,i,j})^2
\end{aligned} \tag{31}$$

Notice that the matrices (24) and (25) are here replaced by band matrices.

4. Numerical results

We conducted experiments to compare the following iteration schemes:

- $(L + U)$ not accelerated,
- $(L + U)$ + Nonlinear VPAStab,
- (LU) + Nonlinear VPAStab,

and to compare how different choices of the biological parameters affect the various schemes. After performing an extensive experimentation in dimension one, we concentrated our attention on the two-dimensional case,the results of which are here presented.

We also considered different boundary conditions: *Dirichlet BCs* and *Neumann BCs*. The computations were done on a workstation running an AMD Athlon XP 1500+ with 576 MB of RAM. The iteration is initialized with

$$\rho_{0,i,j} = p_0(1 - y_j), \qquad i, j = 1, 2, \ldots N,$$

where $p_0 = 275$, and the iterations were stopped when $||r^{(k)}||_2 <$ tol, where tol is a given tolerance. As a failure criterion we used either too slow convergence, or a maximum number of nonlinear iterations per timestep (250). In the experiments the numerical solutions were computed with timestep $k = 10^{-2}$.

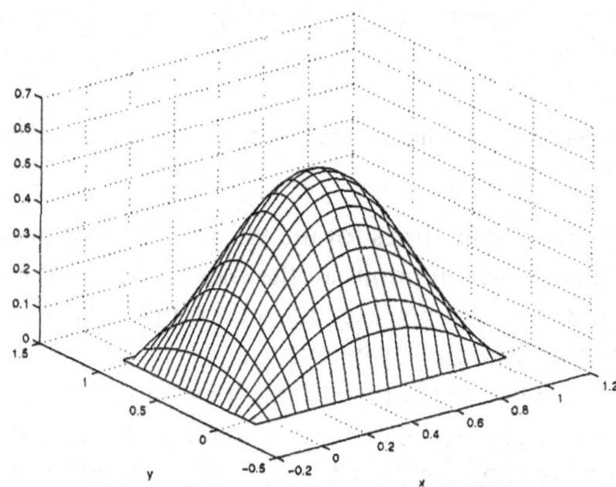

Figure 1. Sample of a numerical solution with Dirichlet BCs, $r = 10$, $K = 100$, $\epsilon = 1$ and $N = 17$.

Since we are interested in evaluation and comparison of the computational cost of the various methods, the following counters are given in tables and figures:

- NI, the number of nonlinear iterations;
- FC, the number of function calls to the residual;
- ET, the execution time.

If we look at the number of nonlinear iterations, we note that acceleration reduces this number. Consequently, the execution times are also greatly reduced in all cases to which we apply acceleration.

In Figures 2 and 3, we respectively show how the number of nonlinear iterations and the execution time vary versus to the required accuracy (specified tolerance) expressed as $\log_{10}(tol)$.

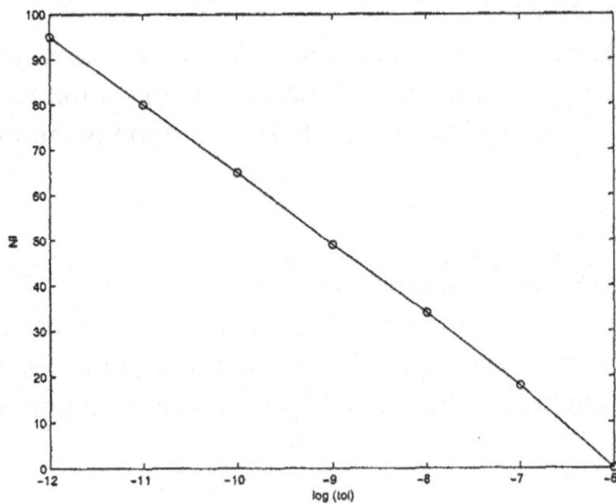

Figure 2. Plot of the number of nonlinear iterations against $\log_{10}(tol)$ for the accelerated LU scheme.

In Figure 3 (left), execution times for the various methods are shown, versus the number N of discretization points in the mesh. We point out that the LU iteration scheme requires a higher execution time than the $L+U$ one, while, in general, it needs fewer nonlinear iterations.

We compare the number of nonlinear iterations for different chosen values of the biological parameters, the reproduction rate r and the carrying capacity K. Results are shown in Figure 4, in Table 1 for different reproduction rates (r) and in Table 2 for different carrying capacities (K).

References

[1] G. A. Baker, P. R. Graves Morris. *Padé Approximants*. Cambridge Univ. Press, Cambridge, 1997.

[2] S. C. Eisenstat, H. F. Walker. Globally convergent inexact Newton methods. *SIAM J. Optimization* 4:393-422, 1994.

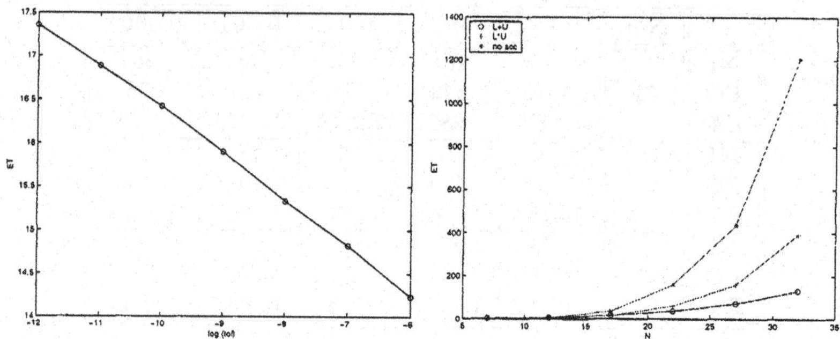

Figure 3. Execution times in seconds against $\log_{10}(tol)$ for the accelerated LU scheme (left) and for the various methods as functions of N (right).

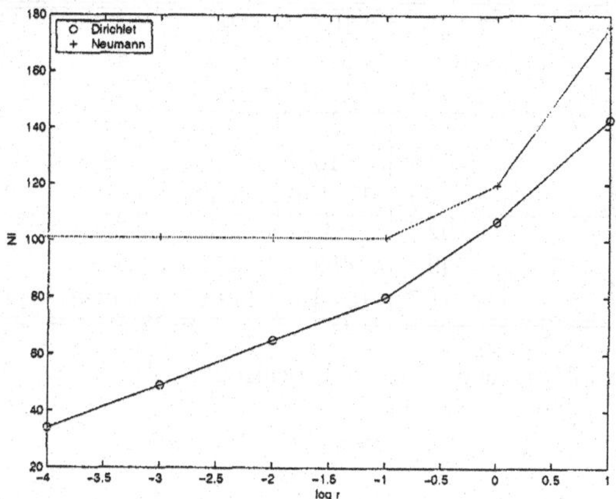

Figure 4. Nonlinear iterations for different values of $\log_{10} r$ for the accelerated LU scheme with tol $= 10^{-8}$.

[3] D. R. Fokkema, G. L. J. Sleijpen, H. A. Van der Vorst. Accelerated inexact Newton schemes for large systems of nonlinear equations. *SIAM J. Sci. Computing* 19:657-674, 1998.

[4] R. Glowinski, H. B. Keller, L. Reinhart. Continuation-conjugate gradient methods for the least squares solution of nonlinear boundary value problems. *SIAM J. Sci. Stat. Comput.* 4:793-832, 1985.

[5] W. B. Gragg. Matrix interpretations and applications of the continued fraction algorithm. *Rocky Mountain J. Math.* 4:213-225, 1974.

[6] P. R. Graves–Morris. VPAStab: stabilized vector-Padé approximation with application to linear systems. *Numer. Algorithms* 33:293-304, 2003.

[7] P. R. Graves-Morris. BiCGStab, VPAStab and an adaptation to mildly nonlinear systems. *Submitted.*

	N\r	1.	0.1	0.01	0.001	0.0001
NI		122	107	95	76	58
FC	11	616	549	496	430	376
ET		1.4020	1.2820	1.2820	1.1320	1.2310
NI		122	94	74	58	44
FC	21	616	510	436	376	334
ET		16.153	15.232	14.321	13.940	13.439
NI		109	81	64	52	38
FC	31	571	469	404	358	316
ET		161.11	154.50	150.85	150.80	145.40

Table 1. Accelerated *LU* iteration scheme - Dirichlet BCs - $\epsilon = 1$, $K = 10$, tol$= 10^{-8}$

	N\K	1	10	100	1000	10000
NI		121	107	95	77	59
FC	11	610	549	496	433	379
ET		1.8730	1.2010	1.1620	1.1220	1.0610
NI		119	94	74	60	44
FC	21	603	510	436	382	334
ET		16.153	15.082	14.300	13.820	13.419
NI		102	81	64	52	39
FC	31	549	469	404	358	319
ET		160.97	154.15	149.95	146.83	144.51

Table 2. Accelerated *LU* iteration scheme - Dirichlet BCs - $\epsilon = 1$, $r = 0.1$, tol$= 10^{-8}$

[8] M. E. Gurtin, R. C. MacCamy. On the diffusion of biological populations. *Math. Biosci.* 33:35-49, 1977.

[9] M. E. Gurtin, R. C. MacCamy. Diffusion models for age-structured populations. *Math. Biosci.* 54:49-59, 1981.

[10] M. H. Gutknecht. Lanczos-type solvers for non-symmetric linear systems of equations. *Acta Numerica.* 6:271-397, 1997.

[11] C. T. Kelley. *Iterative methods for linear and nonlinear equations.* Frontiers in Applied Mathematics 16, SIAM, Philadelphia, PA, 1995.

[12] G. D. Smith. *Numerical Solution of Partial Differential Equations: Finite Difference Methods*, 2nd Ed., Clarendon Press, Oxford, 1979.

[13] H. A. Van der Vorst. BiCGStab: A fast and smoothly convergent variant of BiCG for the solution of non-symmetric linear systems. *SIAM J. Sci. Statist. Comput.* 13:631-644, 1992.

NUMERICAL APPROXIMATION OF A CONTROL PROBLEM FOR ADVECTION–DIFFUSION PROCESSES

A. Quarteroni,[1,2] G. Rozza,[1] L. Dedè[2] and A. Quaini[1]

[1]*École Polytechnique Fédérale de Lausanne (EPFL), FSB, Chaire de Modelisation et Calcul Scientifique (CMCS), Station 8, 1015, Lausanne, Switzerland, {alfio.quarteroni,gianluigi.rozza, annalisa.quaini}@epfl.ch* [2]*MOX–Dipartimento di Matematica "F. Brioschi", Politecnico di Milano, 20133, Milano, Italy, luca.dede@mate.polimi.it*

Abstract Two different approaches are proposed to enhance the efficiency of the numerical resolution of optimal control problems governed by a linear advection–diffusion equation. In the framework of the Galerkin–Finite Element (FE) method, we adopt a novel a posteriori error estimate of the discretization error on the cost functional; this estimate is used in the course of a numerical adaptive strategy for the generation of efficient grids for the resolution of the optimal control problem. Moreover, we propose to solve the control problem by adopting a reduced basis (RB) technique, hence ensuring rapid, reliable and repeated evaluations of input–output relationship. Our numerical tests show that by this technique a substantial saving of computational costs can be achieved.

keywords: optimal control problems; partial differential equations; finite element approximation; reduced basis techniques; advection–diffusion equations; stabilized Lagrangian; numerical adaptivity.

1. Introduction

Many physical processes, which involve diffusion and transport of scalar quantities, can be modelled by linear advection–diffusion partial differential equations. These phenomena are studied, e.g., in Environmental Sciences, to investigate the distribution forecast of pollutants in water or in atmosphere. In this context it might be of interest to regulate the source term of the advection–diffusion equation so that the solution is as near as possible to a desired one, e.g., to operate the emission rates of industrial plants to keep the concentration of pollutants near (or below) a desired level.

This problem can be conveniently accommodated in the optimal control framework for PDEs, where we consider the Lagrangian functional formulation [3], as complementary to the classical approach of J.L. Lions [6].

Please use the following format when citing this chapter:
Author(s) [insert Last name, First-name initial(s)], 2006, in IFIP International Federation for Information Processing, Volume 199, System Modeling and Optimization, eds. Ceragioli F., Dontchev A., Furuta H., Marti K., Pandolfi L., (Boston: Springer), pp. [insert page numbers].

To avoid numerical instabilities that arise in a transport dominating regime, we propose a stabilization on the Lagrangian functional [5]. We consider two numerical approaches that allow an efficient resolution of the optimal control problem, in the context of an iterative optimization procedure. In the first case we solve the equations governing the control problem by means of the Galerkin–FE method. Grid adaptivity is driven by *a posteriori* error estimate on the cost functional, which we assume as an indicator of the whole error on the control problem [3]. Moreover, we propose a separation of the *iteration* and *discretization* error [5], for which we define a posteriori error estimate. As soon as the iteration error is brought below a desired threshold by means of the iterative optimization method, we operate the adaptive strategy to reduce the discretization error [5]. Then we solve numerically the equations governing the control problem by means of the *reduced basis* (RB) method [8], which leads to a large saving of computational costs. In fact the RB method permits a rapid, reliable and repeated evaluation of the input–output relationship [7]; in the case of the control problem the inputs are the control function for the state equation, and the observation for the adjoint one, while the outputs are respectively the state variable and the adjoint one.

At the end we report some numerical tests to validate the methods here presented, referring in particular to a pollution control problem in atmosphere.

2. Mathematical Model of the Control Problem

In this section we recall the Lagrangian functional approach for optimal control problems and the associated iterative optimization method, in a general setting [3]; then we specialize it to an advection–diffusion control problem.

2.1 The general setting

Let us consider the following control problem:

$$find \quad u \in \mathcal{U} \quad : \quad J(w,u) \quad minimum, \quad with \quad Aw = f + Bu, \tag{1}$$

where $w \in \mathcal{V}$ is the state variable, u the control function, A is an elliptic operator defined on $\mathcal{V}$ with values in $\mathcal{V}'$, B is an operator defined on $\mathcal{U}$ and valued in $\mathcal{V}'$, f is a source term, $\mathcal{V}$ and $\mathcal{U}$ are two Hilbert spaces. We write the state equation $Aw = f + Bu$ in weak form: $find\, w \in \mathcal{V} \,:\, a(w,\varphi) = (f,\varphi) + b(u,\varphi),\ \forall \varphi \in \mathcal{V}$. The associated Lagrangian functional reads:

$$\mathcal{L}(w,p,u) := J(w,u) + b(u,p) + (f,p) - a(w,p), \tag{2}$$

where $a(\cdot,\cdot)$ and $b(\cdot,\cdot)$ are the bilinear forms associated with A and B, respectively, $(\cdot,\cdot)$ is the L^2–inner product, while $p \in \mathcal{V}$ is the adjoint variable. Should there exist, the solution of the control problem (w^*, p^*, u^*) is the stationary point of $\mathcal{L}(w,p,u)$. By differentiating the Lagrangian functional, we obtain the Euler–Lagrange system governing the optimal control problem:

$$\begin{cases} \mathcal{L}_{,w}[\phi] = 0 \longrightarrow \textit{find } p \in \mathcal{V} \ : \ a(\phi,p) = J_{,w}(w,u)[\phi], \ \forall \phi \in \mathcal{V}, \\ \mathcal{L}_{,p}[\varphi] = 0 \longrightarrow \textit{find } w \in \mathcal{V} \ : \ a(w,\varphi) = (f,\varphi) + b(u,\varphi), \ \forall \varphi \in \mathcal{V}, \\ \mathcal{L}_{,u}[\psi] = 0 \longrightarrow J_{,u}(w,u)[\psi] + b(\psi,p) = 0, \ \forall \psi \in \mathcal{U}. \end{cases} \tag{3}$$

The first equation in (3) is the adjoint equation, the second one is the state equation, while, by the Riesz theorem, from the third one we can extract the sensitivity of the cost functional δu with respect to the control function u ($\mathcal{L}_{,u}[\psi] = (\delta u(p,u),\psi)$). The control problem can be solved by means of an iterative method [1]. At each step j we solve sequentially the state and the adjoint equation and we compute the sensitivity $\delta u(p^j,u^j)$; then we evaluate the latter in an appropriate norm, which we compare with a prescribed tolerance. If this stopping criterium is not fulfilled, we adopt an optimization iteration on the control function u, such as the *steepest–descent* method, $u^{j+1} = u^j - \tau^j \delta u(p^j,u^j)$, where τ^j is a relaxation parameter.

2.2 The case of an advection–diffusion problem

Let us consider now the specific case of a linear advection–diffusion state equation, referring to a $2D$–domain Ω:

$$\begin{cases} L(w) := -\nabla\cdot(\nu\nabla w) + \mathbf{V}\cdot\nabla w = u, \quad \textit{in } \Omega, \\ w = 0, \quad \textit{on } \Gamma_D, \\ \nu\frac{\partial w}{\partial n} = 0, \quad \textit{on } \Gamma_N, \end{cases} \tag{4}$$

Γ_D and Γ_N are two disjoint portions of the domain boundary $\partial\Omega$ such that $\Gamma_D \cup \Gamma_N = \partial\Omega$, $u \in L^2(\Omega)$ is the control variable, while ν and $\mathbf{V}$ are given functions. We assume homogeneous Dirichlet condition on the inflow boundary $\Gamma_D := \{\mathbf{x} \in \partial\Omega : \mathbf{V}(\mathbf{x})\cdot\mathbf{n}(\mathbf{x}) < 0\}$, being $\mathbf{n}(\mathbf{x})$ the unit vector directed outward, and homogeneous Neumann condition on the outflow boundary $\Gamma_N := \partial\Omega \setminus \Gamma_D$. We consider the observation on a part $D \subseteq \Omega$ of the domain, for which the control problem reads:

$$\textit{find } u \ : \ J(w,u) := \frac{1}{2}\int_D (g\,w(u) - z_d)^2 \, dD \ \ \textit{minimum}, \tag{5}$$

where $g \in C^\infty(\Omega)$ projects w in the observation space and z_d is the desired observation function. Adopting the formalism of the previous section and assuming $\mathcal{V} = H^1_{\Gamma_D} := \{v \in H^1(\Omega) : v_{|\Gamma_D} = 0\}$ and $\mathcal{U} = L^2(\Omega)$, the Lagrangian functional becomes:

$$\mathcal{L}(w,p,u) := J(w,u) + F(p;u) - a(w,p). \tag{6}$$

where:

$$a(w,\varphi) := \int_\Omega \nu \nabla w \cdot \nabla \varphi \, d\Omega + \int_\Omega \mathbf{V} \cdot \nabla w \, \varphi \, d\Omega, \tag{7}$$

$$F(\varphi; u) := \int_\Omega u\varphi \, d\Omega. \tag{8}$$

By differentiating $\mathcal{L}$ with respect to the state variable, we obtain the adjoint equation in weak form:

$$find \, p \in \mathcal{V} \; : \; a^{ad}(p,\phi) = F^{ad}(\phi; w), \;\; \forall \phi \in \mathcal{V}, \tag{9}$$

with:

$$a^{ad}(p,\phi) := \int_\Omega \nu \nabla p \cdot \nabla \phi \, d\Omega + \int_\Omega \mathbf{V} \cdot \nabla \phi \, p \, d\Omega, \tag{10}$$

$$F^{ad}(\phi; w) = \int_D (g\, w - z_d)\, g\, \phi \, dD. \tag{11}$$

In the distributional sense this yields:

$$\begin{cases} L^{ad}(p) := -\nabla \cdot (\nu \nabla p + \mathbf{V} p) = \chi_D g\, (g\, w - z_d), & in \; \Omega, \\ p = 0, \quad on \; \Gamma_D, \\ \nu \frac{\partial p}{\partial n} + \mathbf{V} \cdot \mathbf{n} \, p = 0, \quad on \; \Gamma_N, \end{cases} \tag{12}$$

being χ_D the characteristic function of the subdomain D. Finally, by differentiating $\mathcal{L}$ with respect to the control function u, we have the optimal control constraint, from which we define the cost functional sensitivity: $\delta u(p) = p$.

3. Numerical Approximation and Stabilization

For the numerical resolution of both the state and adjoint equations, we adopt the Galerkin–FE method with linear elements on unstructured triangular meshes. Both equations are of advection–diffusion type with a transport term that can dominate the diffusive one; when it happens an appropriate stabilization is mandatory to avoid numerical instabilities and their propagation in the course of the optimization iterative procedure ([9]).

3.1 "Optimize–then–discretize" and "discretize–then–optimize" approaches

From a numerical point of view, the algorithm outlined in Sec.2.1 (or in Sec.2.2 for the specific advection–diffusion case) requires, at each iterative step, the approximation of the state and adjoint equations. This approximation can be based, e.g., on a suitable FE subspace $\mathcal{X}_h \subset \mathcal{V}$ and the GLS (Galerkin–Least–Squares) method [9], obtaining respectively:

$$find \, w_h \in \mathcal{X}_h \; : \; a(w_h, \varphi_h) + \overline{s}_h(w_h, \varphi_h) = F(\varphi_h; u_h), \;\; \forall \varphi_h \in \mathcal{X}_h, \tag{13}$$

$$\overline{s}_h(w_h, \varphi_h) := \sum_{K \in \mathcal{T}_h} \delta_K \int_K R(w_h; u_h)\, L(\varphi_h)\, dK, \tag{14}$$

$$find\, p_h \in \mathcal{X}_h \; : \; a^{ad}(p_h, \phi_h) + \overline{s}_h^{ad}(p_h, \phi_h) = F^{ad}(\phi_h; w_h), \;\; \forall \phi_h \in \mathcal{X}_h, \tag{15}$$

$$\overline{s}_h^{ad}(p_h, \phi_h) := \sum_{K \in \mathcal{T}_h} \delta_K \int_K R^{ad}(p_h; w_h)\, L^{ad}(\phi_h)\, dK, \tag{16}$$

where δ_K is a stabilization parameter, $R(w; u) := L(w) - u$, $R^{ad}(p; w) := L^{ad}(p) - G(w)$, with $G(w) := \chi_D g\,(g\, w - z_d)$. This paradigm is resumed in the slogan "optimize–then–discretize" [2, 4]. An alternative approach is "discretize–then–optimize", for which first we discretize and stabilize the state equation, e.g. still by the GLS method (Eq.(13) and (14)), then we define the discrete Lagrangian functional:

$$\mathcal{L}_h(w_h, p_h, u_h) := J(w_h, u_h) + F(p_h; u_h) - a(w_h, p_h) - \overline{s}_h(w_h, p_h), \tag{17}$$

from which, by differentiation with respect to w_h, we obtain the discrete adjoint equation (15), however with the following stabilization term:

$$\overline{s}_h^{ad}(p_h, \phi_h) = \overline{\overline{s}}_h^{ad}(p_h, \phi_h) := \sum_{K \in \mathcal{T}_h} \delta_K \int_K L(\phi_h)\, L(p_h)\, dK. \tag{18}$$

Differentiating $\mathcal{L}_h$ with respect to u_h and applying the Riesz theorem, being $u_h \in \mathcal{X}_h$, we obtain: $\delta u_h = p_h + \sum_{K \in \mathcal{T}_h} \delta_K \int_K L(p_h)\, dK$.

3.2 The stabilized Lagrangian approach

We consider a stabilization on the Lagrangian functional itself [5], for which our stabilized Lagrangian functional is:

$$\mathcal{L}_h^s(w_h, p_h, u_h) := \mathcal{L}(w_h, p_h, u_h) + S_h(w_h, p_h, u_h), \tag{19}$$

with:

$$S_h(w, p, u) := \sum_{K \in \mathcal{T}_h} \delta_K \int_K R(w; u)\, R^{ad}(p; w)\, dK. \tag{20}$$

This approach can be regarded as a particular case of the "discretize–then–optimize" one if we identify $\overline{s}_h(w_h, p_h)$ with $-S_h(w_h, p_h, u_h)$. By differentiating $\mathcal{L}_h^s$ we obtain the (stabilized) approximate state and adjoint equations (13) and (15), assuming $\overline{s}(w_h, \varphi_h) = s_h(w_h, \varphi_h; u_h)$ and $\overline{s}^{ad}(p_h, \phi_h) = s_h^{ad}(p_h, \phi_h; w_h)$, where:

$$s_h(w_h, \varphi_h; u_h) := -\sum_{K \in \mathcal{T}_h} \delta_K \int_K R(w_h; u_h)\, L^{ad}(\varphi_h)\, dK, \tag{21}$$

$$s_h^{ad}(p_h, \phi_h; w_h) \tag{22}$$
$$:= - \sum_{K \in \mathcal{T}_h} \delta_K \int_K \Big(R^{ad}(p_h; w_h)\, L(\phi_h) - R(w_h; u_h)\, G'(\phi_h) \Big)\, dK,$$

having set $G'(w) := \chi_D g^2 w$. Finally, the cost functional sensitivity reads:

$$\delta u_h(p_h, w_h) = p_h - \sum_{K \in \mathcal{T}_h} \delta_K\, R^{ad}(p_h; w_h). \tag{23}$$

4. A Posteriori Error Estimate

For the definition of an appropriate error estimate for the optimal control problem, we identify the error on the control problem as being the error on the cost functional, as proposed in [3]. Moreover, we propose to separate this error in two parts: the iteration and the discretization error. For the latter we define a suitable estimate according with the duality principles [3], adopted in the course of mesh adaptive strategy.

4.1 Iteration and discretization errors

At each iterative step j of the optimization procedure we consider the following error:

$$|\varepsilon^{(j)}| = |J(w^*, u^*) - J(w_h^j, u_h^j)|, \tag{24}$$

where $*$ indicates optimal variables, while w_h^j stands for the discrete variable evaluated at the step j. If we refine the mesh, according with an adaptive procedure, we certainly reduce the component of the full error $\varepsilon^{(j)}$ (Eq.(24)) related to the numerical approximation at the step j, which we call the *discretization error* $\varepsilon_D^{(j)}$. On the other hand, the part of $\varepsilon^{(j)}$ expressing the difference between the cost functional computed on continuous variables at the step j and the optimal cost functional, which we call the *iteration error* $\varepsilon_{IT}^{(j)}$, can generally increase [5]. From Eq.(24):

$$\varepsilon^{(j)} = \Big(J(w^*, u^*) - J(w^j, u^j) \Big) + \Big(J(w^j, u^j) - J(w_h^j, u_h^j) \Big) = \varepsilon_{IT}^{(j)} + \varepsilon_D^{(j)}; \tag{25}$$

then we will define a posteriori error estimate only for $\varepsilon_D^{(j)}$, the only part of $\varepsilon^{(j)}$ which can be reduced by mesh refinement. Since $\nabla \mathcal{L}(\mathbf{x})$ is linear in $\mathbf{x}$, the iteration error $\varepsilon_{IT}^{(j)}$ becomes $\varepsilon_{IT}^{(j)} = \frac{1}{2}\ (\ \delta u(p^j, u^j)\ ,\ u^* - u^j\ )$, which, in the case of our advection–diffusion control problem (see Sec.2.2 and [5]), can be written as:

$$\varepsilon_{IT}^{(j)} = -\frac{1}{2}\tau \|p^j\|^2_{L^2(\Omega)} - \frac{1}{2}\tau \sum_{r=j+1}^{\infty} (\ p^j, p^r\ )_{L^2(\Omega)}. \tag{26}$$

Since the iteration error can not be correctly evaluated by means of this expression, we can assume that $|\varepsilon_{IT}^{(j)}| \approx \frac{1}{2}\tau\|p^j\|_{L^2(\Omega)}^2$, or, more simply $|\varepsilon_{IT}^{(j)}| \approx \|p^j\|_{L^2(\Omega)}^2$, which leads to the usual criterium $|\varepsilon_{IT}^{(j)}| \approx \|\delta u(p^j)\|$ (L^2–norm).

4.2 A posteriori error estimate and adaptive strategy

We define the a posteriori error estimate for the discretization error only, based on the following theorem ([5]).

THEOREM 1 *For a linear control problem with the stabilized Lagrangian $\mathcal{L}_h^s$ (Eq.(19) and Eq.(20)), the discretization error at the j–th iteration reads:*

$$\varepsilon_D^{(j)} = \frac{1}{2}(\ \delta u(p^j, u^j), u^j - u_h^j\ ) + \frac{1}{2}\nabla\mathcal{L}_h^s(\mathbf{x}_h^j)\cdot(\mathbf{x}^j - \mathbf{x}_h^j) + \Lambda_h(\mathbf{x}_h^j), \quad (27)$$

where $\mathbf{x}_h^j := (w_h^j, p_h^j, u_h^j)$ is the Galerkin–FE approximation and $\Lambda_h(\mathbf{x}_h^j) := S_h(\mathbf{x}_h^j) + s_h(w_h^j, p_h^j; u_h^j)$, being $s_h(w_h^j, p_h^j; u_h^j)$ the stabilization term (21).

Applying (27) to our advection–diffusion control problem and highlighting the contributions on the elements of the mesh $K \in \mathcal{T}_h$ ([3]), we obtain the following estimate:

$$|\varepsilon_D^{(j)}| \le \eta_D^{(j)} := \frac{1}{2}\sum_{K\in\mathcal{T}_h} \{\ (\omega_K^p\rho_K^w + \omega_K^w\rho_K^p + \omega_K^u\rho_K^u) + \lambda_K\ \}, \quad (28)$$

where, according with the symbol definitions given in Sec.3:

$$\begin{aligned}
\rho_K^w &:= \|R(w_h^j; u_h^j)\|_K + h_K^{-\frac{1}{2}}\|r(w_h^j)\|_{\partial K},\\
\omega_K^p &:= \|(p^j - p_h^j) - \delta_K L^{ad}(p^j - p_h^j) + \delta_K G'(w^j - w_h^j)\|_K + h_K^{\frac{1}{2}}\|p^j - p_h^j\|_{\partial K},\\
\rho_K^p &:= \|R^{ad}(p_h^j; w_h^j)\|_K + h_K^{-\frac{1}{2}}\|r^{ad}(p_h^j)\|_{\partial K},\\
\omega_K^w &:= \|(w^j - w_h^j) - \delta_K L(w^j - w_h^j)\|_K + h_K^{\frac{1}{2}}\|w^j - w_h^j\|_{\partial K},\\
\rho_K^u &:= \|\delta u_h(p_h^j, w_h^j) + \delta u(p^j)\|_K = \|p^j + p_h^j - \delta_K R^{ad}(p_h^j; w_h^j)\|_K,\\
\omega_K^u &:= \|u^j - u_h^j\|_K,\\
\lambda_K &:= 2\delta_K\|R(w_h^j; u_h^j)\|_K\ \|G(w_h^j)\|_K,\\
r(w_h^j) &:= \begin{cases} -\frac{1}{2}\left[\nu\frac{\partial w_h^j}{\partial n}\right], & on\ \partial K\backslash\partial\Omega,\\ -\nu\frac{\partial w_h^j}{\partial n}, & on\ \partial K \in \Gamma_N,\end{cases}\\
r^{ad}(p_h^j) &:= \begin{cases} -\frac{1}{2}\left[\nu\frac{\partial p_h^j}{\partial n} + \mathbf{V}\cdot\mathbf{n}\,p_h^j\right], & on\ \partial K\backslash\partial\Omega,\\ -\left(\nu\frac{\partial p_h^j}{\partial n} + \mathbf{V}\cdot\mathbf{n}\,p_h^j\right), & on\ \partial K \in \Gamma_N;\end{cases}
\end{aligned} \quad (29)$$

∂K indicates the boundary of $K \in \mathcal{T}_h$, while $[\cdot]$ stands for the jump of the embraced quantity across ∂K.

To use the estimate (28), we need to evaluate w^j, p^j and u^j. Indeed, we replace w^j and p^j by the respective quadratic reconstructions, $(w_h^j)^q$ and $(p_h^j)^q$, and u^j by $(u_h^j)^q := u_h^j - \tau(\delta u_h((p_h^j)^q, (w_h^j)^q) - \delta u_h(p_h^j, w_h^j))$, according to the steepest-descent iterative method with $\tau^j = \tau$. The following adaptive strategy is then adopted to allow an efficient generation of adapted meshes:

1 we adopt the optimization iterative method till convergence to the iteration error tolerance Tol_{IT}, assuming an initial coarse mesh;

2 we adapt the mesh, balancing the error on the elements $K \in \mathcal{T}_h$, according with the error estimate $\eta_D^{(j)}$ (28), till convergence to the discretization error tolerance Tol_D;

3 we re-evaluate the variables and $\varepsilon_{IT}^{(j)}$ on the adapted mesh: if $\varepsilon_{IT}^{(j)} \geq Tol_{IT}$, we return to point 1 and we repeat the procedure, while if $\varepsilon_{IT}^{(j)}$ is inferior to Tol_{IT}, we stop.

5. A Numerical Test: Pollution Control Problem

We apply the a posteriori error estimates $\eta_D^{(j)}$ (28) for the discretization error and the strategy presented in Sec.4.2 to a numerical test, which can be regarded as a pollution control problem in atmosphere. Our goal consists in regulating the emissions of industrial chimneys to keep the pollutant concentration below a desired threshold in an observation area (a town).

To this aim we consider a simple advection–diffusion model [5, 8], which can be regarded as a *quasi–3D* model: the pollutant concentration w at the emissive height H is described by the advection–diffusion equation introduced in Sec.2.2, while the concentration at soil is obtained by projection by means of the function $g(\mathrm{x}, \mathrm{y})$ described in Sec.2.2. The values assumed by the diffusion coefficient $\nu(\mathrm{x}, \mathrm{y})$ and the function $g(\mathrm{x}, \mathrm{y})$ depend on the distance from the emission sources and the atmospherical stability class (stable, neutral or unstable). In particular, we consider the case of neutral atmospherical conditions and, referring to the domain reported in Fig.1, we assume $\mathbf{V} = V_{\mathrm{x}}\hat{\mathrm{x}} + V_{\mathrm{y}}\hat{\mathrm{y}}$, with $V_{\mathrm{x}} = V\cos(\frac{\pi}{30})$ and $V_{\mathrm{y}} = V\sin(\frac{\pi}{30})$, being $V = 2.5\ m/s$. Moreover we consider that the chimneys maximum rate of emission is $u_{max} = 800\ g/s$ at the emission height $H = 100\ m$, for which the pollutant concentration (we consider SO_2) is higher than the desired level $z_d = 100\ \mu g/m^3$. In Sec.2.2 we have considered the case of u distributed over all the domain Ω, while here we deal with a particular case which can be accommodated in the general case assuming $u = \sum_{i=1}^{N} u_i \chi_i$, where χ_i is the characteristic function of the chimney U_i.

In Fig.2a we report the pollutant concentration at the ground corresponding to the maximum emission rates; in Fig.2b we plot the concentration at ground at the completion of the optimization strategy; we observe that the "optimal" emission rates become $u_1 = 0.0837 \cdot u_{max}$, $u_2 = 0.0908 \cdot u_{max}$ and $u_3 = 1.00 \cdot u_{max}$. In Fig.3 we report a comparison among adapted meshes; in Fig.3a that obtained

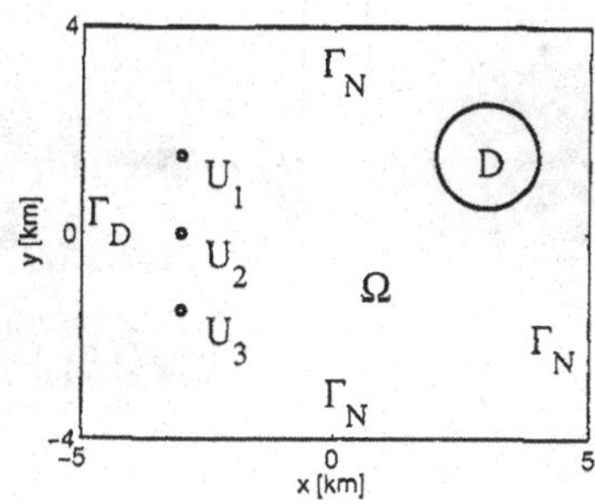

Figure 1. Domain for the pollution problem.

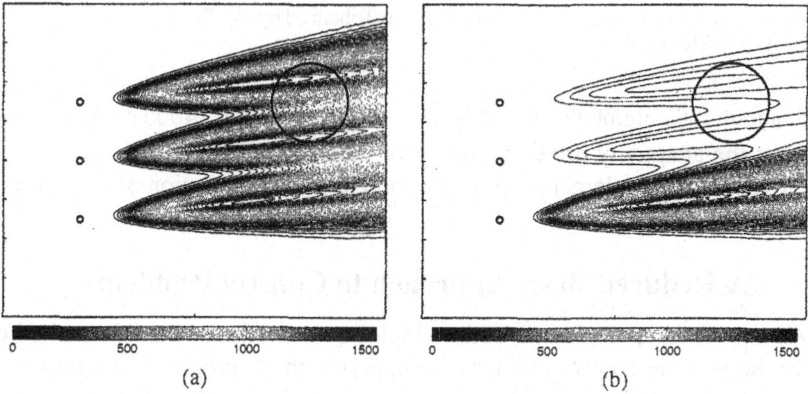

Figure 2. Pollutant concentration $[\mu g/m^3]$ at the ground before (a) and after (b) the regulation of the sources.

by our estimator $\eta_D^{(j)}$ and in Fig.3b by the following estimators [5] (which lead to analogous results):

1 the *energy norm* indicator $(\eta_E^w)^{(j)} := \sum_{K \in \mathcal{T}_h} h_K \, \rho_K^w$;

2 the indicator $(\eta_E^{wpu})^{(j)} := \sum_{K \in \mathcal{T}_h} h_K \, \{(\rho_K^w)^2 + (\rho_K^p)^2 + (\rho_K^u)^2\}^{\frac{1}{2}}$.

For symbols definitions see Eq.(29); the results are compared with those obtained with a fine mesh with about 80000 elements. The adaptivity driven by the error indicator $\eta_D^{(j)}$ leads to concentrate elements in those areas that are more relevant for the optimal control problem. This fact is underlined by comparing the errors on the cost functional and other interesting quantities for the meshes obtained with the different error indicators, but with the same number of elements. E.g., the indicator $\eta_D^{(j)}$ provides an error on the optimal cost functional J of about 20% against the 55% obtained by $(\eta_E^w)^{(j)}$ and $(\eta_E^{wpu})^{(j)}$ with meshes

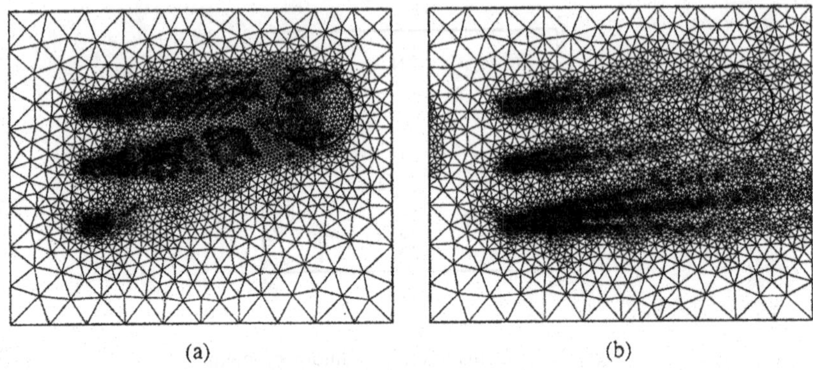

(a) (b)

Figure 3. Adapted meshes (about 14000 elements) obtained by $\eta_D^{(j)}$ Eq.(28) (a) and $(\eta_E^{wpu})^{(j)}$ (b) (and similarly $(\eta_E)^{(j)}$).

with about 4000 elements, and of 6% vs. 15% with about 14000 elements. We see that the adaptivity driven by the error indicator $\eta_D^{(j)}$ permits large savings of number of mesh elements, allowing more efficient resolution of the optimal control problem.

6. A Reduced Basis Approach to Control Problems

As a second approach to improve efficiency, we consider the RB method to solve the optimal control problem, by applying the approach to the state and adjoint equations. For a review on the use of the RB method and for optimal control problems, see [7, 8, 10].

6.1 Reduced basis: abstract formulation

The RB method allows the evaluation of input–output relationships by means of a precise and efficient procedure. The goal consists in calculating a quantity (the output) $s(\mu) = l(w(\mu);\mu)$ depending on the solution of the following parametrized equation:

$$find \quad w(\mu) \in \mathcal{X} \quad : \quad a(w(\mu), v; \mu) = f(v;\mu), \quad \forall v \in \mathcal{X}, \tag{30}$$

where $\mu \in \mathcal{D}$ and $\mathcal{D}$ is a set of parameters, $\mathcal{X}$ is a Hilbert space, the form $a(\cdot,\cdot;\mu)$ is bilinear, continuous and coercive and the forms $f(\cdot;\mu)$ and $l(\cdot;\mu)$ are linear and continuous, for all μ. Moreover, we assume that the form $a(\cdot,\cdot;\mu)$ is affine parameter dependent, that is:

$$a(w(\mu), v; \mu) = \sum_{q=1}^{Q} \sigma^q(\mu) a^q(w, v), \quad \forall w, v \in \mathcal{X}, \quad \forall \mu \in \mathcal{D}, \quad q = 1, \ldots, Q, \tag{31}$$

where $\sigma^q : \mathcal{D} \to R$ are parameter–dependent functions, while $a^q : \mathcal{X} \times \mathcal{X} \to R$ are parameter–independent forms; affine parameter dependence is required also for $f(\cdot;\mu)$ and $l(\cdot;\mu)$. To build the RB space we need to introduce a finite dimensional subspace $\mathcal{X}_h$ of $\mathcal{X}$, which we identify with a Galerkin–FE space associated with a very fine triangulation of the domain Ω. The Galerkin–FE element method consists in solving the following $\mathcal{N}$–dimensional problem $find\ w_h(\mu) \in \mathcal{X}_h\ :\ a(w_h(\mu), v; \mu) = f(v),\ \forall v \in \mathcal{X}_h$, which, if $\mathcal{N}$ is large, leads to computational expensive evaluations of the output $s_h(\mu) = l(w_h(\mu); \mu)$ for several values of the input μ. We consider a set of samples $S_N^\mu = \{\mu^i \in \mathcal{D}, i = 1, \ldots, N\}$ and we define the N–dimensional RB space as $W_N = span\{\zeta^i, i = 1, \ldots, N\}$, where $\zeta^n = w_h(\mu^n)$, with $n = 1, \ldots, N$. The RB method consists in evaluating the output $s_N(\mu) = l(w_N(\mu); \mu)$, where $w_N(\mu)$ is given by the following problem (of dimension N):

$$find\quad w_N(\mu) \in W_N \quad : \quad a(w_N(\mu), v; \mu) = f(v), \quad \forall v \in W_N. \tag{32}$$

Then we write $w_N(\mu)$ as $w_N(\mu) = \sum_{j=1}^N w_{N_j}(\mu)\zeta^j$, being $\underline{w}_N(\mu) = \{w_{N_1}(\mu),$
$\ldots, w_{N_N}(\mu)\}^T$ the solution of the following linear system of order N:

$$\underline{A}_N(\mu)\underline{w}_N(\mu) = \underline{F}_N(\mu), \tag{33}$$

where $A_{N_{i,j}}(\mu) = a(\zeta^j, \zeta^i; \mu)$ and $F_{N_i}(\mu) = f(\zeta^i, \mu)$, with $i, j = 1, \ldots, N$; the output $s_N(\mu)$ is calculated as $s_N(\mu) = \underline{L}_N(\mu)^T w_N(\mu)$, where $L_{N_i}(\mu) = l(\zeta^i; \mu)$. From the affine dependence property, we can split the matrix $\underline{A}_N(\mu)$ and the vectors $\underline{F}_N(\mu)$ and $\underline{L}_N(\mu)$ into a parameter–dependent part and a parameter–independent part. In the case of matrix $\underline{A}_N(\mu)$, this means that $\underline{A}_N(\mu) = \sum_{q=1}^Q \sigma^q(\mu)\underline{A}_N^q$, where $\underline{A}_{N_{i,j}}^q = a^q(\zeta^j, \zeta^i)$ is a parameter–independent matrix. Similar expressions hold for $\underline{F}_N(\mu)$ and $\underline{L}_N(\mu)$. These decompositions allow a very convenient computational procedure composed by off–line and on–line stages. In the off–line stage we afford the larger computational costs, computing the basis ζ^n of W_N and assembling the matrices $\underline{A}_N^q$ and the vectors $\underline{F}_N^q$ and $\underline{L}_N^q$, which require N FE solutions and inner products. In the on–line stage, given μ, we assemble $\underline{A}_N(\mu)$, $\underline{F}_N(\mu)$ and $\underline{L}_N(\mu)$, we solve the system (33) and we compute $s_N(\mu)$. Let us notice that N is usually very low with respect to $\mathcal{N}$ according to the precision required on $s_N(\mu)$; this leads to large computational costs savings in case of recursive evaluation of $s_N(\mu)$ for different parameters μ.

6.2 The reduced basis method applied to control problems

The resolution of optimal control problems by an iterative method leads to a recursive resolution of the state and adjoint equations. The computational cost of the whole procedure can therefore be quite relevant, especially if great precision is required. In this context, see Sec.2.1, the input of the state equation can be regarded as the control function u, while the output as the state variable w. Similarly, the adjoint equation input is the observation on the system, related to w, while the output is the adjoint variable p itself, which, by means of the iterative method adopted (e.g., the steepest–descent method), becomes the input for the state equation. The iterative optimization method can be seen as a recursive evaluation of an input–output relationship. Moreover, it can be interesting to perform an optimization for different values of some physical or geometrical parameters [8]: e.g., referring to the pollution control problem introduced in Sec.5, the diffusivity, the velocity of the wind, or the reciprocal distance among the chimneys. The main idea consists in parametrizing both the state and adjoint equations by the parameters of interest and the source terms (related respectively to u and w); then we solve these equations by means of the RB method in the course of the optimization iterative procedure. The strategy allows large savings of computational costs with respect to the conventional FE iterative optimization method. The RB space for the adjoint equation does not need to be coherent with that for the state equation.

6.3 Numerical tests

We report two numerical tests, which refer to the pollution control problem considered in Sec.5.

In the first case we assume that at the initial step the emissions of the chimneys are respectively the 45%, 0% and 55% of the total emission $u_{tot} = 2700\ g/s$. The optimization procedure leads to the following distribution of emissions on the three chimneys: 3.49%, 0% and 55.02% of u_{tot}. The saving in computational costs with respect to the resolution by the FE method is about the 73%, having chosen $\tau = 800$ and an error tolerance on J of 10^{-8}.

The second test considers as parameters the emissions of the chimneys and the wind velocity field, i.e. $\mu = \{u_1, u_2, u_3, V_x, V_y\}$, where $\mathbf{V} = V_x\hat{\mathrm{x}} + V_y\hat{\mathrm{y}}$, with $\sqrt{V_x^2 + V_y^2} = 1$. We assume $V_x = \cos(\frac{\pi}{4})$ and $V_y = \sin(\frac{\pi}{4})$ and we start with the following initial emissions 30%, 40% and 30% of u_{tot}. The optimization procedure provides the following optimal emissions, respectively for the three chimneys, the 30%, 38.8% and 7.3% of u_{tot}. The RB strategy, with $\tau = 800$ and the RB dimension $N = 81$, allows a 55% time saving with respect to the FE method, with a final error on J and u respectively about $3.9 \cdot 10^{-9}$ and $1.1 \cdot 10^{-6}$. The original grid is made of 1700 nodes.

7. Conclusions

We have proposed two strategies to improve the efficiency of the numerical resolution of optimal control problems governed by linear advection–diffusion equations, in the context of an iterative optimization procedure. In particular, having identified the error on the control problem as error on the cost functional, we have separated it into the iteration and discretization errors. For the latter we have proposed an a posteriori error estimate, which is adopted in a strategy of grid adaptivity. Then we have considered the RB method, applied to the equations governing the control problem itself, in order to save computational costs by adopting a reliable method. The efficiency of these approaches is proved by numerical tests, that are concerned with a pollution control problem in atmosphere.

References

[1] V. Agoshkov. *Optimal Control Methods and Adjoint Equations in Mathematical Physics Problems.* Institute of Numerical Mathematics, Russian Academy of Science, 2003.

[2] R. Becker. Mesh adaptation for stationary flow control. *J. Math. Fluid Mech.* 3:317-341, 2001.

[3] R. Becker, H. Kapp, R. Rannacher. Adaptive finite element methods for optimal control of partial differential equations: basic concepts. *SIAM J. Control Optim.* 39:113-132, 2000.

[4] S.S. Collis, M. Heinkenschloss. Analysis of the Streamline Upwind/Petrov Galerkin method applied to the solution of optimal control problems. *CAAM report* TR02-01, `http://www.caam.rice.edu`

[5] L. Dedè, A. Quarteroni. Optimal control and numerical adaptivity for advection–diffusion equations. *MOX report* 55.2005, `http://mox.polimi.it`, *M^2AN*, to appear, 2005.

[6] J.L. Lions. *Optimal Control of Systems Governed by Partial Differential Equations.* Springer–Verlag, New York, 1971.

[7] C. Prud'homme, D. Rovas, K. Veroy, Y. Maday, A.T. Patera, G. Turinici. Reliable real–time solution of parametrized partial differential equations: reduced–basis output bound methods. *J. Fluids Eng.* 124:70-80, 2002.

[8] A. Quaini, G. Rozza. Reduced basis methods for advection–diffusion optimal control problems. In progress. 2005.

[9] A. Quarteroni, A. Valli. *Numerical Approximation of Partial Differential Equations.* Springer, Berlin and Heidelberg, 1994.

[10] G. Rozza. Optimal flow control and reduced basis techniques in shape design with applications in haemodynamics. PhD Thesis, EPFL, 2005.

A NEW LOW RANK QUASI-NEWTON UPDATE SCHEME FOR NONLINEAR PROGRAMMING

R. Fletcher,[1]

[1] *Department of Mathematics, University of Dundee, Dundee DD1 4HN, Scotland, UK, fletcher@maths.dundee.ac.uk*

Abstract

A new quasi-Newton scheme for updating a low rank positive semi-definite Hessian approximation is described, primarily for use in sequential quadratic programming methods for nonlinear programming. Where possible the symmetric rank one update formula is used, but when this is not possible a new rank two update is used, which is not in the Broyden family, although invariance under linear transformations of the variables is preserved. The representation provides a limited memory capability, and there is an ordering scheme which enables 'old' information to be deleted when the memory is full. Hereditary and conjugacy properties are preserved to the maximum extent when minimizing a quadratic function subject to linear constraints. Practical experience is described on small (and some larger) CUTE test problems, and is reasonably encouraging, although there is some evidence of slow convergence on large problems with large null spaces.

keywords: nonlinear programming, filter, SQP, quasi-Newton, symmetric rank one, limited memory.

1. Introduction

This work arises as part of a project to provide effective codes for finding a local solution $\mathbf{x}^*$ of a nonlinear programming (NLP) problem, which for convenience we express in the form

$$\begin{array}{ll} \underset{\mathbf{x}\in\mathbb{R}^n}{\text{minimize}} & f(\mathbf{x}) \\ \text{subject to} & c_i(\mathbf{x}) \geq 0 \quad i = 1, 2, \ldots, m, \end{array} \tag{1.1}$$

although in practice a more detailed formulation would be appropriate, admitting also equations, linear constraints and simple bounds. In particular we aim to develop a new trust-region filter SQP (sequential quadratic programming) code which only uses first derivatives of the problem functions $f(\mathbf{x})$ and $c_i(\mathbf{x})$.

Please use the following format when citing this chapter:
Author(s) [insert Last name, First-name initial(s)], 2006, in IFIP International Federation for Information Processing, Volume 199, System Modeling and Optimization, eds. Ceragioli F., Dontchev A., Furuta H., Marti K., Pandolfi L., (Boston: Springer), pp. [insert page numbers].

Filter methods for NLP were first introduced by Fletcher and Leyffer [6], and a production code `filterSQP` has been shown to be reliable and reasonably efficient. This code requires second derivatives of the problem functions to be made available by the user. The code has been hooked up to the AMPL modelling language, which includes a facility for automatically providing second derivatives, and is available for use under NEOS. More recently, convergence proofs for different types of filter method have been developed, and a code `filter2` has been written to implement the method considered in the paper of Fletcher, Leyffer and Toint [7]. This code also requires second derivatives to be made available. The practical performance of `filter2` is similar to that of `filterSQP`. An early version of the new quasi-Newton filter SQP code, referred to as `filterQN`, has already been tried on a range of problems with some success.

In view of the ready availability of second derivatives through the AMPL modelling language, one might question whether there is a need for NLP algorithms that use only first derivatives. To answer this, one should first point to the success of the NLP solver SNOPT (Gill, Murray and Saunders, [8]), based on an augmented Lagrangian formulation, which only requires first derivatives to be available. This is one of the most effective existing codes for NLP. Other reasons include the fact that Hessian matrices are often indefinite, which in an SQP context might render some QP solvers inapplicable. Even if the QP solver can handle indefinite matrices, there is usually no guarantee that a global (or even local) solution is found to the QP subproblems. (Although it has to be said that there is little evidence that this is a serious difficulty in practice.) Another argument is that NLP problems often have small or even empty null spaces, in which case only a small part of the Hessian is in a sense useful.

There are certain types of problem however in which Hessian calculations can be seriously time consuming and hence impracticable. Such an example is the optimal design of a Yagi-Uda antenna, shown to me by Martijn van Beurden (see [1] for details). The antenna is constructed from a number of wires along an axis, and there are two design variables (length and position along the axis) for each wire. Also there are 31 complex variables on each wire to model the current (62 real variables if complex arithmetic is not available). These variables satisfy a complex dense nonsingular system of linear equations. When modelled in AMPL, both the design and current variables appear explicity in the model, and the second derivative calculation is seriously time consuming. The largest problem that could be handled by AMPL via NEOS, with various solvers, had 5 wires, and hence 10 design variables and 310 real current variables. For this problem, filterSQP took about 2 hours to solve the problem whereas SNOPT, which only requires first derivatives, took about 15 minutes. For filterSQP, the memory usage was 610MB.

An much more effective procedure is not to use AMPL at all, and to use the complex linear equations to eliminate the complex variables, leaving a much smaller problem in just the design variables. Factors derived from the complex equations can be used efficiently to compute the gradient of the reduced problem, whereas computing the Hessian of the reduced problem remains very expensive. When posed in this way, various QN-SQP solvers, such as DONLP2, NPSOL and an early version of the filterQN code, were able to solve the 5-wire problem in around one minute. In fact even the 20-wire problem, with 40 design variables and 2480 real current variables could be solved in reasonable time.

This paper describes a new quasi-Newton scheme for updating a low rank positive semi-definite Hessian approximation, primarily for use in SQP methods for NLP. The paper is organised as follows. Section 2 reviews existing quasi-Newton methodology, and gives two results relating to hereditary conditions and quadratic termination for the symmetric rank one update, one of which may not be well known. Section 3 considers the implications for NLP, and describes the form UU^T of the representation. In Section 4, the interpretation as a limited memory approximation is discussed, and an it is shown how to update the representation so that the most recent information is contained in the leftmost columns of U. Section 5 focusses on how the projection part of the BFGS update might be implemented in this context, and Section 6 describes a new scheme which combines this update with the symmetric rank one update, for use when the latter alone is inapplicable. The outcome is a new rank two update which is not in the Broyden family, although invariance under linear transformations of the variables is preserved. The underlying motivation is seen to be the preservation of hereditary properties to the maximum extent. Conjugacy properties of the update in the quadratic case are brought out in Section 7 and a result somewhat akin to a hereditary property is shown to hold. Preliminary practical experience is described in Section 8 on small (and some larger) CUTE test problems, and is reasonably encouraging, although there is some evidence of slow convergence on large problems with large null spaces. Some conclusions are drawn in Section 9.

2. Quasi-Newton methodology

In this section we review existing quasi-Newton (QN) methodology in the context of uncontrained optimization ($m = 0$ in (1.1)). A QN method is based on updating symmetric matrices $B^{(k)}$ that approximate the Hessian matrix $\nabla^2 f$ of the objective function. These matrices are then used on iteration k of a Newton-like line search or trust region method. The initial matrix $B^{(1)}$ is arbitrary and is usually chosen to be positive definite, for example the unit matrix. At the completion of iteration k of the QN method, difference vectors

$$\delta^{(k)} = \mathbf{x}^{(k+1)} - \mathbf{x}^{(k)} \tag{2.1}$$

in the variables, and

$$\gamma^{(k)} = \nabla f(\mathbf{x}^{(k+1)}) - \nabla f(\mathbf{x}^{(k)}) \tag{2.2}$$

in the gradients are available, and an updated matrix $B^{(k+1)}$ is computed, usually so as to satisfy the *secant condition*

$$B^{(k+1)}\delta^{(k)} = \gamma^{(k)} \tag{2.3}$$

which would be satisfied to first order by the true Hessian $\nabla^2 f(\mathbf{x}^{(k)})$.

There are many ways to satisfy (2.3), but there are two well known QN updating formulae which have featured in many applications. These are the *Symmetric Rank 1 (SR1)* formula

$$B^{(k+1)} = B + \frac{(\gamma - B\delta)(\gamma - B\delta)^T}{(\gamma - B\delta)^T\delta}, \tag{2.4}$$

suggested independently by various authors in 1968-69, and the *BFGS formula*

$$B^{(k+1)} = B - \frac{B\delta\delta^T B}{\delta^T B\delta} + \frac{\gamma\gamma^T}{\delta^T\gamma} \tag{2.5}$$

suggested independently by various authors in 1970. Superscript (k) has been suppressed on all vectors and matrices on the right hand sides of these formulae, and also elsewhere in the subsequent presentation, so as to avoid over-complicating the notation. More details and references may be found in Fletcher [5] for example.

An important property of the BFGS formula is that if $B^{(k)}$ is positive definite and $\delta^T\gamma > 0$, then $B^{(k+1)}$ is positive definite. Since $\nabla^2 f(\mathbf{x}^*)$ is positive semi-definite and usually positive definite, it is desirable that the approximating matrices $B^{(k)}$ also satisfy this property. It then follows that the Newton direction $-(B^{(k)})^{-1}\nabla f(x^{(k)})$ is a descent direction, and a line search along this direction enables $f(\mathbf{x})$ to be reduced. It is also possible to implement the line search in such a way that $\delta^T\gamma > 0$ always holds. Because of these properties, the BFGS formula has been the method of choice in most cases. On the other hand the denominator $(\gamma - B\delta)^T\delta$ in the SR1 formula may be negative, so that $B^{(k+1)}$ is not positive semi-definite, or even zero, in which case the formula breaks down. However the SR1 formula has been used, particularly in the context of trust region methods, with some safeguards. Indeed there is some evidence (Conn, Gould and Toint [4]) that the matrices $B^{(k)}$ converge more rapidly to $\nabla^2 f(\mathbf{x}^*)$ when the SR1 update is used.

Both formulae usually generate dense matrices $B^{(k)}$, even when the true Hessian $\nabla^2 f$ is sparse, and so are only suitable for solving small to medium size problems. Special purpose methods have been developed for solving large systems, for example the limited memory BFGS (L-BFGS) method (Nocedal, [9]), the sparse Hessian update scheme of Powell and Toint [11], and the use of the SR1 update for partially separable functions (Conn, Gould and Toint, [3]).

Another pointer to the effectiveness of a QN update, albeit somewhat indirect, is whether the property of *quadratic termination* can be proved. That is to say, can the associated QN method find the minimizer of a quadratic function in a finite number of steps. This property usually holds for the BFGS method only if exact line searches along the Newton direction are carried out. A stronger termination property holds for the SR1 method in which the differences $\delta^{(k)}$ can be defined in an almost arbitrary manner. This may be a pointer to the effectiveness of the SR1 update in a trust region context. This result is established in the following well known theorem.

THEOREM 1 *Consider n SR1 updates using difference vectors $\delta^{(k)}$ and $\gamma^{(k)}$ for $k = 1, 2, \ldots, n$, where $\gamma^{(k)} = W\delta^{(k)}$ and W is symmetric. If $B^{(1)}$ is symmetric, and if for $k = 1, 2, \ldots, n$ the denominators in (2.4) are non-zero, and the vectors $\delta^{(k)}$ are linearly independent, then $B^{(n+1)} = W$.*

Proof Clearly the SR1 update preserves the symmetry of the matrices $B^{(k)}$. It is shown by induction that

$$B^{(k)}\delta^{(j)} = \gamma^{(j)} \qquad j = 1, 2, \ldots, k-1, \tag{2.6}$$

where $1 \leq k \leq n+1$. For $k = 1$ the condition is vacuous and hence true. Now let it be true for some k such that $1 \leq k \leq n$. The definition of $B^{(k+1)}$ gives

$$B^{(k+1)}\delta^{(j)} = B^{(k)}\delta^{(j)} + \frac{\mathbf{u}^{(k)}\mathbf{u}^{(k)T}\delta^{(j)}}{\mathbf{u}^{(k)T}\delta^{(k)}} \tag{2.7}$$

where $\mathbf{u}^{(k)} = \gamma^{(k)} - B^{(k)}\delta^{(k)}$. For $j = k$ the right hand side is $B^{(k)}\delta^{(k)} + \mathbf{u}^{(k)}$ which is equal to $\gamma^{(k)}$ by definition of $\mathbf{u}^{(k)}$. For $j < k$ it follows from (2.6) that $B^{(k)}\delta^{(j)} = \gamma^{(j)}$, and also using the definition of $\mathbf{u}^{(k)}$ and symmetry of $B^{(k)}$ that

$$\mathbf{u}^{(k)T}\delta^{(j)} = (\gamma^{(k)} - B^{(k)}\delta^{(k)})^T\delta^{(j)} = \gamma^{(k)T}\delta^{(j)} - \delta^{(k)T}\gamma^{(j)}.$$

Because $\gamma^{(j)} = W\delta^{(j)}$ for all $j = 1, 2, \ldots, n$ it follows for $j < k$ that $\mathbf{u}^{(k)T}\gamma^{(j)} = 0$. Thus for both $j = k$ and $j < k$ it has been shown that $B^{(k+1)}\delta^{(j)} = \gamma^{(j)}$ and hence (2.6) has been established with $k+1$ replacing k. Hence by induction, (2.6) is true for all $k = 1, 2, \ldots, n+1$.

For $k = n + 1$, and using $\gamma^{(j)} = W\delta^{(j)}$, (2.6) can be written as

$$B^{(n+1)}\delta^{(j)} = W\delta^{(j)} \qquad j = 1, 2, \ldots, n, \tag{2.8}$$

or as

$$B^{(n+1)}\Delta = W\Delta, \tag{2.9}$$

where Δ is an $n \times n$ matrix with columns $\delta^{(j)}$, $j = 1, 2, \ldots, n$. But Δ is nonsingular by the linear independence assumption, so it follows that $B^{(n+1)} = W$. **QED**

A consequence of the therorem is that if the SR1 method is applied to minimize a quadratic function with positive definite Hessian W, then a Newton iteration on iteration $n + 1$ will locate the minimizer exactly. A key feature of the proof is the establishment of so-called *hereditary conditions* (2.6), in which secant conditions (2.3) from previous iterations remain satisfied by subsequent $B^{(k)}$ matrices. In other words, when the correct behaviour $B^{(k+1)}\delta^{(k)} = \gamma^{(k)} = W\delta^{(k)}$ is introduced, it persists in subsequent $B^{(k)}$ matrices.

A less well known result in the quadratic case is that if W is positive definite, and $B^{(1)} = 0$ is chosen, then the denominators in the SR1 update are all positive, and the matrices $B^{(k)}$ are positive semi-definite.

THEOREM 2 *Consider n SR1 updates using difference vectors $\delta^{(k)}$ and $\gamma^{(k)}$ for $k = 1, 2, \ldots, n$, where $\gamma^{(k)} = W\delta^{(k)}$ and W is symmetric positive definite. If $B^{(1)} = 0$, and the vectors $\delta^{(k)}$ $k = 1, 2, \ldots, n$ are linearly independent, then the SR1 updates are well defined, the matrices $B^{(k)}$ are positive semi-definite of rank $k - 1$, and $B^{(n+1)} = W$.*

Proof Without loss of generality we can take $W = I$ since the SR1 update is independent under linear transformations ([5], Theorem 3.3.1). It is shown by induction that

$$B^{(k)}\delta^{(j)} = \delta^{(j)} \qquad j = 1, 2, \ldots, k - 1, \tag{2.10}$$

and

$$B^{(k)}\mathbf{v} = 0 \quad \forall\, \mathbf{v} \in \{\mathbf{v} \,|\, \mathbf{v}^T\delta^{(j)} = 0,\ j = 1, 2, \ldots, k - 1\}, \tag{2.11}$$

so that $B^{(k)}$ is an orthogonal projector of rank $k - 1$. This is true for $k = 1$ because $B^{(1)} = 0$. Because $\gamma^{(k)} = \delta^{(k)}$, (2.4) may be written

$$B^{(k+1)} = B + \frac{(I - B)\delta\delta^T(I - B)}{\delta^T(I - B)\delta}. \tag{2.12}$$

Because the $\delta^{(k)}$ are linearly independent, it follows that the denominator in (2.12) is positive. As in Theorem 1 $B^{(k+1)}\delta^{(k)} = \delta^{(k)}$, and for $j < k$ it follows using (2.10) that $(I - B^{(k)})\delta^{(j)} = 0$ and hence $B^{(k+1)}\delta^{(j)} = \delta^{(j)}$. Also the rank one correction in (2.12) is in span$\{\delta^{(1)}, \delta^{(2)}, \ldots, \delta^{(k)}\}$, so (2.11) follows for $B^{(k+1)}$. Thus the inductive step has been established. The rest of the theorem follows as for Theorem 1. **QED**

In a non-quadratic context, this theorem suggests that if $B^{(1)} = 0$ is chosen, then there is less likelihood that the SR1 update will break down, or give rise to an indefinite matrix.

3. Quasi-Newton updates in NLP

This section looks at the new issues that arise when QN updates are used in the context of an NLP calculation, particularly when n is large. In this case it is impracticable to update a full dense Hessian. However, it is only the reduced Hessian that needs to be positive semi-definite at a solution, and local and superlinear convergence can be achieved without updating a full Hessian. A low rank Hessian approximation is presented which allows rapid local convergence of SQP, whilst requiring much less storage to implement.

When the NLP problem has nonlinear constraints, it is the Hessian $W = \nabla^2\mathcal{L}(\mathbf{x}, \boldsymbol{\lambda}^*)$ of a Lagrangian function $\mathcal{L}(\mathbf{x}, \boldsymbol{\lambda}^*) = f(\mathbf{x}) - \mathbf{c}(\mathbf{x})^T\boldsymbol{\lambda}^*$ that determines the local convergence properties of an SQP method, where $\boldsymbol{\lambda}^*$ is the vector of KT multipliers at the solution (see, for example [5]). In this case $\gamma^{(k)}$ should be computed from differences in the Lagrangian gradients, and Nocedal and Overton [10] recommend

$$\gamma^{(k)} = \nabla\mathcal{L}(\mathbf{x}^{(k+1)}, \boldsymbol{\lambda}^{(k+1)}) - \nabla\mathcal{L}(\mathbf{x}^{(k)}, \boldsymbol{\lambda}^{(k+1)}) \tag{3.1}$$

as an effective choice (amongst others), where $\boldsymbol{\lambda}^{(k+1)}$ is the most recently available estimate of the KT multipliers.

In general, the Lagrangian Hessian matrix $W^* = \nabla^2\mathcal{L}(\mathbf{x}^*, \boldsymbol{\lambda}^*)$ at the solution may not be positive semi-definite, in contrast to the unconstrained case. Only the $d \times d$ reduced Hessian matrix $Z^T W^* Z$ is positive semi-definite (and usually positive definite), where columns of the matrix Z are a basis for the null space $\mathcal{N}^* = \{\mathbf{z} \mid A^{*T}\mathbf{z} = \mathbf{0}\}$, where A^* denotes the matrix of active constraint gradients at the solution (see [5] for example). Quite often the dimension of $\mathcal{N}^*$ much smaller than n. A related consequence of this is that the denominator $\delta^T\gamma$ that arises in the BFGS method cannot be assumed to be positive, again in contrast to the unconstrained case.

Sufficient conditions for the Q-superlinear convergence of SQP methods under mild assumptions are that

$$B^{(k)}Z \sim W^*Z, \tag{3.2}$$

(see Boggs, Tolle and Wang [2]). That is to say, $B^{(k)}$ should map the null space correctly in the limit, but $B^{(k)} \sim W^*$ is not necessary. In this paper we aim to achieve something akin to (3.2) based on the quadratic termination properties of the SR1 update.

Quadratic termination for an NLP solver relates to how the solver performs when applied to solve the equality constrained QP problem

$$\begin{array}{ll} \underset{\mathbf{x}\in\mathbb{R}^n}{\text{minimize}} & q(\mathbf{x}) = \frac{1}{2}\mathbf{x}^T W \mathbf{x} + \mathbf{c}^T\mathbf{x} \\ \text{subject to} & A^T\mathbf{x} = \mathbf{b}, \end{array} \tag{3.3}$$

where $A \in \mathbb{R}^{n\times m}$, $m < n$, and $\text{rank}(A) = m$. We let $Z \in \mathbb{R}^{n\times d}$ be a matrix whose columns are a basis for the null space $\mathcal{N}(A^T) = \{\mathbf{z} \mid A^T\mathbf{z} = \mathbf{0}\}$ of dimension $d = n - m$. The QP problem (3.3) has a unique solution if and only if the reduced Hessian Z^TWZ is positive definite. In this case, if

$$B^{(k)}Z = WZ, \tag{3.4}$$

if $\mathbf{x}^{(k)}$ is a feasible point, and if iteration k is an SQP iteration

$$Z^TB^{(k)}Z\mathbf{t}^{(k)} = -Z^T\nabla f(x^{(k)}) \tag{3.5}$$

$$\mathbf{x}^{(k+1)} = \mathbf{x}^{(k)} + Z\mathbf{t}^{(k)}, \tag{3.6}$$

then $\mathbf{x}^{(k+1)}$ solves (3.3). If the SR1 update is used, and if d consecutive and linearly independent steps $\delta^{(k-1)}, \delta^{(k-2)}, \ldots, \delta^{(k-d)}$ in $\mathcal{N}(A^T)$ can be completed, then these vectors can form the columns of Z, and (3.4) follows by the hereditary properties of the SR1 update. Thus quadratic termination is obtained under these conditions.

These results do not require W to be positive definite. However, if necessary a related QP problem with the same solution can be defined by adding a squared penalty $\frac{1}{2}\sigma(A^T\mathbf{x}-\mathbf{b})^T(A^T\mathbf{x}-\mathbf{b})$ into $q(\mathbf{x})$. If Z^TWZ is positive definite and σ is sufficiently large, then the Hessian $W + \sigma AA^T$ of the modified objective function is positive definite.

Quadratic termination for an inequality constraint QP problem is less obvious, because it is uncertain what will happen to $B^{(k)}$ before the correct active set is located. However tests with `filterQN` on some inequality QP problems from the CUTE test set did exhibit termination.

In this paper we represent $B^{(k)}$ by the low rank positive semi-definite approximation

$$B^{(k)} = U^{(k)}U^{(k)T}, \tag{3.7}$$

where $U^{(k)}$ is a dense $n \times r$ matrix. Usually $U^{(k)}$ has rank r but the current implementation does not guarantee that this is so. Clearly $B^{(k)}$ is positive semi-definite and has the same rank as $U^{(k)}$. Of course only $U^{(k)}$ is stored, and $B^{(k)}$

is recovered implicitly from (3.7). We shall use the SR1 formula to update $U^{(k)}$ whenever possible. When this is not possible, we shall arrange matters so as to retain as many of the most recent hereditary properties as possible in $B^{(k)}$. By this means we hope that, once the correct active set is located by the NLP solver, we shall then build up hereditary properties in the correct null space, and hence obtain rapid convergence.

In using (3.7), it follows from the remarks in Section 2 that there may be some advantage to be gained by initializing $B^{(1)} = 0$. We do this simply by setting $r = 0$. In general, a trust region constraint $\|\delta\| \leq \rho$ will ensure that the SQP subproblem is bounded, so that no difficulty arises from the rank deficiency of $B^{(k)}$.

In passing, we note that even less storage is needed if an approximating matrix $M^{(k)} \sim Z^T W Z$ is used, and $B^{(k)} = Z^{(k)} M^{(k)} Z^{(k)T}$, where $Z^{(k)}$ is a current approximation to Z^*, obtained from the current QP subproblem. However, the active set in the QP subproblem can change considerably from iteration to iteration, and it is not easy to suggest a robust strategy for updating $M^{(k)}$.

4. Updating the representation $B^{(k)} = U^{(k)}U^{(k)T}$

In this section we consider some issues relating the use of an update formula in conjunction with (3.7). The SR1 update is seen to be most suitable, if it is applicable. It is shown how to implement the update so as order the columns of U in such a way that the most recent information is contained in the leftmost columns. This provides a useful limited memory capability.

The SR1 update (2.4) can only be used to update $U^{(k)}$ if $(\gamma - B\delta)^T\delta > 0$. In this case, one way to update $U^{(k)}$ would be simply to append the column vector

$$\mathrm{u} = \frac{\gamma - B\delta}{((\gamma - B\delta)^T\delta)^{1/2}} \tag{4.1}$$

to $U^{(k)}$. Unless u is in the range space of $U^{(k)}$, we would obtain a matrix $U^{(k+1)}$ with rank $r^{(k+1)} = r^{(k)} + 1$. Thus the SR1 update provides a way of building up information in U and is used whenever possible. For the BFGS update, the first two terms on the right hand side of (2.5) perform a projection operation which usually reduces the rank of B by one. The final rank one term restores the rank, so that usually $r^{(k+1)} = r^{(k)}$. Thus the BFGS update is unable to build up information in U.

The low rank representation (3.7) gives the method the flavour of a limited memory method, and indeed we shall introduce a *memory limit*

$$r \leq r_{max}. \tag{4.2}$$

Ideally r_{max} should be greater or equal to the dimension d of the null space at the solution. In this case we would hope for local superlinear convergence of our SQP method. When $r_{max} < d$ only linear convergence can be expected, and it is not clear how slow this might be. However some hope might be derived from the fact that methods such as conjugate gradients and some new gradient methods are able to solve instances of very large problems in relatively few iterations.

When the memory is full, the SR1 update would usually cause $r^{(k+1)}$ to be greater than r_{max}, so that we are faced with the need to delete a column of $U^{(k+1)}$ from the memory. We would like to ensure that when this occurs, it is the *oldest* information that is deleted, in a certain sense. We can exploit the fact that the representation (3.7) is not unique, to the extent that

$$B = UU^T = UQQ^TU^T = (UQ)(UQ)^T$$

where Q is any orthogonal matrix. We shall choose Q in such a way that the most recent information is contained in the *leftmost* columns of U, when the SR1 update is being used. Then we delete the rightmost column of U when the memory is overfull. We shall refer to this as the *priority ordering* of U.

We therefore express

$$U^{(k+1)} = [\mathbf{u} \quad U]\,Q \tag{4.3}$$

(suppressing superscript (k)), in which

$$\mathbf{u} = \frac{\gamma - B\delta}{((\gamma - B\delta)^T\delta)^{1/2}} = \frac{\gamma - U\mathbf{v}}{\alpha}, \tag{4.4}$$

where $\mathbf{v} = U^T\delta$ and $\alpha = (\delta^T\gamma - \mathbf{v}^T\mathbf{v})^{1/2}$. Using (4.4) we may write

$$[\mathbf{u} \quad U] = [\gamma \quad U]\begin{bmatrix} 1/\alpha & & & & \\ -v_1/\alpha & 1 & & & \\ -v_2/\alpha & & 1 & & \\ \vdots & & & \ddots & \\ -v_r/\alpha & & & & 1 \end{bmatrix}. \tag{4.5}$$

Next we implicitly transform the spike matrix to an upper triangular matrix, R say, by postmultiplying by a product of plane rotation matrices in columns $(1, j)$ for $j = r+1, r, \ldots, 2$, each rotation being chosen so as to eliminate the entry in row j of column 1. These rotations are explicitly applied to the columns of the left hand side matrix in (4.5).

Denoting the product of plane rotations by Q, the resulting matrix may be expressed as

$$U^{(k+1)} = [\mathbf{u} \quad U]\,Q = [\gamma \quad U]\,R. \tag{4.6}$$

It follows by a simple induction argument that

- column 1 of $U^{(k+1)}$ depends only on $\gamma^{(k)}$
- column 2 of $U^{(k+1)}$ depends only on $\gamma^{(k)}, \gamma^{(k-1)}$
- column 3 of $U^{(k+1)}$ depends only on $\gamma^{(k)}, \gamma^{(k-1)}, \gamma^{(k-2)}$

etc., over the range of previous iterations on which SR1 updates have been used, by virtue of R being upper triangular.

We refer to this calculation as

$$U^{(k+1)} = \mathrm{sr1}(U, \delta, \gamma). \tag{4.7}$$

Its cost is $O(nr)$ arithmetic operations, which is the same order of magnitude as the cost of a matrix product with U or U^T, and hence is readily affordable.

It is possible that (4.7) may give rise to a matrix $U^{(k+1)}$ whose columns are rank deficient. An example is given by

$$U = \begin{bmatrix} 1 & 1 \\ 1 & 0 \\ 2 & 0 \end{bmatrix} \qquad \delta = \begin{pmatrix} 1 \\ 0 \\ 0 \end{pmatrix} \qquad \gamma = \begin{pmatrix} 3 \\ 2 \\ 4 \end{pmatrix}. \tag{4.8}$$

It is clear that γ is in the range of U, and $\alpha = 1$, so that the SR1 update does not break down. The matrix $U^{(k+1)}$ has 3 non-trivial columns but has rank 2. At present, there is no evidence to suggest that this possibility is causing any practical disadvantages, although if it were so, it would not be difficult to suggest modifications to ensure that $U^{(k)}$ always has full rank. Indeed, it does seem more appropriate in these circumstances to use an update that keeps the same the same number of columns in U, such as is described in Section 6 below.

5. The BFGS projection update

Although we have argued that the SR1 update will often be well defined, this will not always be so. Thus we have to decide how to update $U^{(k)}$ when the SR1 denominator is non-positive.

An extreme case is when $\delta^T\gamma \leq 0$. In the quadratic case (3.3), this suggests that δ is not in the null space spanned by columns of Z, so that γ provides no useful information for updating B. Now the curvature estimate of the current B matrix along δ is $\delta^T B\delta$. Thus, if $\delta^T B\delta > 0$, the curvature estimate is seen to be incorrect, and we use the part of the BFGS update

$$B^{(k+1)} = B - \frac{B\delta\delta^T B}{\delta^T B\delta}. \tag{5.1}$$

which projects out the existing information along δ, and has the correct invariance properties. This update reduces the rank of B by one. If $r = 0$ or $B\delta = 0$

we just set $B^{(k+1)} = B$. We implement (5.1) in such a way as to reduce the number of columns in U by one, which ensures that $U^{(k+1)}$ has full rank if $U^{(k)}$ has. We may express

$$B^{(k+1)} = U\left(I - \frac{\mathbf{v}\mathbf{v}^T}{\mathbf{v}^T\mathbf{v}}\right)U^T = UQ\left(I - \frac{Q^T\mathbf{v}\mathbf{v}^TQ}{\mathbf{v}^T\mathbf{v}}\right)Q^TU^T \quad (5.2)$$

where $\mathbf{v} = U^T\boldsymbol{\delta}$ and $\mathbf{v}^TQ$ is an orthogonal transformation with a product of plane rotation matrices in columns (j, r), $j = 1, 2, \ldots, r-1$, so as eliminate successive elements v_j of $\mathbf{v}^T$. That is to say, $\mathbf{v}^TQ = \pm\|\mathbf{v}\|_2\mathbf{e}_r^T$ where $\mathbf{e}_r^T = (0, \ldots, 0, 1)$. Then

$$B^{(k+1)} = UQ\left(I - \mathbf{e}_r\mathbf{e}_r^T\right)Q^TU^T. \quad (5.3)$$

Thus to update U we apply the same rotations to the columns of U, and then delete the last column of the resulting matrix. We may write this as

$$U^{(k+1)} = U\bar{Q} \qquad \text{where} \qquad \bar{Q} = Q\begin{bmatrix} I \\ \mathbf{0}^T \end{bmatrix}. \quad (5.4)$$

It follows that $\boldsymbol{\delta}^TU^{(k+1)} = \mathbf{v}^T\bar{Q} = \mathbf{0}^T$, reflecting the fact that $B^{(k+1)}\boldsymbol{\delta} = \mathbf{0}$. We refer to the entire projection update as

$$U^{(k+1)} = \text{proj}(U, \boldsymbol{\delta}). \quad (5.5)$$

As for (4.7), the cost is $O(nr)$ arithmetic operations. We observe that (5.5) destroys any priority ordering properties, and in the quadratic case, hereditiary properties.

6. A new QN update

We have seen that the SR1 update can be used when $\boldsymbol{\delta}^T\boldsymbol{\gamma} > \mathbf{v}^T\mathbf{v}$, and the projection update when $\boldsymbol{\delta}^T\boldsymbol{\gamma} \leq 0$. When $0 < \boldsymbol{\delta}^T\boldsymbol{\gamma} \leq \mathbf{v}^T\mathbf{v}$, there would appear to be useful information in $\boldsymbol{\delta}$ and $\boldsymbol{\gamma}$, but the SR1 update can no longer be used. In this section we present a new update which maximizes the extent to which hereditary properties, built up using the SR1 update on previous iterations, are preserved.

To do this we partition

$$U^{(k)} = [U_1 \quad U_2], \quad \mathbf{v} = \begin{pmatrix} \mathbf{v}_1 \\ \mathbf{v}_2 \end{pmatrix}, \quad B = B_1 + B_2 = U_1U_1^T + U_2U_2^T \quad (6.1)$$

where U_1 has $r_1 \geq 0$ columns and U_2 has $r_2 \geq 0$ columns. The new QN update applies the SR1 update to U_1 and the projection update to U_2, and $U^{(k+1)}$ is obtained by concatenating the resulting matrices, that is

$$U^{(k+1)} = [\text{sr1}(U_1, \boldsymbol{\delta}, \boldsymbol{\gamma}) \quad \text{proj}(U_2, \boldsymbol{\delta})]. \quad (6.2)$$

By choosing r_1 sufficiently small, it is always possible to ensure that the SR1 denominator $\alpha = (\delta^T\gamma - \mathbf{v}_1^T\mathbf{v}_1)^{1/2}$ in (6.2) exists and is positive. We observe that (6.2) usually leaves the rank of U unchanged.

In our current implementation we have chosen r_1 to be the largest integer for which $\delta^T\gamma - \mathbf{v}_1^T\mathbf{v}_1 > 0$. This choice maximizes the extent to which priority ordering and, in the quadratic case, hereditary conditions in U_1 are preserved in $U^{(k+1)}$. In fact it may well be better to require $\delta^T\gamma - \mathbf{v}_1^T\mathbf{v}_1 \geq \tau$ where $\tau > 0$ is some tolerance, for example $\tau = \varepsilon\delta^T\gamma$ with $\varepsilon = 10^{-6}$ say.

The new update may also be expressed in the form

$$B^{(k+1)} = B_1 + \frac{(\gamma - B_1\delta)(\gamma - B_1\delta)^T}{(\gamma - B_1\delta)^T\delta} + B_2 - \frac{B_2\delta\delta^T B_2}{\delta^T B_2\delta}. \tag{6.3}$$

If $r_1 = 0$ we get the BFGS update (2.5), and if $r_1 = r$ then the SR1 update (2.4). Intermediate values give a new rank two correction formula, but not one that is in the Broyden class. We observe the following properties

- Satisfies the secant condition (2.3).
- Preserves positive semi-definite $B^{(k)}$ matrices.
- Is invariant under a linear transformation of variables (see [5], Section 3.3).
- Any priority ordering or hereditary conditions in U_1 are preserved.

To summarize, if $\delta^T\gamma \leq 0$, then the BFGS projection update is used and the rank of U decreases by one (usually). If $0 < \delta^T\gamma \leq \mathbf{v}^T\mathbf{v}$, then the new update is used, choosing r_1 as described, and the rank of U is unchanged. The choice $r_1 = 0$ gives rise to a BFGS update. If $\delta^T\gamma > \mathbf{v}^T\mathbf{v}$ then an SR1 update is used (the rank of U usually increasing by one), except in the case that $r = r_{max}$ and the memory is full. In this case we apply the SR1 update and then delete column $r_{max} + 1$ of the resulting matrix. We note that this procedure still preserves the secant condition (2.3). This follows from (4.6) and (4.4) by virtue of

$$\delta^T U^{(k+1)} = \delta^T \begin{bmatrix} \mathbf{u} & U \end{bmatrix} Q = \begin{bmatrix} \alpha & \mathbf{v}^T \end{bmatrix} Q = (\eta, 0, \ldots, 0)$$

where $\eta^2 = \delta^T\gamma$, by virtue of the way in which Q is chosen. Since the last element on the right hand side is zero, the secant condition $\delta^T U^{(k+1)} U^{(k+1)T} = \gamma^T$ is unaffected by deleting the last column of $U^{(k+1)}$. We also observe that any priority ordering in U is preserved, and any hereditary conditions up to a maximum of $r - 1$.

7. Conjugacy conditions

The use of the BFGS projection operation can destroy hereditary properties that have been built up in the U_2 matrix, in the quadratic case. In this section we show that this is not as unfavourable as it might seem, and that something akin to a hereditary property holds, even when the BFGS projection operation is used to update U_2. Also some conjugacy properties of the new update are shown. The results of this section apply when $B^{(1)} = 0$ and we are investigating the quadratic case in which the relationship $\gamma^{(k)} = W\delta^{(k)}$ holds for all k, where W is a fixed matrix. We shall also assume that W is nonsingular, which is a minor requirement that can always be achieved if necessary with an arbitrarily small perturbation using a quadratic penalty (see the paragraph following (3.6)).

When $B^{(1)} = 0$, it follows easily for both (5.5) and (6.2) by induction that the columns of $U^{(k)}$ are in span$(\gamma^{(1)}, \gamma^{(2)}, \ldots, \gamma^{(k-1)})$. It follows that U has an image

$$\Delta = W^{-1}U \tag{7.1}$$

whose columns are correspondingly in span$(\delta^{(1)}, \delta^{(2)}, \ldots, \delta^{(k-1)})$. In Theorem 3 below it is proved that normalised conjugacy coditions

$$U^T W^{-1} U = I \tag{7.2}$$

are satisfied by U (that is, $U^{(k)}$) for all k, a consequence of which is that $U^T\Delta = I$. Likewise, conjugacy conditions

$$\Delta^T W \Delta = I \tag{7.3}$$

are satisfied by Δ. A consequence is that

$$W\Delta = U = UU^T W^{-1} U = UU^T \Delta = B\Delta. \tag{7.4}$$

Hence B maps the subspace range(Δ) in the 'correct' way, that is $B\Delta = W\Delta = U$.

The implication of this in the quadratic case is that although the use of some projection operations may destroy hereditary conditions, it does not introduce any 'wrong' information. That is to to say, the image Δ of U has columns in span$(\delta^{(1)}, \delta^{(2)}, \ldots, \delta^{(k-1)})$, and the current matrix $B = UU^T$ maps Δ correctly into U (which is $W\Delta$). Thus, although hereditary conditions may not be satisfied, the exists a set of directions (columns of Δ) which satisfy a similar condition $B\Delta = U$ to hereditary conditions.

We now prove the theorem on which (7.2) depends.

THEOREM 3 *Let $B^{(1)} = 0$, and let there exist a nonsingular symmetric matrix W such that the difference vectors are related by $\gamma^{(k)} = W\delta^{(k)}$ for all k. Then conjugacy conditions (7.2) are preserved by the updating scheme described in Section 6.*

Proof We prove the result by induction. The result is trivially true when $k = 1$. Now we assume that (7.2) is true for some value of $k \geq 1$ and consider the calculation of $U^{(k+1)}$. If $\delta^T\gamma > 0$, then $U^{(k+1)}$ is defined by (6.2), which may be expressed as

$$U^{(k+1)} = [\mathbf{u} \quad U_1 \quad U_2] \begin{bmatrix} Q_1 & \\ & \bar{Q}_2 \end{bmatrix}, \tag{7.5}$$

using the notation of (4.6) and (5.4) with subscripts 1 and 2 to indicate matrices derived from the SR1 and projection updates respectively. We note that Q_1 has $r_1 + 1$ rows and columns, and $\bar{Q}_2$ has r_2 rows and $r_2 - 1$ columns. It follows that

$$\begin{aligned} &U^{(k+1)T} W^{-1} U^{(k+1)} \\ &= \begin{bmatrix} Q_1^T & \\ & \bar{Q}_2^T \end{bmatrix} \begin{bmatrix} \mathbf{u}^T W^{-1}\mathbf{u} & \mathbf{u}^T W^{-1} U \\ U^T W^{-1}\mathbf{u} & I \end{bmatrix} \begin{bmatrix} Q_1 & \\ & \bar{Q}_2 \end{bmatrix}, \end{aligned} \tag{7.6}$$

where $U^T W^{-1} U = I$ has been substituted from the inductive hypothesis. It now follows from (4.4) and $W^{-1}\gamma = \delta$ that

$$\begin{aligned} U^T W^{-1}\mathbf{u} &= U^T W^{-1} \left(\frac{\gamma - B_1\delta}{\alpha} \right) \\ &= \begin{pmatrix} U_1^T \\ U_2^T \end{pmatrix} \left(\frac{\delta - W^{-1} U_1 U_1^T \delta}{\alpha} \right) = \begin{pmatrix} 0 \\ U_2^T\delta/\alpha \end{pmatrix} \end{aligned} \tag{7.7}$$

using $U_1^T W^{-1} U_1 = I$ and $U_2^T W^{-1} U_1 = 0$ from the inductive hypothesis. Also from (4.4)

$$\mathbf{u}^T W^{-1}\mathbf{u} = \frac{(\gamma - B_1\delta)^T W^{-1} (\gamma - B_1\delta)}{\alpha^2} = 1 - \frac{(\gamma - B_1\delta)^T W^{-1} B_1\delta}{\alpha^2}$$

from $\delta = W^{-1}\gamma$ and the definition of α. But

$$\begin{aligned} (\gamma - B_1\delta)^T W^{-1} B_1\delta &= \delta^T((W - U_1U_1^T) W^{-1} U_1 U_1^T \delta \\ &= \delta^T (U_1 U_1^T - U_1 U_1^T W^{-1} U_1 U_1^T)\delta = 0 \end{aligned}$$

from $U_1^T W^{-1} U_1 = I$. Hence $\mathbf{u}^T W^{-1}\mathbf{u} = 1$. Finally we substitute this and (7.7) into (7.6). Then using $Q_1^T Q_1 = I$, $\bar{Q}_2^T \bar{Q}_2 = I$ and $\bar{Q}_2^T U_2^T \delta = 0$ (from (5.4)), it follows that $U^{(k+1)T} W^{-1} U^{(k+1)} = I$ and establishes that the inductive hypothesis holds for $k+1$ in this case.

In the case that $\delta^T\gamma \leq 0$, only the 2,2 partition of the above argument is used and the result follows similarly. In the case that the memory is filled, and a column of $U^{(k+1)}$ is deleted, it is clear that $U^{(k+1)T} W^{-1} U^{(k+1)} = I$ will continue to hold, but for a unit matrix with one fewer row and column. Thus the result is established in all cases used in the update scheme. **QED**

8. Practical experience

In this section, some preliminary practical experience with the new update scheme is described. An experimental code `filterQN` is currently under development, and indeed has been so for some time. It is a trust region filter SQP code, based on the method considered in [7], but using the quasi-Newton update scheme described in this paper, rather than the exact second derivatives as used in the `filter2` code, referred to in Section 1. The delay in finalizing the code is mainly due to uncertainty as to how best to implement feasibility restoration when second derivatives are not available.

The results in this section are sampled from CUTE test problems in which the dimension d of the null space at the solution is a significant proportion of n. For such problems, feasibility restoration often plays a minor role in determining the outcome of the calculation. Thus, although the results cannot yet be taken as definitive, they do give some indication as to what level of performance can be expected from a QN code. The problems are solved to an accuracy of better than 10^{-6} in the KT conditions. The memory limit is $r_{max} = \min(n, 100)$.

Table 1. Performance of `filterQN` on small CUTE problems

	n	m	d	$\#g$	$f2$
HS90	4	1	3	29	6*
HS92	6	1	5	33	6*
HS99	7	2	5	11	6
HS100	7	4	5	14	14
HS100LNP	7	2	5	15	18
HS100MOD	7	4	6	18	13
HS101	7	5	5	230	21
HS102	7	5	4	209	18
HS103	7	5	3	28	25
HS111	10	3	6	45	25
HS111LNP	10	3	6	45	25
HS113	10	8	4	13	6
HS117	10	5	4	19	21
CANTILVR	5	1	4	25	17
DIPIGRI	7	4	5	14	14
ERRINBAR	18	9	3	70	25
MISTAKE	9	13	5	17	14
POLAK3	12	10	9	58	37
ROBOT	14	2	5	19	11
TENBARS1	18	9	5	59	28
TENBARS2	18	8	4	36	28
TENBARS3	18	8	4	76	28
TENBARS4	18	9	5	82	29

* `filter2` finds a locally infeasible point

Table 1 gives results on Hock-Schittkowski test problems in the left hand column, and other small CUTE test problems in the right hand column. Headings give the number of variables n, the number of constraints m (excluding simple bounds), the dimension d of the null space at the solution, the number of gradient calls $\#g$ required by `filterQN`, and the number of gradient calls $f2$ required by `filter2`. One gradient call includes the evaluation of both the gradient of the objective function and the Jacobian of the vector of constraint functions. In the case of `filter2`, it also includes the evaluation of all second derivatives. Both codes require about one QP subproblem to be solved for each gradient evaluation, most often in warm start mode.

Generally the problems are solved reliably and accurately by `filterQN`, and we see rapid local convergence. HS90 and HS92 are successfully solved by `filterQN`, whereas `filter2` can only find a locally infeasible point. For HS101 and HS102, `filterQN` spends about 200 and 180 iterations respectively in feasibility restoration, which accounts for the poor performance. Future work on the feasibility restoration algorithm should resolve this difficulty. Apart from that, we observe that `filterQN` mostly takes more iterations than `filter2`. To some extent this is expected due to the need for `filterQN` to spend something like d extra iterations in building up a matrix $B = UU^T$ with the property that $BZ \sim WZ$.

Next we show some results in Table 2, obtained on some larger CUTE test problems, again chosen to have significant null space dimensions. In fact the exact dimension of the null space is not so easy to determine, because the accuracy required from the QP subproblem can often be obtained without having to build up the full reduced Hessian. In the table an approximate value of d is given based on the size of the QP reduced Hessian on termination of the `filter2` run.

Some of the problems are solved quite efficiently by `filterQN`. The largest problem DTOC1L is the only linearly constrained problem in the set, so feasibility restoration is not an issue in this case. Thus it is very satisfactory that this problem is solved accurately and quickly by `filterQN`, even though the dimension d of the null space is very large. Less satisfactory is the performance on the ORTHREG problems, and also DIXCHLNV, ORTHRGDS and ZAMB2, which mostly have large null spaces with $d \gg r_{max}$. In these problems, slow convergence is observed in the asymptotic phase, and the build up of information in the U matrix is very slow. The restriction on r may also be having an effect. On other problems, the behaviour of `filterQN` is reasonably satisfactory, bearing in mind the need for for extra iterations to build up an effective Hessian approximation.

Table 2. Performance of `filterQN` on some larger CUTE problems

	n	m	$\sim d$	$\#g$	$f2$
AIRPORT	84	42	42	74	12
LAKES	90	78	12	62	421
READING6	102	50	8	40	23
ZAMB2-8	138	48	30	36	7
ZAMB2-9	138	48	10	25	4
ZAMB2-10	270	96	10	19	6
ZAMB2-11	270	96	59	65	6
DIXCHLNV	100	50	35	473	22
DTOC1L	5998	3996	1998	287	8
DTOC5	1999	999	998	24	5
EIGENB2	110	55	52	64	21
OPTCDEG2	1202	800	10	18	6
OPTCTRL6	122	80	39	74	6
ORTHRDM2	203	100	103	49	9
ORTHRDS2	203	100	103	36	32
ORTHREGC	1005	500	501	392	10
ORTHREGD	1005	500	280	216	13
ORTHREGE	36	20	12	111	63
ORTHREGF	1205	400	767	228	22
ORTHRGDS	1003	500	503	458	29
SVANBERG	500	500	19	71	9
TRAINH	808	402	17	39	14
ZAMB2	1326	480	142	349	8

9. Conclusions

We itemize the main features of this paper as follows. A new QN update scheme for low rank Hessian approximation in SQP has been presented. Where possible it uses the SR1 update formula, but when this is not possible a new rank two update is used, which is not in the Broyden family, although invariance under linear transformations of the variables is preserved. The Hessian approximation is a positive semi-definite matrix, which ensures that global solutions of QP subproblems are calculated. It also enables interior point methods to be used to solve the QP subproblems, if required. The representation provides a limited memory capability, and there is a priority ordering scheme which enables 'old' information to be deleted when the memory is full. Hereditary and conjugacy properties are preserved to the maximum extent when minimizing a quadratic function subject to linear constraints. Practical experience is reasonably encouraging on small (and some larger) problems. There is some evidence of slow convergence on some larger problems with large null spaces. It may be that this is to some extent caused by the use of an l_∞ trust region. Future work will continue to develop the `filterQN` code, especially the feasibility restoration part, and will also investigate the use of an l_2 trust region.

References

[1] M.C. van Beurden. Integro-differential equations for electromagnetic scattering: analysis and computation for objects with electric contrast. Ph.D. Thesis, Eindhoven University of Technology, Eindhoven, The Netherlands, 2003.

[2] P.T. Boggs, J.W. Tolle and P. Wang. On the local convergence of quasi-Newton methods for constrained optimization. *SIAM J. Control and Optimization.* 20:161-171, 1982.

[3] A.R. Conn, N.I.M. Gould and Ph.L. Toint. An introduction to the structure of large scale nonlinear optimization problems and the LANCELOT project. In *Computing Methods in Applied Sciences and Engineering.* R. Glowinski and A. Lichnewsky, eds, SIAM Publications, Philadelphia, 1990.

[4] A.R. Conn, N.I.M. Gould and Ph.L. Toint. Convergence of quasi-Newton matrices generated by the symmetric rank one update. *Mathematical Programming.* 50:177-196, 1991.

[5] R. Fletcher. *Practical Methods of Optimization.* Wiley, Chichester, 1987.

[6] R. Fletcher and S. Leyffer. Nonlinear Programming Without a Penalty Function. *Mathematical Programming.* 91:239-269, 2002.

[7] R. Fletcher, S. Leyffer and Ph.L. Toint. On the Global Convergence of a Filter–SQP Algorithm. *SIAM J. Optimization.*, 13:44-59, 2002.

[8] P.E. Gill, W. Murray, M.A. Saunders. SNOPT: An SQP Algorithm for Large-Scale Constrained Optimization. *SIAM J. Optimization.* 12:979-1006, 2002.

[9] J. Nocedal. Updating quasi-Newton matrices with limited storage. *Mathematics of Computation.* 35:773-782, 1980.

[10] J. Nocedal and M.L. Overton. Projected Hessian updating algorithms for nonlinearly constrained optimization. *SIAM J. Numerical Analysis.* 22:821-850, 1985.

[11] M.J.D. Powell and Ph.L. Toint. On the estimation of sparse Hessian matrices. *SIAM J. Numerical Analysis.* 16:1060-1074, 1979.

RELIABILITY IN COMPUTER NETWORKS

S. Minkevicius,[1,2] and G. Kulvietis[2]
[1] *Institute of Mathematics and Informatics, Akademijos 4, 2600 Vilnius, Lithuania, stst@ktl.mii.lt*
[2] *Vilnius Gediminas Technical University, Sauletekio 11, 2040 Vilnius, Lithuania, gk@fm.vtu.lt*

Abstract We use a mathematical model of an open queueing network in heavy traffic. The probability limit theorem for the virtual waiting time of a customer in heavy traffic in open queueing networks has been presented. Finally, we present an application of the theorem - a reliability model from computer network practice.

keywords: mathematical models of technical systems, reliability theory, queueing theory, open queueing network, heavy traffic, the probability limit theorem, virtual waiting time of a customer.

1. Introduction

One can apply the theory of queueing networks to obtain probability characteristics of technical systems (for example, the reliability function of computer networks).

At first we try to present a survey of papers designated to applying the results of the queueing theory in reliability. In [2], it is investigated the reliability of a distributed program in a distributed computing system and it has been showen a probability that a program which runs on multiple processing elements that have to communicate with other processing elements for remote data files will be executed successfully. In [8], a single machine, subject to breakdown, that produces items to inventory, is considered. The main tool employed is a fluid queue model with repair. To analyze the performance of multimedia service systems, which have unreliable resources, and to estimate the capacity requirement of the systems, a capacity planning model using an open queueing network has been developed in [6]. Paper [1] discusses a novel model for a reliable system composed of N unreliable systems, which can hinder or enhance one another's reliability. Paper [10] analyzes the behaviour of a heterogeneous finite-source system with a single server. As applications of this model, some problems in the field of telecommunications and reliability theory are treated. In [7] the management policy of an $M/G/1$ queue with a single removable

Please use the following format when citing this chapter:
Author(s) [insert Last name, First-name initial(s)], 2006, in IFIP International Federation for Information Processing, Volume 199, System Modeling and Optimization, eds. Ceragioli F., Dontchev A., Furuta H., Marti K., Pandolfi L., (Boston: Springer), pp. [insert page numbers].

and non-reliable server is investigated. They use the analytic results of this queueing model and apply an efficient Matlab program to calculate the optimal threshold of management policy and some system characteristics. In [3, 4], using the law of the iterated logarithm for the queue length of customers, the reliability function of computer network is estimated and a theorem similar to Theorem 2.1 is proved.

In this paper, we present the probability limit theorem for the virtual waiting time of a customer in heavy traffic in open queueing networks.

First we consider open queueing networks with the "first come, first served" service discipline at each station and general distributions of interarrival and service time. The basic components of the queueing network are arrival processes, service processes, and routing processes. The service discipline is "first come, first served" (FCFS). We consider open queueing networks with the FCFS service discipline at each station and general distributions of interarrival and service times. The queueing network studied by us has k single server stations, each of which has an associated infinite capacity waiting room. Every station has an arrival stream from outside the network, and the arrival streams are assumed to be mutually independent renewal processes. Customers are served in the order of arrival and after service they are randomly routed to either another station in the network, or out of the network entirely. Service times and routing decisions form mutually independent sequences of independent identically distributed random variables.

The basic components of the queueing network are arrival processes, service processes, and routing processes. In particular, there are mutually independent sequences of independent identically distributed random variables $\left\{z_n^{(j)}, n \geq 1\right\}$, $\left\{S_n^{(j)}, n \geq 1\right\}$ and $\left\{\Phi_n^{(j)}, n \geq 1\right\}$ for $j = 1, 2, \ldots, k$; defined on a probability space. The random variables $z_n^{(j)}$ and $S_n^{(j)}$ are strictly positive, and $\Phi_n^{(j)}$ have support in $\{0, 1, 2, \ldots, k\}$. We define $\mu_j = \left(M\left[S_n^{(j)}\right]\right)^{-1} > 0$, $\sigma_j = D\left(S_n^{(j)}\right) > 0$ and $\lambda_j = \left(M\left[z_n^{(j)}\right]\right)^{-1} > 0$, $a_j = D\left(z_n^{(j)}\right) > 0$, $j = 1, 2, ..., k$; with all of these terms assumed finite. Denote $p_{ij} = P\left(\Phi_n^{(i)} = j\right) > 0$, $j = 1, 2, \ldots, k$. In the context of the queueing network, the random variables $z_n^{(j)}$ function as interarrival times (from outside the network) at the station j, while $S_n^{(j)}$ is the nth service time at the station j, and $\Phi_n^{(j)}$ is a routing indicator for the nth customer served at the station j. If $\Phi_n^{(i)} = j$ (which occurs with probability p_{ij}), then the nth customer served at the station i is routed to the station j. When $\Phi_n^{(i)} = 0$, the associated customer leaves the network. The matrix P is called a routing matrix.

To construct renewal processes generated by the interarrival and service times, we assume the following for $l \geq 1, j = 1, 2, \ldots, k$

$$z_j(0) = 0, z_j(l) = \sum_{m=1}^{l} z_m^{(j)}, S_j(0) = 0, S_j(l) = \sum_{m=1}^{l} S_m^{(j)}.$$

Observe that this system is quite general, encompassing the tandem system, acyclic networks of $GI/G/1$ queues, and networks of $GI/G/1$ queues with feedback.

Let us define $V_j(t)$ as a virtual waiting time of a customer at the jth station of the queueing network in time t,

$$\hat{\beta}_j = \frac{\lambda_j + \sum_{i=1}^{k} \mu_i \cdot p_{ij}}{\mu_j} - 1, \qquad \hat{\sigma}_j^2 = \frac{\sigma_j^2 \cdot \left(\sum_{i=1}^{k} p_{ij}^2 + 1 \right) + \left(\sum_{i=1}^{k} p_{ij}^2 \cdot \sigma_i^2 \right) + a_j^2}{\mu_j}$$

where $j = 1, 2, \ldots, k$.

We suppose that the following condition is fulfilled:

$$\lambda_j + \sum_{i=1}^{k} \mu_i \cdot p_{ij} > \mu_j, \;\; j = 1, 2, \ldots, k. \tag{9.1}$$

Note that this condition quarantees that, with probability one there exists a virtual waiting time of a customer and this virtual waiting time of a customer is constantly growing.

One of the results of the paper is the following theorem on the probability limit theorem for the virtual waiting time of a customer in an open queueing network.

THEOREM 1 *If conditions (1) are fulfilled, then*

$$\lim_{n \to \infty} P\left(\frac{V_j(nt) - \beta_j \cdot n \cdot t}{\hat{\sigma}_j \cdot \sqrt{n}} < x \right) = \int_{-\infty}^{x} \exp(-y^2/2) dy,$$

$0 \leq t \leq 1$ *and* $j = 1, 2, \ldots, k$.

Proof. This theorem is proved on conditions $\lambda_j > \mu_j, \;\; j = 1, 2, \ldots, k$ (see, for example, [9]). Applying the methods of [4], it can be proved that this theorem is true under more general (1) conditions.

The proof of the theorem is complete.

2. Reliability functions of the computer network

Now we present a technical example from the computer network practice. Assume that queues arrive at a computer v_j at the rate λ_j per hour during business hours, $j = 1, 2, \ldots, k$. These queues are served at the rate μ_j per hour in the computer v_j, $j = 1, 2, \ldots, k$. After service in the computer v_j, with probability p_j (usually $p_j \geq 0.9$), they leave the network and with probability p_{ji}, $i \neq j$, $1 \leq i \leq k$ (usually $0 < p_{ji} \leq 0.1$) arrive at the computer v_i, $i = 1, 2, \ldots, k$. Also, we assume the computer v_j fails when the virtual waiting time of queues is more than k_j, $j = 1, 2, \ldots, k$.

In this section, we will prove the following theorem on the reliability function of the computer network (probability of stopping the computer network).

THEOREM 2 *If* $t \geq \max_{1\leq j\leq k} \frac{k_j}{\hat{\beta}_j}$ *and conditions (1) are fulfilled, the computer network becomes unreliable (all computers fail).*

Proof. At first, using Theorem 1 we get for $0 < \varepsilon < 1$ that

$$\lim_{n\to\infty} P\left(\frac{V_j(n) - \beta_j \cdot n}{\hat{\sigma}_j \cdot \sqrt{n}} < x\right) = \int_{-\infty}^{x} \exp(-y^2/2)dy,\ j = 1, 2, \ldots, k. \tag{9.2}$$

Let us investigate a computer network which consists of the elements (computers) v_j that are indicators of stations X_j, $j = 1, 2, \ldots, k$.

Denote

$$X_j = \begin{cases} 1, & \text{if the element } v_j \text{ is reliable,} \\ 0, & \text{if the element } v_j \text{ is unreliable,} \end{cases}$$

$j = 1, 2, \ldots, k$.

Note that $\{X_j = 1\} = \{V_j(t) < k_j\}$, $j = 1, 2, \ldots, k$.

Denote the structural function of the system of elements connected by scheme 1 from k (see, for example, [5]) as follows:

$$\phi(X_1, X_2, \ldots, X_k) = \begin{cases} 1, & \sum_{i=1}^{k} X_i \geq 1 \\ 0, & \sum_{i=1}^{k} X_i < 1. \end{cases}$$

Let us estimate the reliability function of the computer network using the formula of full conditional probability (see [4])

$$h(X_1, X_2, \ldots, X_k, t) \leq \sum_{i=1}^{k} P(X_i = 1)$$

Thus,

$$0 \leq h(X_1, X_2, \ldots, X_k, t) \leq \sum_{i=1}^{k} P(V_i(t) \leq k_i). \tag{9.3}$$

Applying Theorem 1 (when $t = 1$) we obtain that

$$0 \leq \lim_{t\to\infty} P(V_j(t) < k_j) = \lim_{n\to\infty} P(V_j(n) < k_j) =$$

$$\lim_{n\to\infty} P\left(\frac{V_j(n) - \beta_j \cdot n}{\hat{\sigma}_j \cdot \sqrt{n}} < \frac{k_j - \beta_j \cdot n}{\hat{\sigma}_j \cdot \sqrt{n}}\right) = \int_{-\infty}^{-\infty} \exp(-y^2/2)dy = 0. \tag{9.4}$$

Thus (see (4)),

$$\lim_{t\to\infty} P\left(V_j(t) < k_j\right) = 0,\ j = 1, 2, \ldots, k. \tag{9.5}$$

Consequently, $\lim_{t\to\infty} h(X_1, X_2, \ldots, X_k, t) = 0$ (see (3) and (5)).

The proof of the theorem is complete.

Finally, we give an exact expression for $h(X_1, X_2, \ldots, X_k, t),\ t > 0$. We will prove the following theorem on this probability.

THEOREM 3 *$h(X_1, X_2, \ldots, X_k, t)$ is equal to* $\exp(-\sum_{j=1}^{k} P(V_j(t) < k_j))$.

Proof. First denote $\lambda_j,\ j = 1, 2, \ldots, k$ as intensities of structural elements, that form a complex stochastic system. Then probability of stopping this system is equal to $e^{-\sum_{j=1}^{k} \lambda_j}$ (see, for example, [11]).

But

$$\lambda_j = MX_j = P(X_j = 1) = P(V_j(t) < k_j),\ j = 1, 2, \ldots, k. \tag{9.6}$$

Applying (6), we obtain that $h(X_1, X_2, \ldots, X_k, t)$ is equal to

$$e^{-\sum_{j=1}^{k} \lambda_j} = e^{-\sum_{j=1}^{k} P(V_j(t) < k_j)}.$$

The proof is complete.

As one can see, using Theorems 2 and 3, it is possible to estimate the reliability of a complex computer network.

References

[1] E. Gelenbe, J. M. Fourneau. G - networks with reset. *Performance Evaluation.* 49(1-4):179-191, 2002.

[2] M. S. Lin, D. J. Chen. The computational complexity of reliability problem on distributed systems. *Information Processing Letters.* 64(3):143-147, 1997.

[3] S. Minkevivcius, G. Kulvietis. On reliability in computer networks. *Proceedings of International Conference of LAD' 2004.* St. Petersburg, Russia, 223-229, 2004.

[4] S. Minkevivcius, G. Kulvietis. Application of the law of the iterated logarithm in open queueing networks. 2005 (to be published).

[5] J.J. Morder, S. E. Elmaghraby. *(eds.) Handbook of operational research models and applications.* Van Nostrand Reinhold, New York, 1978.

[6] K. Park, S. I. Kim. A capacity planning model of unreliable multimedia service systems. *Journal of Systems and Software.* 63(1):69 -76, 2002.

[7] W. L. Pearn, J. C. Ke, C. Chang. Sensitivity analysis of the optimal management policy for a queueing system with a removable and non-reliable server. *Computers and Industrial Engineering.* 46(1):87-99, 2004.

[8] D. Perry, M. J. Posner. A correlated M/G/1-type queue with randomized server repair and maintenance modes. *Operations Research Letters.* 26(3):137-148, 2000.

[9] L.L. Sakalauskas, S. Minkevivcius. On the law of the iterated logarithm in open queueing networks. *European Journal of Operational Research.* 120:632-640, 2000.

[10] J. Sztrik, C. S. Kim. Markov-modulated finite-source queueing models in evaluation of computer and communication systems. *Mathematical Modelling and Computer.* 38(7-9):961-968, 2003.

[11] V. Zubov. *(eds.) Mathematical theory of reliability of queueing systems.* Radio i sviaz. 1964.